INVENTING CANAD

Early Victorian Science and the Idea
of a Transcontinental Nation

The years of the nineteenth century that saw British North America attempting to establish a transcontinental nation also saw the fruition of scientific ideas that challenged traditional conceptions of man's relationship with nature and the land. Victorian thought was undergoing a radical transition, from the static, orderly world modelled by eighteenth-century mathematical physics to the world-in-process epitomized by the mid-nineteenth-century theory of evolution by natural selection. In British North America this intellectual transition played itself out in a number of contexts: in the reorganization of science from the natural history tradition to a utilitarian ideology promoted by business and professional classes; in the institutionalization of scientific inventories such as the Geological Survey of Canada; and in the expansion from local to transcontinental interests.

Tapping a wide range of archival and published sources, Suzanne Zeller documents the place of Victorian science in British North American thought and society during the era of Confederation. Four prominent Victorian 'inventory' sciences provide a focus for her study: geology, terrestrial magnetism, meteorology, and botany, each set within its wider context. She considers the role of individuals instrumental in these pursuits – Sir William Edmond Logan, Sir John Henry Lefroy, and George Lawson – and of a host of scientists, politicians, educators, journalists, businessmen, and 'improving' farmers who promoted public support of these sciences. Together they formed a community that believed that science not only enhanced the possibilities of Canada's material progress but also provided a fertile ground for a 'new nationality' to take root as a northern variation of the British nation.

Victorian science offered a means to assess and control nature as a rational alternative to retreat from nature's harshness. It also helped develop a sense of Canada's past and brighten the prospects for a transcontinental future.

SUZANNE ZELLER has taught history at the University of Windsor and Trent University and is currently assistant professor of history at Wilfrid Laurier University.

SUZANNE ZELLER

Inventing Canada:
Early Victorian Science
and the Idea of a
Transcontinental Nation

UNIVERSITY OF TORONTO PRESS
Toronto Buffalo London

© University of Toronto Press 1987
Toronto   Buffalo   London
Printed in Canada

ISBN 0-8020-2644-3 (cloth)
ISBN 0-8020-6606-2 (paper)

Printed on acid-free paper

Canadian Cataloguing in Publication Data
Zeller, Suzanne Elizabeth, 1952–
Inventing Canada : early Victorian science and
the idea of a transcontinental nation

Includes bibliographical references and index.
ISBN 0-8020-2644-3 (bound)   ISBN 0-8020-6606-2 (pbk.)

1. Science – Canada – History. 2. Science –
Canada – Societies, etc. – History. 3. Nationalism
– Canada – History. I. Title.

Q127.C3Z44 1987   509'.71   C87-094593-9

COVER ILLUSTRATION Details from the cover of *Le Naturaliste canadien* (1870)

This book has been published with the help of a grant from the Canadian Federation
for the Humanities, using funds provided by the Social Sciences and Humanities
Research Council of Canada.

Contents

PREFACE vi
INTRODUCTION 3

Part I Geology
1 Exposing the Strata 13
2 Montreal Masonry 31
3 Logan's Geological Inventory: 'Construction and Extension,' 1842–1850 51
4 'Grandeur and Historical Renown,' 1851–1856 78
5 'Permanence,' 1857–1869 94

Part II Terrestrial Magnetism and Meteorology
6 The Spirit of the Method 115
7 Mutual Attractions, 1845–1850 131
8 Science as a Cultural Adhesive, 1850–1853 145
9 Encompassing the North 161

Part III Botany
10 Adventitious Roots 183
11 The Metamorphosed Leaf 204
12 Fragile Stems, 1857–1863 218
13 Flower and Fruit: The Nation as Variation 240

CONCLUSION 269
NOTES 275
NOTE ON SOURCES 337
INDEX 341

Preface

The history of science in Canada is still largely unexplored, and its place in the formation of Canadian culture has not been fully assessed. This study is concerned with the way in which ideas of science informed the outlooks of British North Americans during the Victorian age. The approach is through a few individuals who helped to organize science and who knew and communicated with one another in many instances. Yet the ideas intertwined, cross-fertilizing one another in the larger community. The focus has been broadened to include politicians, businessmen, educators, newspaper editors, and farmers who shared in these ideas and supported the work of practitioners of science. In appealing to science and technology to overcome material difficulties, members of this larger community assumed that science could also provide a meaningful interpretation of their experience, a premise that deeply influenced the course of British North American history from the nineteenth century onward.

It is a pleasure to be able to thank so many people without whose assistance this study could not have been completed. These include archivists at several repositories in Great Britain: the Royal Botanical Gardens, Kew; the Royal Botanic Garden, Edinburgh; the National Museum of Wales, Cardiff; the Public Record Office, the Imperial College of Science and Technology, the Institute of Geological Science, and the Geological Society, all in London. Closer to home, I received much assistance from Jane Lynch in the Interlibrary Loans Office, Robarts Library, and from the staff of the Fisher Rare Book Library, both at the University of Toronto. I owe thanks as well to D'Arcy Beckstead, Energy, Mines and Resources Canada; Tom Hacking, Atmospheric Environment Services, Downsview, Ont.; and the staffs of the following archives and libraries: the Public Archives of Canada; Baldwin Room, Metropolitian Toronto Reference Library; Public Archives of Ontario; Queen's University Archives; McGill

University Archives; the McCord Museum, Montreal; the Molson Company, Montreal; the Archive du Séminaire de Chicoutimi; Archive du Séminaire de Québec; Public Archives of Manitoba; Dalhousie University Archives; Public Archives of Nova Scotia; and the Centre for Newfoundland Studies at Memorial University. My stay in Great Britain was made memorable by the hospitality of Krystyna Phillips; my time in Montreal, by the family of Sol and Anita Shore, who welcomed a stranger as family. Financial support from the Social Sciences and Humanities Research Council, the University of Toronto, and Trent University is gratefully acknowledged.

Professors J.M. Bliss and T.H. Levere of the University of Toronto supervised the PHD thesis from which this study is derived. This work benefited also from the careful reading and thoughtful comments of Dr George Urbaniak, John H. Noble, and the manuscript assessors, and from the guidance of Gerry Hallowell at the University of Toronto Press. Don McLeod gave invaluable help with the index. I alone am responsible for any errors that remain.

I wish to dedicate this book to my family, for their support throughout.

Sir William Logan. Public Archives Canada (PAC) C10459

Robert Bell. PAC C6941

Sir J.W. Dawson. PAC C49822

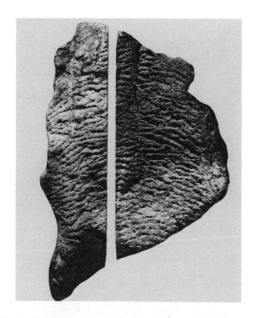

Specimen of the 'fossil' *Eozoon canadense*. Geological Survey of Canada, Ottawa (GSC)

Logan's collecting basket. GSC

'My Tent,' drawing by Sir William Logan, from his Journal, 1843. GSC

Buggy odometer used by Logan. GSC

Logan's wheel odometer. GSC

Sir J.H. Lefroy. Atmospheric Environment Services, Environment Canada (AES)

Sir Charles James Riddell. AES

Frederick Templeman Kingston. AES

Toronto Observatory, 1852, on the site of Convocation Hall, University of Toronto.
Metropolitan Toronto Library Board

OPPOSITE LEFT: Charles Smallwood's Observatory at St Martin's, Isle Jésus, near Montreal. *Canadian Naturalist and Geologist* 3/5 (October 1858): 361-3
External view of the Observatory
A   Thermometer for solar radiation
B   Screen of Venetian blinds
C   Thermometer
D   Opening in ridge of the roof, closed with shutters, to allow use of transit instrument
E   Rain gauge with conducting pipe through the roof
F   Velocity shaft of the anemometer
G   Mast for elevating apparatus for collecting electricity
H   Cord for hoisting the collecting apparatus
I   Direction shaft of the anemometer
J   Copper wire for conducting the electricity into the building

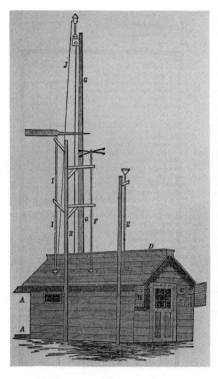
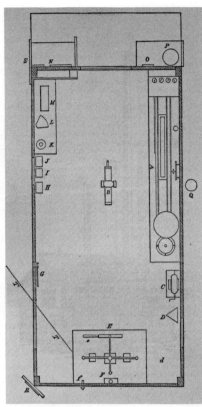

Plan of the Observatory (TOP RIGHT)
A Anemometer
B Small transit for correcting time
C Electrical machine for charging the distinguisher
D Peltier's electrometer
d Space occupied by drosometer, polariscope, etc.
E Electrometer
e Discharger
F Distinguisher
f Small stove – sometimes used in damp weather
G Thermometer placed in the prismatic spectrum for investigations on light
H Nigretti & Zambra's barometers and cisterns, 118 feet above the level of the sea
I Small-tube barometer
J Newman's barometer
K Aneroid barometer
L Quadrant and artificial horizon
M Microscope and apparatus for ascertaining the forms of snow crystals
N Thermometer, psychometer, etc, 4 feet high. A space is left between the two walls to ensure insulation and prevent radiation.
O Ozonometer
P Evaporator – removed in winter and replaced by scales for showing the amount of evaporation from the surface of ice
Q Post sunk in the ground, and 40 feet high, to carry the arms of support for the anemometer
R Solar radiation
S Venetian blinds
T Iron rod beneath the surface of the ground connected with the discharger to ensure safety

Map of the north-west part of Canada, Indian Territories, and Hudson Bay, Crown Lands Department, March 1857, showing geological formations and isotherms. Ontario Archives, Toronto

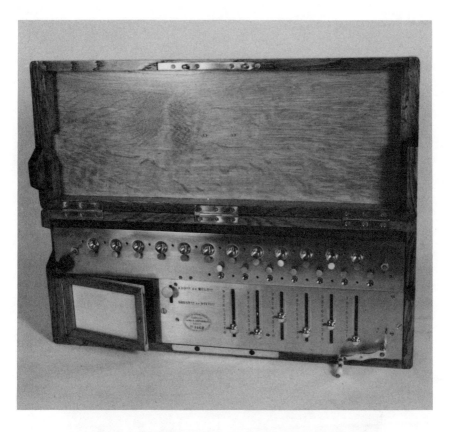

Calculating machine used in the Toronto Observatory. AES

Long shaft windmill vane used by Kingston in the Toronto Observatory. AES

Long shaft anemometer used by Kingston in the Toronto Observatory. AES

George Lawson. PAC C37834

William Hincks. University of Toronto Archives

Introduction

> I went into science a great deal myself at one time; but I saw it would not do. It leads to everything; you can let nothing alone.
>
> George Eliot, *Middlemarch* (1871–2)

It was difficult for immigrants to the backwoods during the early nineteenth century to see Canada as anything but a harsh, unyielding wilderness. Many were dismayed by the apparent barrenness of the indomitable rock formations, by the extremities of heat and cold, and by the unrelenting tangle of primordial forest. Success, it seemed, would be mere survival in such formidable circumstances. But could one ever really feel at home in such rudimentary surroundings? Would it be possible in this wilderness to cultivate garden provinces like England and New England? Retreat from nature's unpredictable and brutal force became a natural defence in the face of such basic uncertainty and is said to have congealed to form a fundamental trait in Canadian culture.[1]

Yet retreat constituted only one response to nature in the more complex process of 'cultivating' the British North American wilderness. It found its counter-thrust in science, the rational study of nature. Science penetrated British culture from the late eighteenth century, displacing art as the 'dominant cultural mode' of Britain's new industrial civilization.[2] The authority of science as a spearhead of the age of progress grew dramatically throughout Victoria's domains, mainly because of its apparent power to promote utilitarian ends. Proffering solutions to settlers' most basic problems, science held out the promise of a means to locate good soils for agriculture and valuable mineral deposits for mining and industry, to cope with climate, and to make commercial use of plants and other natural products. Science offered a chance for real prosperity, more than mere survival. Victorian Canadians, like Victorians elsewhere, marvelled at the technological signs of material progress and economic development – canals, railways, and electric telegraphs – and tended to identify these with science.

Organized science in Victorian Canada maintained strong ties not only to the domestic advance of technology but also to its own seventeenth-century Baconian and Newtonian heritage. 'Baconianism' assured that new scientific concepts and theories would arise through the amassing of facts by observation and experiment; 'Newtonianism' provided a mathematical model for the mechanical view of nature as an orderly machine subject to universal laws. Both identified the natural with the rational, and both promoted an emphasis on the usefulness of science which emerged more clearly during the eighteenth and nineteenth centuries. But Victorians also witnessed an unsettling challenge to the traditional view of nature as static and mechanical; by the 1840s this challenge was marked by a growing sense of the world as an organism in the process of historical change, a view that reached its zenith in the publication of Charles Darwin's *The Origin of Species* in 1859.[3]

Early Victorian science as it was practised in British North America consisted of various branches of both natural history (natural science), which also touched on agricultural chemistry, and natural philosophy (physical science), especially geophysics. Together these two main disciplines made up a major scientific tradition in Western culture. Sometimes called the 'geographical' tradition, its purpose was to explore and to exploit new lands all over the world. More broadly, the geographical tradition expressed the 'incremental spirit' which pervaded modern science and whose purpose was to increase and diffuse knowledge. Any individual could add piecemeal to the stock of knowledge and was encouraged to do so.[4]

A common inventorial purpose linked early Victorian scientific pursuits in British North America, and they are here collectively called 'inventory science' to highlight the mapping and cataloguing of resources and other natural phenomena which preoccupied the colonists. Inventory science in early Victorian Canada included geology, terrestrial magnetism, meteorology, botany, and to a lesser extent entomology, zoology, and anthropology. Its institutionalization was best exemplified by the preliminary geological surveys of New Brunswick and Newfoundland during the 1830s and by the founding of the more permanent Magnetic and Meteorological Observatory at Toronto in 1839 and of the Geological Survey of Canada at Montreal in 1842.

The inventory tradition derived not only from the incremental spirit of the Scientific Revolution but also from the more recent Scottish Enlightenment. Its foremost example was the *Statistical Account of Scotland* compiled by Sir John Sinclair in 1791–2. Sinclair (1754–1835) surveyed Scottish parishes for details of their local environs and published the results to demonstrate the usefulness of 'statistics,' a term borrowed from the German tradition of statecraft under Frederick the Great. Sinclair heralded the statistical movement of early Victorian Britain, which aimed to improve the quality of life through detailed fact-gathering.

Sinclair's achievement was not so much that he invented the statistical survey as that he carried one through on a large scale. The basic idea of soliciting from local 'Gentlemen and Clergy' information on 'the Antiquities and natural History of their respective Parishes,' for the purpose of 'exciting them to favour the World with a fuller and more satisfactory Account of their Country, than it is in the Power of a Stranger and transient Visitor to give,' dates back in Britain at least to Thomas Pennant's popular A *Tour in Scotland 1769*, published in 1772. Pennant (1726–98), an English naturalist, circulated questionnaires in order to gain as much local information as possible for his collections.[5]

Inventory in the tradition of Pennant's natural history questionnaire ranked among the first scientific tasks undertaken by inhabitants of Canada. In 1827 the newly founded Natural History Society of Montreal appointed an 'Indian Committee' to survey the Indian tribes, the physical geography, and the natural history of the interior of the British North American continent, with an eye to assessing its agricultural and commercial potential. The committee prepared 253 detailed questions for distribution even to the farthest northern and western outposts of the Hudson's Bay Company. Such a scientific inventory was necessary, the society had agreed, since even the partially settled provinces of Upper and Lower Canada remained a 'very extensive' yet 'almost unknown portion of the empire.' The society enlisted the co-operation of inhabitants of the interior in this inventory, 'for the benefit of science and our country.'[6] As will be seen, a similar questionnaire was issued to members of the Legislative Assembly, concerning local details of their constituencies, by William Logan when he began his geological survey of Canada in 1843.

Of crucial importance in justifying inventory science was the doctrine of utilitarianism. Victorian science in British North America both reflected and reinforced the criterion of practical value or usefulness. But beneath these strongly pragmatic overtones, the utilitarian element in Victorian thought masked subjective social, economic, and political interests of its proponents. With roots set deep in the British experience, utilitarianism lent a sense of purpose and meaning to the arduous task of settling British North America. It drew science into a value system which had emerged out of the industrial and agricultural changes of late eighteenth-century Britain. Utilitarianism encouraged the belief that even social problems were manageable through quantification and the statistical accumulation of facts. Science in the utilitarian sense was a tool, not merely to locate sources of material wealth but also to construct an ordered society. Victorians saw science emerge from a peripheral leisure-class activity to become the fundamental basis of industrial society.[7]

Science became the gauge by which Canadians assessed what their country and, through it, they themselves could one day become. Nor was this view limited to English Canadians. In 1846 Étienne Parent warned his fellow French Canadians that their collective national survival demanded pragmatic responses to the scientific and technological challenges of the modern age:

Nous sommes dans un monde ou tout se meut, s'agit, tourbillone. Nous serons usés, broyés, si nous ne remuons aussi ... Voilà, messieurs, l'image du laisser-aller et du mouvement industriel. Que cette révolution qui s'est operée de nos jours, sous nos yeux, ne soit pas perdue pour nous, et qu'elle nous apprenne que l'empire du monde moderne a été donné au mouvement, à l'activité, à l'action vive, constate de l'homme sur la matière.[8]

Science provided nineteenth-century colonists, both English Canadians and some French Canadians, with not only the practical means to dominate their physical surroundings but also an ideological framework within which to comprehend the experience of doing so. It enabled Canadians to make sense of the New World by translating from the experience of the Old and even, conversely, to clarify the familiar in the light of a fresh perspective. It refocused Canadians' vision of the land they inhabited, broadening their conceptual horizons and encouraging them to carve out a place for their developing society in the larger world. In its inventorial capacity, nineteenth-century science was a promissory note on which to draw self-knowledge and self-confidence. In general, this dual function of science was so important to the ideology of nineteenth-century nation-building that, in a Swedish historian's view, it remains surprising that no Nobel Prize for inventory science was ever instituted.[9] In particular, science gave credence and respectability to the very idea of a transcontinental Canadian nation, and to the conviction that, with science, the idea would become reality.

This conceptual relationship between science and nation-building found expression in the form of scientific metaphors applied to issues of the day. Canadians who debated the issue of Confederation in 1865 described a sense of being drawn together by a centripetal force that made the unification of British North America inevitable in the long run. Remarkable recent scientific and technological advances were widely held to be responsible for this situation because of their tendency to 'annihilate time' and 'devour space.' Members of the Legislative Assembly of the Province of Canada accordingly devised scientific and technological analogies to convey their ideas about Confederation. Sir

E.P. Taché invoked the Newtonian tradition when he imbued the possibility of American annexation with the power of a natural law: 'We are,' he declared, 'in our present position, small, isolated bodies, and it may probably be with us, as in the physical world, where a large body attracts to itself the smaller bodies within the sphere of its influence.'[10]

Mechanical metaphors were also used to conjure up negative images of past political failures: Confederation was felt to be necessary because the 'wheels' of the deadlocked government of the Province of Canada were 'stuck fast.' Canadians were also shocked to witness in the Civil War the collapse of the looser American federation, where 'instead of those institutions, framed with such mathematical precision, and that mechanism so finished and so regular in its course, there are to be seen but violent and jerking motions, overturnings, and the collision and smashing of the component parts of the disconnected machinery of state.'[11]

Visions of the future were more often expressed through organic rather than mechanical analogies, and sometimes formulated by naturalists themselves. In this view, exemplified in an unsigned book review published in the *Canadian Naturalist and Geologist* in 1858, the destiny of British North America appeared subject to the same natural laws which guided every living organism:

Physically considered, British America is a noble territory, grand in its natural features, rich in its varied resources. Politically, it is a loosely united aggregate of petty states, separated by barriers of race, local interest, distance, and insufficient means of communication. As naturalists, we hold its natural features as fixing its future destiny, and indicating its present interests, and regard its local subdivisions as arbitrary and artificial.[12]

Natural history, then, seemed to provide clues to the unfolding of Canada's political history and appeared to point to the natural growth of a larger British North American nation. The reviewer felt sure that 'nothing more enlarges men's minds than the belief that they form units, however small, in a great nationality. Nothing more dwarfs them than exclusive devotion to the interests of class, a coterie, or a limited nationality.' Thomas D'Arcy McGee pursued this arresting theme in 1863 when he observed that the very idea of a British North American nationality 'begets a whole progeny, kindred to itself, – such as ideas of extension, construction, permanence, grandeur, and historical renown.' It opened up what he discerned as 'long, gleaming perspectives, into both time and space.'[13]

Science also tempted British North Americans with an organizing principle to marshal their common assumptions about progress and development. The verdicts of science sharpened the arguments of politicians who supported Confederation by singling out the Province of Canada's practical need for Nova Scotia coal if it expected ever to acquire wealth and prosperity through industrialization: 'Perhaps we shall find coal in Canada. No, says Sir Wm. Logan, our learned geologist – impossible; science tells us that it does not exist. (Hear, hear.)'[14]

In an age of railroads and lightning telegraphs it seemed that increasing speed in travel and communication had 'a great deal to do in quickening the perceptions of mankind.'[15] These quickening perceptions were expressed nowhere more clearly than in the idea of a 'new nationality,' which, it was hoped, Confederation would create. The idea of creating a nationality out of several colonial peoples dispersed over a vast territory began to appear feasible only in the light of the scientific progress of the age, which suggests the influence and authority with which Victorian science informed Canadian thought by the 1860s.

These examples demonstrate the importance of historical context to any discussion of the idea of a nation as well as the futility of insisting on a universally applicable definition of nationhood. They also illustrate what the proponents of Confederation understood by the 'new nationality' they thought they were creating: one based upon common interests rather than common culture. As one member of the Legislative Assembly optimistically predicted, 'Give the people of Upper and Lower Canada a common objective to pursue, and common interests to sustain, and all questions of origin, and creed, and institutions and language will vanish in the superior end to be attained and by their closer union among ourselves, or by their wider union with other colonies under the proposed scheme of Confederation.'[16] George Étienne Cartier was more pointed when he declared, 'in ancient times the manner in which a nation grew up was different from that of the present day ... Nations were now formed by the agglomeration of communities having kindred interests and sympathies. Such was our case at the present moment.' For, Cartier continued, 'The idea of a unity of races was utopian – it was impossible. Distinctions of this kind would always exist. Dissimilarity, in fact, appeared to be the order of the physical world and of the moral world, as well as in the political world.'[17]

The vision held by D'Arcy McGee in 1863 might also be ascribed to the contributions of Victorian inventory science to the idea of a transcontinental Canadian nation: the territorial expansionism inherent in the practice, or 'extension,' of inventory science; the eventual development, or 'construction,' of an independent scientific tradition which grew from impatience with the slow and often irrelevant responses of experts in the British scientific metropolis; the use of science to bridge cultural and political divisions which threatened the 'permanence' of Canadian society and the future stability of the North Atlantic triangle; the sense of 'grandeur' enjoyed in public pride, confidence, and prestige resulting from Canadian scientific achievements; and finally the notion that the development of a transcontinental Canadian nation not only could be compatible with both its 'historical' British heritage and the influence of its North American environment but would also be natural and perhaps even inevitable.

The links between Victorian science and the growth of a transcontinental nation in Canada highlight a historical stage of organization, an attempt to muster both human and natural resources to deal with the perceived pre-eminent tasks of the Victorian age. As the Scientific Revolution had paved the way to the Industrial Revolution, inventory science laid a conceptual and practical foundation for the reorganization of British North America. The creation of the Dominion of Canada was a response to the challenges of modernization and industrialization. It reflected a determination to overcome perceived economic backwardness and to justify peaceful imperial expansion. In this light, it was no mere coincidence that the aggressive north-westward-looking Canadian expansionist movement of the mid-nineteenth century found nourishment in those same backwoods of central Canada which also inspired insecurity and retreat. By the same token, an intellectual bridge of common assumptions and aspirations seemed to supporters of science to tie the Atlantic colonies increasingly to central Canada during the early Victorian age.

The rise of science in early Victorian Canada can best be understood by looking beyond its material usefulness. The tasks of identification, inventory, and mapmaking gave form to the idea of a transcontinental national existence; they imparted to Canadians a sense of direction, stability, and certainty for the future. A corollary of Victorian science as it was practised in British North America was the spirit of possession which grew from the inventory process; nationalism provided a vocabulary for justifying and ennobling that possessive spirit. 'Cultivators' of Victorian science, growing numbers of practitioners and supporters in all the British North American colonies, drew verbal, visual, and symbolic representations from the natural and physical world in which they lived. But this transfer of ideas worked both ways, for they also contributed an intellectual mode in which to 'cultivate' that same world, in helping to invent the idea of Canada as we know it today.

PART I
GEOLOGY

1

Exposing the Strata

There is a kind of freemasonry in the study or pursuit of Natural History.

Edward P. Thompson, *The Note-Book of a Naturalist* (1845)

Of all the sciences emerging from the older traditions of natural history and natural philosophy, geology held out the most obvious possibilities for Canada. As early as the sixteenth century, the first European explorers had noted the climatic extremes and rocky terrain characteristic of the northern interior of the New World; the agricultural potential of New France seemed far less promising than the likelihood of unearthing valuable minerals. Discoveries dating from the days of Intendant Jean Talon included coal at Cape Breton, lead in the Gaspé, iron in the St Maurice Valley, and copper at Lake Superior. These discoveries in turn fuelled recurring rumours of an Eldorado on the shores of the upper Great Lakes, especially during bleak economic times. But there was no demand for a geological survey of Canada before 1815. Only after the War of 1812 in America and the Napoleonic Wars in Europe did a series of political, economic, and social changes nurture such a demand. Combined with new theories of the earth and intensified applications of science in exploration, these changes informed perceptions of what Canada was and, more important, what it could become.

The idea of a geological survey, a systematic inventory of the earth's structure and mineral resources, had gained currency by the late eighteenth century in the work of the Saxon mining inspector Abraham Gottlob Werner (1750–1817), the Scottish theorist James Hutton (1726–97), and the English stratigrapher William ('Strata') Smith (1769–1839). By the 1820s several American states, including the Carolinas, Massachusetts, and Tennessee, were organizing such surveys, and other states following suit. The Geological Survey of Great Britain was launched under the direction of Henry De la Beche (later Sir Henry) in 1835.

In British North America these examples, and the proximity of the rich coalfields of Nova Scotia, inspired the New Brunswick and Newfoundland governments to initiate colonial geological reconnaissances, the first by Abraham Gesner in 1838 and the second by Joseph Beete Jukes in 1839. Farther inland, the need for salt and other minerals in Loyalist settlements west of Montreal had motivated exploratory expeditions during the 1790s. But inventory science was most successfully institutionalized in the public funding of a geological survey of Canada in 1841 and the appointment of William Edmond Logan as its director some months later, in 1842.[1]

Interest in geology spread to British North America with the influx of British immigrants who had been infected by a popular revival of the amateur naturalist tradition in the mother country. An interest in nature and a knowledge of natural history formed an important component of English culture, especially among the middle and upper classes. The strength of the amateur naturalist tradition was epitomized by Gilbert White's *The Natural History of Selborne* (1788-9), which alloyed familiar parochial scenes with rational impersonal observation to produce one of the most widely read and most frequently reprinted works in the English language. Interest in geology grew from the natural history tradition in the early Victorian outlook, promoting discipline and useful activity as well as gratifying a growing mania for collecting specimens. Moreover, geological knowledge seemed to reflect the popular tenets of William Paley's *Natural Theology* (1802). Paley stood fully in the orthodox religious tradition, encouraging the collection of specimens as evidence of God's design in nature. Geology confirmed religious faith and attracted ministers like William Buckland (1784-1856) and Adam Sedgwick (1785-1873), prominent geologists in their own right. But by the same token it attained a certain degree of elegance and prestige by attracting wealthy and influential supporters like Roderick Murchison (1792-1871) and Charles Lyell (1797-1875). It fit a growing tendency to define science in terms of an entrepreneurial ideology, a utilitarian outlook suggested to Victorians by the seventeenth-century anti-Scholasticism of Francis Bacon. Just as Bacon's writings took on a new relevance with the spread of industrialization in Britain, immigrants eager to apply new methods and ideas to their pristine chosen land anticipated geology's promise of economic returns in exchange for following the scientific method.[2]

A surprising number of immigrants to Canada in the early nineteenth century had had firsthand contact with exciting theoretical controversies in British geology, particularly in Scotland. Scottish geology formed part of a broader natural history and contrasted with the independent local observations empha-

15 Exposing the Strata

sized in English geology. It was included in the popular education of surgeon-naturalists and explorers who served on British government expeditions in British North America. Moreover, settlers too arrived preconditioned to interpret the rocky Canadian landscape in ways determined by current issues in geology. A theoretical revolution in British geology during the late eighteenth and early nineteenth centuries centred on a debate between followers of Abraham Gottlob Werner and those of James Hutton over the aqueous or igneous origins of the earth's geological structures. This conflict culminated in the 1830s with a new theoretical synthesis concerning the history of the earth.

The Wernerian school, or 'Neptunists', dominated the Scottish geology taught at the University of Edinburgh, so it closely paralleled the thought of the Scottish Enlightenment. Werner theorized that all the rocks of the earth's crust were precipitated out of water into distinct layers. His views encouraged the development of new techniques in mineralogy that led to the emergence of geology as an earth science. Wernerians classified rocks on the basis of their location in the ordered strata of the earth's crust, a useful concept for finding minerals and for identifying those located *in situ*, where they had been formed. These exact criteria inspired confidence in the order of the strata over wide expanses of land and challenged those trained in Wernerian methods to hone their skills among the complex strata of the Scottish terrain.[3] Such training proved especially relevant in Canada, where early inland explorations tended to penetrate the continent along the St Lawrence waterway, a setting ideal for confirming Wernerian assumptions about the agency of water in the origin of geological formations.

Popular open-air lectures by Wernerians such as Robert Jameson and Thomas Charles Hope at Edinburgh gave further impetus to the dissemination of Scottish geological knowledge as useful knowledge. This overriding concern with utility resulted from a perceived need to modernize the backward Scottish economy. Both Scotland's shift to a coal-based economy and the proximity of iron ore deposits were seen as instrumental in the country's recent rise to industrial and intellectual leadership.[4] The desire to overcome a relative state of economic backwardness was easily transposed from the Scottish to the Canadian milieu, where economic backwardness imperilled an already precarious political existence throughout the nineteenth century. It became the major premise in a powerful argument for economic development that not even the French-Canadian élite, by the 1840s, would deny outright.

The conceptual development of British geology was accompanied by an institutional one. The Geological Society of London, founded in 1807, provided a major focus and a stimulus for active research in mineralogy and stratigraphy. By the 1820s, the society had organized a palaeontological section to encourage the collection of fossils on as wide a geographical basis as possible. In so doing, the Geological Society was recognized as an active scientific arm of the expanding British Empire.[5]

The intellectual, political, economic, and social roots of the Geological Survey of Canada involved a network, or 'freemasonry,' of like-minded supporters in the colonies who responded to these developments in Britain. The task of digging out the roots of this fellowship requires a view of exploration not as a series of isolated events but as integral to the process of colonization. In general, demand for exploration depended upon perceptions of the availability of land to explore, of geographical limits to economic growth, and of the image of the unexplored areas.[6] Once interest in a geological survey did become constant, it still surfaced as a public issue only sporadically. It appeared to rise recurrently, phoenixlike, from its own ashes. Traces of the ideas of those who advocated a geological survey of Canada seem scattered through time and space like fossils in the geological record. But as with those fossils, the historical context in which those ideas lie stratified is the key to understanding the origins of some of our own ideas.

I

Among the first to recognize Canada as a field for geological study were the Royal Engineers, whose ranks swelled after 1815 as part of the Duke of Wellington's British army reforms. The royal corps under the Ordnance Department – the artillery, the sappers and miners, and the engineers – ranked among Britain's leading engineers and scientists. Ordnance soldiers were assigned to safeguard Canada's frontiers and incidentally to aid civilian development. The Royal Engineers' surveys for canals and roads offered unique opportunities to observe the geology of the country. Two of the officers, Richard Henry Bonnycastle and Frederick Henry Baddeley, were ardent and experienced naturalists who included geological observations in their regular reports. Occasional geological descriptions also appeared in the reports of Royal Engineers assigned to survey a strategic navigable route along the Ottawa River system to Lake Huron. Their failure in this primary task provoked a rap on the knuckles from Wellington himself, who complained that 'they go wandering over the Country they don't know where, and report upon anything except what they were sent to examine and report upon.' Later Royal Engineers thought it best not to irritate the Iron Duke's sensibilities any further and so curtailed their written observations. The subsequent failure of Colonel John By's engineering party to note any geological observations along the Rideau Canal route was regretted by those who valued such information.[7]

17 Exposing the Strata

The first regional geological surveys in Canada were made in the Ottawa and Rideau valleys. Strategically important for the settlement of discharged military officers, this area received immigrants who had benefited from a Scottish education. Military settlements grew up near the exposed strata of the Rideau Canal excavations, where interesting geological specimens abounded. One Scottish settlement was Perth, long the military, judicial, political, and social centre for most of the Ottawa Valley. One of the Scottish immigrants to Perth in 1821 was James Wilson, a surgeon trained at Edinburgh and a dedicated amateur geologist. Over the years, Wilson accumulated so complete a reference collection of local specimens that William Logan consulted Wilson before beginning his more extensive geological survey of Canada.[8]

The material richness of Wilson's legacy was supplemented by other contributions from the Ottawa-Rideau region. Another prominent settler, A. J. Christie (1787–1843), founded and edited the *Bytown Gazette* in 1836. Christie, a self-styled surgeon allegedly educated at Edinburgh, often reminded his readers of the immense sources of mineral wealth and great field for enterprise in Canada and acquainted them with current geological terminology. Christie was not himself a naturalist, but another prominent surgeon at Bytown, Edward Van Cortlandt (1805–75), pursued an active interest in geology and mineralogy after his arrival there from Quebec in 1832 and helped to organize several successive natural history societies at Bytown. It was not surprising therefore that William Logan noted years later that the residents of the Ottawa and Rideau valleys seemed better prepared than most to aid him in his geological survey of the province.[9]

By the 1820s Wernerian investigations of the geology of Canada moved farther west. One man who attracted public attention was John Jeremiah Bigsby (1792–1881), an assistant staff surgeon in the medical corps of the British army when he arrived in British North America in 1818. Bigsby had studied medicine at Edinburgh, where part of his training in natural history included Wernerian geology. In 1819 he was instructed by the army's Medical Department to explore the geology of Upper Canada, where he collected specimens as far north as Lake Nipissing. In 1820 he became medical officer to the International Boundary Commission, and for three years he geologized along the Great Lakes, substantiating Neptunist theories about the sedimentary origins of Canada's geological formations.[10]

Bigsby's major contribution was his published writings on Canadian geology. In 1823 he presented his paper 'Notes on the Geography and Geology of Lake Huron' to the Geological Society of London. He is also credited with the first scholarly work on geology published in Canada, the essay 'On the Utility and Design of the Science of Geology,' with hints on the best method of acquiring a knowledge of the subject. Bigsby focused his general remarks with examples and even sketches of the geology of Canada. His purpose was to awaken interest in the benefits of a subject as interesting, instructive, and profitable as it was neglected. In Bigsby's view, geology was 'not merely a recreation for the inquisitive; it exercise[d] a prodigious and immediate influence on the civilization and prosperity of a people.' A recurring theme was that financial assistance from the state was needed to support geologists who intended to investigate more fully the 'distant and savage districts' Bigsby had visited in Canada.[11]

Bigsby's writings on the Huron region reverberated in the settled parts of Canada, sparking particular attention in the commercial centre of Montreal. In a detailed review, David Chisholme, the editor of the *Canadian Review and Magazine* in 1826, wrote – erroneously – that Bigsby's paper foreshadowed a forthcoming 'magnificent work' on the geology and mineralogy of all the British provinces in North America. Although the 'Notes' were only local and partial in their scope, Chisholme considered them 'highly interesting and important.' He shared Bigsby's fascination with the shores of Lake Huron and abhorred 'our gross ignorance of the mineralogical and geological treasures of so vast an extent of country as Canada.' Chisholme's rhetoric embellished Bigsby's achievements, and he hoped to inspire Canadians to action. 'Like the settler,' Chisholme predicted, this work 'will open a path to extensive regions of the country which have formerly been known, but as the dreary and lonesome haunts of wild beasts and savage men; and reconcile us to a country, which, though at first waste and barbarous, must in a few years surrender its treasures, and its fertility to the irresistible power of civilization.' In 'exploring the hidden riches of nature,' Bigsby was 'the true benefactor of his race, and the guide to all those proud and eminent perfections which the mind of man has attained in the arts and sciences, as well as in moral and political happiness.'[12]

Chisholme's review continued with a premature articulation of the need for a government-supported geological survey of Canada. He believed that while Bigsby's individual undertaking was beneficial to society, such advantages would be far surpassed if the state were to underwrite a wider geological survey of the country's inland regions. Finally, he lamented with Bigsby that the science of geology was still 'in its infancy in this country' and, stimulating public interest in collecting minerals, invited readers to submit descriptions of unidentified specimens to the offices of the *Canadian Review*, to be classified by 'experts.'[13]

Throughout his career David Chisholme (1796–1842) fuelled interest in Bigsby's themes as he himself grew in stature and influence. A Scot educated in law who had emigrated to Montreal in 1822, he edited the *Canadian Magazine and Literary Repository* (1823–5) through its various metamorphoses as the *Canadian*

19 Exposing the Strata

Review and Literary and Historical Journal (1824–6) and the *Canadian Review and Magazine* shortly before its demise in 1826. Chisholme was to exert a real influence on the history of the Geological Survey of Canada again in 1836, when he began to promulgate his statist ideas as editor of the Montreal *Gazette*. He believed that, by offering new fields for exploration and discovery, colonies advanced science as much as science advanced colonies. This belief formed part of a larger political ideal that colonies should be recognized as an 'integral part' of the mother country in every way: 'Except what nature and climate may conspire to constitute, let there be no political or moral distinction betwixt the colonist and the natural born subject.' When the British government approved the charter of the Canada Company in 1826 to open up over a million acres of the land bordering on Lake Huron, Chisholme was delighted at the prospect of his dreams becoming reality.[14] Indeed, it was on the Canada Company's Huron Tract that the idea of a geological survey of Canada now found fertile soil, in the Wernerian geology implicit in Scottish culture.

II

The Canada Company (also known as the Canada Land Company) was organized at the peak of a financial bubble in Britain that lasted from 1823 to 1826 and saw the formation of many joint-stock companies to invest in mining and colonization schemes in both North and South America. The more than one millions acres it acquired east of Lake Huron was part of an even larger area purchased by the British government from the Chippewa (Ojibway) nation in 1818; it was known thereafter as the Huron Tract. In return, the company was to make annual payments to the government of Upper Canada. The company's court of directors hoped to realize great financial gains in a land and colonization venture by diverting immigrants arriving on the eastern seaboard to the Huron Tract. The court's sixteen directors, all London businessmen, included at least three with strong connections in British North America: Simon McGillivray and Edward Ellice were involved in the fur trade, and Hart Logan was a 'merchant prince' of one of the principal wholesale firms in Montreal. His nephew, James Logan, was appointed an agent to the company.[15] James was the brother of William Logan, also employed by Hart Logan at the time, in London. The court of directors of the Canada Company functioned as its major decision-making body.

Representing the company in the Huron Tract was its superintendent, John Galt, who went to Canada in October 1826. Galt was to act as the eyes and ears of the company's court of directors, but time and distance permitted him considerable leeway to prepare the tract for settlement as he saw fit. Galt was to be assisted by a 'Warden of the Woods and Forests' appointed by the company in September 1826. William Dunlop was to be a kind of roving caretaker in the Huron Tract, assessing the quality and condition of the land with an eye for its timber, soil, waterways, and whatever else might affect its value for settlement. Dunlop was ordered to keep a daily record of 'every material fact likely to be serviceable to the interests of the Company' and to report these periodically to Galt.[16]

William ('Tiger') Dunlop (1792–1848) had lived in Canada before and knew exactly what he was getting into. A Scot educated in medicine at the University of Glasgow, Dunlop had served as an assistant surgeon in the War of 1812. After the war's end in 1814, he stayed on to help survey a military road from Lake Simcoe to Penetanguishene on Georgian Bay. Dunlop, who described himself as a 'strong broad-shouldered, carroty-haired, slovenly, coarse-looking Scotchman,' was known as 'a character singularly eccentric and almost bizarre,' remarkable for his 'inexhaustible sense of fun' and adventure. He had made a name for himself by embarking on a voyage to India and shooting tigers there while helping to survey a road, this time through the jungle. Dunlop also sojourned for a short time in Edinburgh, where he lectured on medical jurisprudence at the university and wrote for *Blackwood's Edinburgh Magazine*. During the investment mania of 1825, Dunlop worked for various firms with interests in brick, iron, and salt; he also superintended the Cheshire salt-works. His disdain for mediocrity and his own declaration that 'nothing pleases me so much as mad projects' encouraged a general tendency to emphasize this rather mercurial side of his character.[17] But Dunlop's Scottish education, his Edinburgh connections, his experience in surveying rough terrain, and his interest in mineral resources combined to reveal through his activities in the Huron Tract his much more sober aspect.

Dunlop spent several seasons in 1827 and 1828 surveying the hundreds of square miles of forest over which the Canada Company's lots were scattered. He submitted two reports of these researches to Galt before 1829. These reports recorded not simply what Dunlop happened to observe but what his education and experience had trained him to look for. His Wernerian background in geology led Dunlop to conclude:

The whole of the Tract is what in geology would be considered a recent formation, and could be supposed to have taken its origin from the subsiding of an enormous mass of waters which at no very distant period must have covered this part of the continent.

It is all more or less interspersed with petrifactions most of which are of a marine or at least aquatic character.

... All these strata lie perfectly parallel to the plane of the Horizon demonstrating that

21 Exposing the Strata

they must have been deposited by water; and so far as I have observed I have never seen a single instance of a perceptible dip in the strata, along the whole course of Lake Huron or any of the Rivers which empty themselves into it.[18]

Dunlop's Wernerian frame of reference hindered an entirely accurate analysis of large masses of granite, an unstratified igneous rock, in many parts of the tract. He confessed his dissatisfaction at being unable 'to contrive even a feasible conjecture' to explain this apparent anomaly.

But moreover, Warden Dunlop concerned himself with the mineral resources of the tract, even though the rights to gold and silver mines were reserved by the Crown. He felt the loss of a splendid opportunity to geologize more deeply when Galt denied him permission to join Col John By's engineering party, whom he encountered by chance during his rambles. As a result, Dunlop drew back from firm conclusions on the presence of workable minerals, since he had been taught that 'anything like a thorough knowledge of the bowels of the Earth, must be attained by boring on purpose or watching excavations made for other purposes.' What he did notice closely paralleled his previous work in the joint-stock companies: building materials, such as clay for bricks, and deer-licks and salt-springs 'too weak to hold out any rational hope of remuneration from working them' and 'almost uniformly combined with Iron, Sulphur, or carbonated hydrogen gas or perhaps with all three.' In 1828 he reported the discovery of iron but noted he had had insufficient time to ascertain its quantity.[19]

Most important, Dunlop confided to Galt a long-standing plan to direct a mineralogical survey of the whole Huron Tract. Dunlop described the project as 'too obviously necessary to require comment' and blamed 'more urgent business' for pre-empting its realization earlier. He asked Galt to request the court of directors 'when they do send out medical men to let them be good naturalists and mineralogists' and to make it a part of their agreement that they assist him in such enquiries. For by this means alone, he concluded, the company would be 'amply compensated for any expense such Gentlemen may cause them.'[20] Whatever his superiors may have thought of these suggestions in the long run, they held to their primary concern of settling the tract as more immediately remunerative. Galt was soon dismissed from the company's services, and Dunlop was recalled as warden. But Dunlop was retained on a fixed salary, and he soon recognized new opportunities to present his plan for a mineralogical survey to wider and more attentive audiences at York, the capital of Upper Canada.

There Dunlop helped to found a Mechanics' Institution in 1830. After several false starts the institution attracted members by presenting a series of lectures on 'Scientific and Mechanical' subjects. Dunlop delivered one of the first of these lectures, on the typical Scottish theme of education as the 'Grand Source of National Greatness.' 'What,' he asked, 'constitutes the power, the wealth, the greatness, the happiness of nations?' It was not the soil, the climate, or the population in themselves, for they formed but 'the inert body of national greatness.' Rather, it was knowledge, for 'the Nation that possesses the greatest quantity of knowledge, from that very fact, is the most powerful.' However, Dunlop's call for popular education and public inquiry drew resistance from the more élitist John Strachan and the Family Compact.[21]

A second opportunity appeared, more closely related to Dunlop's interest in geology. In July 1831 he joined two colleagues to found the Literary and Philosophical Society of Upper Canada, also known as the York Literary and Philosophical Society. The purpose of the society was to promote a scientific inventory of 'the natural and civil history' of the whole interior of British North America 'as far as the Pacific and Polar Seas.' Such an investigation of the animal, vegetable, and mineral kingdoms would also promote the further study of 'Natural History, Natural Philosophy, and the Fine Arts.' The organizers confessed themselves 'anxious to procure every existing and probable record of the Aborigines and their language, Minerals, Fossils, Animals, Plants, Geological and Topographical Sketches,' interests which combined those of the three founders. Besides Dunlop, they were Dr William Rees, who specified an interest in medical botany, and Charles Fothergill, an amateur naturalist attempting to garner support for a scientific expedition into the British North American interior. The York society suffered organizational difficulties, but eventually found legitimacy by electing Bishop John Strachan as its first president. Its short life was not over before Dunlop used the society to lobby for government support of a geological survey of the province. Dunlop insisted that not only material benefits would result, for in addition Canada's newness offered science a unique opportunity to observe the work of nature in progress.[22]

On 10 December 1832 a petition of the society was presented in the House of Assembly, requesting public funds to appoint 'persons duly qualified' to investigate, 'thoroughly and scientifically' the geology, mineralogy, and general natural history of the province. Such persons would 'procure and report every kind of information tending to promote science, and an acquaintance with the characteristics of the country, such as the more prominent features of land and water, and the capabilities of communication between the different parts.' Dunlop's active interest in such a survey and his continued promotion of geology within the York society suggest that he had a firm hand in the project. However, the petition never emerged from the Committee on Supply, where funding would have to be approved.[23] While Dunlop regrouped his forces, the idea of a geological survey gained ground in quite another quarter.

23 Exposing the Strata

III

On 18 January 1832 Lieutenant-Governor John Colborne, the patron of the York Literary and Philosophical Society, recommended to the favourable consideration of the House of Assembly a memorial from Dr John Rae. Rae (1796-1872) was born near Aberdeen and educated in medicine at the University of Edinburgh. A keen student of natural science and comparative studies, Rae had spent some time travelling before emigrating to Canada in 1822. He taught school at Williamstown, Upper Canada, while in his spare time investigating the colony's geological and economic prospects. His memorial requested financial aid to enable him to continue these investigations. Erroneous ideas about the colony prevailed, Rae argued, because it was 'as yet imperfectly known.' Rae was interested in describing 'the leading features in the Geological Structure,' along with soils, mineral productions, land-forms, climate, and diseases, from north-east of Lake Huron to the rear of the Midland District. He was, he said, 'led to believe that the mineralogical details of this region' were 'very interesting' and that there was 'a probability of valuable minerals being there to be found.' Still another tract, around Nottawasaga Bay and Lake Huron north of the Canada Company's Huron Tract, interested him 'for similar reasons.' Finally, Rae urged that his efforts would help to 'develope the resources of the country, and still further to direct public attention towards it.'[24]

The next day Rae's petition was referred to the Committee on Supply, too late in the session to gain full consideration, since the completed bill of supply was read on 20 January. Rae's memorial may never have been discussed in the committee of the whole House, but its existence and Colborne's support indicated a growing public interest in the geology and mineralogy of the regions he named. At the same time, several tenuous threads began to intertwine in the political sphere.

One of these threads ran through the Canada Company. When John Galt lost the confidence of the company's court of directors in 1829, he was replaced by two commissioners, William Allan and Thomas Mercer Jones. In 1833 Jones commenced a series of letters to Colborne on behalf of the Canada Company, but ostensibly on his own initiative, advancing a claim for lands north of the Huron Tract. The basis of the claim stemmed from the company's charter of 1826, which promised compensation for lands unfit for cultivation. In 1827 Tiger Dunlop had described the summit level of the tract as a massive swamp, and the court of directors was now interested in determining the exact extent of this 'Great Swamp,' which no one had ever seen in its entirety. The company held that despite previous compensation there were still deficiencies in the Huron Tract. It is not clear why Jones chose to revive these claims when he did, but he was well aware both of Rae's convictions about the value of mineral resources north of the tract and of Colborne's agreement in these convictions. Colborne adamantly refused to concede any more of these lands to the company.[25]

Interest stimulated by those who hoped to gain from lands north of the Huron Tract had spread to the government itself. It was for this reason that Colborne was 'very averse' to granting more land to the company. In his Speech from the Throne on 19 November 1833, Colborne spelled out the widely held belief that

the information we possess of the Statistical Changes rapidly taking place, and of the energy displayed by the Inhabitants of several Districts in the improvement of their interior water communications, enables us to judge of the extensive resources of the Colony, and we concur ... that, *were they sufficiently known and appreciated*, the Parent State would be encouraged to regard this fertile country as an asylum for a larger portion of her present redundant population.[26]

Colborne's statement was widely taken to mean, as the editor of the Toronto *Patriot and Farmer's Monitor* put it, that the condition of Upper Canada was like that of 'a young heir, jumping into a new estate.' Perhaps at this juncture, he added, it would be appropriate to 'employ an exploring party to study and make known the Geology of our country; the expense would be but trifling, and the result might be most important ... Surely,' the editor hoped, 'some spirited Member will move this in the House.' As far as he was concerned, public support for a geological survey had been a partisan issue for too long. What was now called for was not a 'petty smallbeer economy,' but rather a search into 'the multifarious resources of our Country' which would 'effectually dispel the gloomy fears of national bankruptcy' and inspire a new outlook, 'liberal and calculated to reinforce us physically and exalt us morally.'[27]

On the initiative of Col Mahlon Burwell, an address to Colborne was accordingly passed by the House of Assembly on 3 and 4 January 1834. Burwell, the member for London, was a provincial surveyor with extensive practical experience in the western part of Upper Canada who had accompanied Tiger Dunlop in his survey of the Huron Tract. The address requested 'an exploring party to penetrate from given points on the north shore of Lake Huron, in continuous right lines, some 50 or 60 miles into the heart of the country.' It anticipated that 'if a practical Surveyor were sent out to produce the lines, and take field notes of the soil, timber, waters, &c., and a gentleman of science were to accompany him, and report upon the geology and mineralogy of the interior, as well as on the borders of the Lake, the result could not fail to be highly beneficial to the interests of this Province, as well as those of the Empire at large.' Therefore, it reasoned, the expenses of such a survey should be paid out of the territorial revenue. Not surprisingly, Colborne responded within two days that he would forward the address to Britain and had no doubt that it would receive attention.[28]

The editor of the Toronto *Patriot and Farmer's Monitor* had hardly been clairvoyant. Thomas Dalton (1781 or 1782–1840) had emigrated in about 1804 from Birmingham to Newfoundland; about 1816 he moved to Kingston, where he established both a brewery and a newspaper. In 1832 he sold the brewery to John Molson and moved the newspaper to York. Well known as one of the first colonial writers to urge a federation of the British North American provinces, Dalton linked this issue to the development of the coal trade. He took great pride in an editorial he wrote in 1832, highlighting the arrival in Canada of a cargo of Cape Breton coal for John Molson's steamship company in Quebec. It was, Dalton believed, the first Canadian instance of coal being imported from Nova Scotia. He outlined the advantages of coal as a fuel in many enterprises, including brewing. A symbol of progress, the trade in coal, he pointed out, 'approaches toward us.' Moreover, he wrote, connections between Nova Scotia and Canada 'of so reciprocal and extensive a benefit' represented a 'desirable bond of Union.' A vehement anti-republican, Dalton predicted in 1834 that internal instability would soon bring about the collapse of the United States, probably within five years. This, he urged, could pave the way for several states (Maine, Ohio, Michigan, Indiana, and Illinois) to join the British North American territories in 'an empire of impregnable strength and inexhaustible wealth and resources.' It was, he insisted, 'all pointed out by nature,' and would 'ensue in the process of time.'[29]

The initiative taken by Mahlon Burwell and his supporters reaped immediate results. In reply to Colborne's request for an expedition north of the Huron Tract, the Colonial Office deemed it a more 'natural course of enquiry' to commence with districts contiguous to settled lands. In order to save the project, Colborne revealed his reasons for the proposed survey of the lands north of Matchadash Bay and between the Ottawa River, Lake Nipissing, and Sault Ste Marie, still without mentioning the Canada Company by name: 'Should the projected survey be sanctioned,' he proposed, 'the result of the labors of the exploring party will prove highly interesting to His Majesty's Government, and afford much detailed information to the local authorities as will enable them to judge how far it may be expedient to limit in future, the sales of the Waste Lands of the Crown.' The dispatch made no direct mention of geology, but once the survey was authorized Colborne insisted that a geologist be appointed as one of its officers. The man in charge of assessing the land's fitness for settlement, Lt John Carthew of the Royal Navy, was also ordered to report on the geological and mineralogical structures encountered by the party.[30]

Lt Frederick H. Baddeley of the Royal Engineers carried out the geological survey as part of the general assessment of the lands to be explored. Baddeley (1794–1879) was well versed in geological surveying. Educated at the Royal Military Academy at Woolwich, he had served in Europe during the Napoleonic Wars and was posted at Quebec in 1821. One of the original members of the Literary and Historical Society of Quebec founded in 1824, he served as its president in 1829. Considered by R.H. Bonnycastle to be 'the most active and best geologist then in Canada,' Baddeley had published a number of papers reporting his observations on the geology of various regions in Canada.[31]

The survey was conducted between 15 July and 8 November 1835, and Baddeley found 'the whole belt of shore on Lake Huron, a mass of granite rock.' Many of the districts seemed unfit for settlement. The formal reports of the exploring party were not submitted to the lieutenant-governor until March 1836 and they were published months later, by which time Colborne had been replaced by Sir Francis Bond Head. It was generally hoped that Carthew and Baddeley had opened the way for deeper investigations into 'what will prove to be the richest mineral country in the province[,] perhaps on this continent,' and there were some who urged, in addition, the construction of water and rail transportation systems to make these resources accessible. Thomas Roy, a civil engineer from Toronto, believed that 'the whole, when brought into operation, would firmly unite the British North American possessions, and tend more than any other measure which could be adopted, to nurse up a great, a rich, and a prosperous people.'[32] But the path towards a geological survey of the province was still not cleared of pitfalls, as developments conspired to alter the light in which Baddeley's geological reports were received.

IV

It was becoming clear by 1836 that the issue of a geological survey was not really a partisan issue in the House of Assembly of Upper Canada; it tended to cross Reform and Constitutionalist (Tory) lines. On 3 February William Lyon Mackenzie gave notice in the Reform-dominated House that he would move for a select committee to plan for a provincial geological survey. On 17 February the House responded by naming four members, including three radical Reformers: David Gibson and John Macintosh (York) and Charles Duncombe (Oxford). In reply Tory newspapers accused the Reformers of making political hay out of issues they knew nothing about. Disparaging the suggestion by the *Albion of Upper Canada* that 'there were not ten men in Upper Canada, who have made Geology their study' and that a more sound preliminary step would be to spend the money on a library and a museum, the Hamilton *Gazette* insisted that there were 'hundreds of *sound* Geologians and mineralogists in Upper Canada.' Mackenzie reiterated that he was interested only in knowing 'how far it was practicable to institute a geological survey of Upper Canada, in order more fully to ascertain its vast internal resources.' In fact, he pointed out, the chairman of the select committee was no Reformer and was named 'because of his extensive knowledge of the subject.'[33]

The chairman of the committee to 'consider and report a plan for the geological survey of the province' was Robert Graham Dunlop (1789-1841), member for the newly created county of Huron and the brother of Tiger Dunlop. A captain in the Royal Navy who had fought in Europe during the Napoleonic Wars, Dunlop had commanded the ship that took his brother to India in 1817. Before his arrival in Canada in 1833, Robert Dunlop had studied science, including mineralogy and geology, and had spent three years at the universities of Glasgow and Edinburgh. Weakened physically by his past exploits, he was convinced by his brother's enthusiastic accounts to 'retire' to Canada. And much like his madcap brother, Robert Dunlop was remembered by those who knew him as 'always full of schemes.' In 1835 he had entered Canadian politics, representing the 'Colborne Clique' in Huron County. His political leanings have been termed 'anti-Compact Constitutionalist,' emphasizing his concern with the practical needs of Upper Canada. Dunlop shared with his brother and others the view that 'money was not the wealth of the country, that the wealth of a country consisted in labor, in agriculture and mineralogical productions, in her forests and fishing.' Only internal improvements could unlock these treasures, and it was the duty of the legislature to encourage their development.[34]

On 1 March 1836 Dunlop's committee submitted its report, which was ordered printed and sent to the Committee on Supply. The select committee put forward 'some of the strongest reasons' in favour of a geological survey. First, since manufacturing was out of the question for such a new country, one had to turn to agriculture and mining. Land in the province was abundant, and productive mines might well be located. These were worth seeking because they entailed further advantages: agriculture and mining could be mutually beneficial, since 'by calling the mineral riches of our country to the aid of her agricultural resources, you at once give an impulse to her commercial prosperity, increase the means of internal improvement, and greatly extend those for the moral and religious instruction of the people.'

Second, there was every indication of the existence of coal in several districts of Upper Canada. Not only would coal be 'of incalculable value' for processing 'our various minerals' and generally for fuel, it was also a symbol of the 'power and conveniences' enjoyed by older countries. Third, salt was known to abound in Canada. If abundant sources could be located, they would be 'of the utmost importance, more especially to those great sources of our national wealth and prosperity, the fisheries.' The final reason reiterated the national importance of scientific inventories: 'As statistics consist in a knowledge of the means and resources of a country, a professional man might as well be supposed to pursue a vocation without his instruments, a merchant to carry on business without his books, as a nation or community to govern itself without statistics.'

In short, the preliminary report beamed with Dunlop enthusiasm. The committee strongly recommended a geological survey of the province, with a report of its natural wealth and resources to be tabled at the next session of Parliament.[35] The report was not implemented, however, for in the meantime the relationship between the Reform House of Assembly and Lieutenant-Governor Francis Bond Head had soured. In April 1836 the House refused to vote supply, and in May Parliament was dissolved. Like the phoenix, though, the geological survey promised to rise again from its ashes.

V

Kingston, at the nether end of the Ottawa-Rideau system, was another centre of interest in a geological survey. In October 1836 the Kingston *Chronicle and Gazette* began a lengthy series of 'Hints' on a geological survey of the province. The author, 'A.B.,' was well acquainted with both geological theory and the geological structure of the province. More important, he was well versed in the formal duties of a geological surveyor in mapping and in composing field reports. In his view, the United States was well worth emulating in its generous support of geological surveys. Since the assembly was about to reconvene, he stressed that it was 'high time' that members be directed to consider a systematic investigation into the mineral resources of Upper Canada. A proper survey could not 'be effected in one session,' he cautioned, but would have to 'embrace many.'[36]

Internal and external evidence suggest that A.B. was Capt. F.H. Baddeley of the Royal Engineers. A.B. was familiar not only with the geological contributions of Bigsby, Bayfield of the Royal Navy, and Bonnycastle of the Royal Engineers but also with details of the Carthew expedition of 1835. He apologized that the geological surveyor of that group 'had to follow rapidly, with little deviation, the track of his party, and no opportunity of selecting his ground of observation was afforded.' 'Chance alone' therefore determined whether 'anything of value geologically or mineralogically interesting was met with.' For A.B. this complication suggested a conflict of interests in the mandate of such exploring expeditions: for while on the one hand the geologist sought out bare rocks on which to base his judgments, on the other hand the rest of the party sought out deep and rich soils in order to measure the potential for settlement.[37] These admissions shed a powerful light on developments about to unfold in the newly elected House of Assembly.

On 9 November 1836 Robert Dunlop gave notice that he would move to introduce a bill for a provincial geological survey. Changing his tactic only two days later, he decided instead to move the question into the committee of the whole House. When the committee convened on 17 November, Mahlon Burwell recommended an address to the Lieutenant-Governor to appropriate funds for the survey. The address failed to elicit a positive response from Bond Head, but Dunlop persisted. On 14 December the House requested that the British government place at the disposal of the lieutenant-governor enough Crown lands to finance 'a correct Geological Survey of the Province.'[38]

Meanwhile A.B. strove to keep the public abreast of these developments. Although he doubted that bituminous coal would be found in Upper Canada, he emphasized that the only avenue to such crucial answers was science. Mineralogy and modern geology, he explained, derived 'a vast collection of facts' from 'close and laborious examination' and arranged by 'men of the highest authority.'

Without this knowledge, the mere practical miner must constantly be liable to fail in discriminating minerals, and in recognizing the rocks they are likely to be found associating with, even those minerals to which he exclusively gives his attention ... He has of course no chance of finding out the useful properties of minerals ... previously known to him.[39]

Despite assurances that Dunlop and his cohorts would not relinquish their political efforts, A.B. was perplexed by their failures to date. He discounted the expense of a geological survey, as well as popular fears that the imperial government would seize property on which valuable ores were discovered. Some months passed before Baddeley laid the blame on his preliminary report of the 1835 expedition. Printed perhaps coincidentally on the same page in the House of Assembly *Journals* as the report by Dunlop's select committee of 1836, Baddeley's report might have convinced Canadians that he had already conducted a geological survey of the country traversed. Instead, the report was merely 'a summary of facts ... drawn up in a hurried manner in the field, in anticipation of the questions which it was supposed Sir John Colborne would put to me on my return.'[40]

There may have been some basis for Baddeley's concern, but responsibility lay also with Lieutenant-Governor Sir Francis Bond Head. Head (1793–1875) was a Royal Engineer who had served in the garrison at Edinburgh. But he had suffered a discomfiting experience during the investment boom of 1825, when he left the army to become a mining supervisor with the Rio Plata Mining Association in Argentina. The company eventually collapsed, after which Head began a literary career describing his adventures on the Pampas. He also recorded his analysis of the errors which had burst the financial bubble. The main problem, he argued, was twofold: British investors knew nothing of the countries in which 'our money lies buried'; nor had workers from Britain been trained in 'scientific principles' to help alleviate this problem. Head transposed to North America his conviction that 'nature has formed the vast continent of America on a scale very different from that of the Old World.' He concluded from past experience that the popular estimate of the value of New World mines had been, 'all along, greatly exaggerated.'[41] The additional fact that the motion for a geological survey was put forward by his own political nemesis, William Lyon Mackenzie, certainly did not help to persuade Bond Head otherwise.

Despite Bond Head's personal reservations, increasing numbers of social conservatives in Upper Canadian society saw economic development as the surest way to build an orderly and contented society. They continued to call for a geological survey because it seemed consistent with these beliefs. As the geographical limits to expansion by settlement in Upper Canada became apparent, these same people had all the more reason to wonder about the mining potential of the country. A significant member of this economically progressive Tory 'freemasonry' in the Kingston area was James Macfarlane, the editor of the *Chronicle and Gazette*, which had published A.B.'s letters. Macfarlane was an enterprising publisher who bought the paper in 1822; by 1834 he had built in into one of the leading papers of the province. With his reasoned editorials and pleas for moderation and public improvement, Macfarlane continued the policy of the paper's earlier proprietor, John Macaulay, the surveyor general of Upper Canada. Like David Chisholme and Thomas Dalton, Macfarlane and Macaulay and others like them believed not only that science could benefit from expanded fields for research and investigation but, moreover, that sciences like geology offered Canada new dimensions for expansion in turn.[42] The practice of geology was initiated in Canada by a combination of the formal British imperial tie and the Scottish tradition of utilitarianism and Wernerian geology. By the late 1830s the impetus for a geological survey became a dynamic movement among the professional and business classes, supported by equally dynamic modifications in the science itself.

2
Montreal Masonry

David Chisholme's writings in the *Canadian Review and Magazine* during the 1820s alluded to pressure for a publicly funded geological survey, a movement by no means limited to Upper Canada. Newspapers in Lower Canada had followed events in Upper Canada as well as progress by various American state surveys.[1] But the main locus of support for a geological survey in Lower Canada was Montreal, where strong bonds cemented leading members of Montreal society in a perceived common interest. This social 'freemasonry' buttressed the possibility not only of attaining state support for a geological survey of Canada but also of appointing one of their own to direct it.

I

The most obvious channel through which to press this dual case was the Natural History Society of Montreal (NHSM). Founded in 1827 to take inventory of 'the natural productions of the country,' to collect specimens in a museum, and to 'foster a general spirit of scientific and literary research' in Montreal, the society drew its early membership from the city's professional, commercial, and largely (but not exclusively) English-speaking middle classes. The small group of physicians who convoked the first meeting singled out 'the want of scientific pursuits' which they agreed was 'forcibly felt' in Montreal. Although interested in all branches of natural history, the society anticipated that mineralogy would become its strongest department, if only because, as it reported in 1828, rocks and minerals were more easily transported than botanical or zoological specimens. Since the museum was to be 'a visible sign of the existence and utility' of the society and a rallying point for members, it seemed important to build up the collections. This line of thinking inspired the society's 'Indian Committee' in 1828 to undertake its extensive scientific inventory of the interior of British North America.²

From its inception, the NHSM promoted public interest in geology. In 1831 it held a successful course of geological lectures, utilizing a 'handsome and extensive' collection of minerals purchased with a grant from the provincial legislature. Biding its time, the society consciously worked more for posterity than for its own generation in 'laying up materials for instruction, ready for use whenever the state of society' in Canada demanded more advanced knowledge. Several outbreaks of cholera suspended meetings in 1832 and 1834, but the society regained its momentum by 1836. Recent donations, it reported, included a copy of the *Report on the Geology of Massachusetts* from that state's highly regarded geological surveyor, Edward Hitchcock. Hitchcock's survey of Massachusetts was an inspiration, the first of its kind in America to be completed. In addition, the society's mineral cabinet had now been inventoried in an updated *catalogue raisonnée*; dissatisfaction lingered, however, that the collection of Canadian minerals was not yet as extensive as might be hoped.³

Yet the society felt sure that recent discoveries in Canadian localities afforded 'the most pleasing anticipations' as to the mineralogical riches of the country which might be revealed by more thorough exploration. New materials lay waiting to be discovered in Canada, some already known to be valuable, and others 'more or less rare.' The society hoped that such promise of 'valuable consequences,' especially in the light of American successes, would stimulate the Canadian legislatures to dedicate a small part of their revenues to the cause of a mineralogical inventory. Nor would staffing the survey present insurmountable problems. For although it was preferable to look to the local ranks, if resident mineralogists could not be found to undertake the task, the society would not hesitate to look to Europe and the United States for competent individuals. Echoing David Chisholme's remarks a decade earlier, the society insisted on the greater need for government assistance in Canada than in other lands, where 'great progress in science, literature and arts' was 'more highly estimated, and where the accumulation of wealth allows a powerful individual effort.' Besides, it urged, Canadians had already conceded the principle in granting public money for educational purposes.⁴ In May 1837 the society resolved to petition the legislature of Lower Canada for a geological survey, in the hope that if Robert Dunlop's efforts finally succeeded in Upper Canada, then both provinces could join in the commission.⁵

Interest in a joint geological survey was promoted by the NHSM's leading members. These included Dr Andrew Fernando Holmes, a founder of the society and a member of the faculty of medicine at McGill College. Holmes (1797-1860) was born in Spain and brought to Canada as a child in 1801. He studied medicine at Montreal and Edinburgh, deepening a lifelong interest in botany and geology. Membership in the Lyceum of Natural History of New York and the Connecticut Academy of Arts and Sciences enhanced his admiration of American geological surveyors such as Edward Hitchcock and Charles Upham Shepard.[6] Hitchcock's donation of his published reports cemented the appreciation of the entire society.

Another influential member was Robert Armour, Jr, secretary of the society's Indian committee. Armour (1809–45) was the son of Robert Armour (1781–1857), a Scottish bookseller and publisher who had immigrated in 1798. The elder Armour purchased the Montreal *Gazette* in 1832, and Robert, Jr edited the paper until David Chisholme succeeded him in 1836. Robert, Jr, then joined with Hew Ramsay to form the bookselling and publishing firm of Armour and Ramsay, which took over publication of the *Gazette*. Both Ramsay and Andrew H. Armour, the brother of Robert, Jr, served on the council of the NHSM during the 1830s, and Andrew replaced Robert as Ramsay's business partner. Other members of the society included Adam Thom, the editor of the Montreal *Herald*; J.S. McCord and William Badgley, both prominent Montreal lawyers; and John T. Brondgeest, secretary to the Montreal Committee of Trade. These connections suggest the social and business network underpinning David Chisholme's editorial arguments for a geological survey.[7]

By the autumn of 1837 the efforts of R.G. Dunlop in Upper Canada and the NHSM in Lower Canada to initiate a geological survey had come a long way. But both projects were once again frustrated when rebellion broke out in both provinces. In Upper Canada the select committee of 1836 dissolved into its Reform and Constitutionalist factions when Gibson and Macintosh joined Mackenzie's rebels in Toronto and Duncombe led his own revolt in the southwest. The Dunlop brothers, in contrast, organized the Huron County militia against the insurgents.[8]

In Lower Canada divisions were complicated by the largely ethnic dichotomy which informed long-standing social, political, and economic grievances. French-Canadian attitudes towards a government-supported geological survey were coloured by the nature of reformist thought in 1837. Reform newspapers in Upper Canada gave much less space to the issue than did Tory papers. Yet even the less xenophobic *Le Canadien*, edited by Étienne Parent, made no mention of the growing case for a survey, although he regularly reported the activities of the Literary and Historical Society of Quebec as well as other developments in science. It was science entailing government involvement that Parent could not bring himself to support.[9]

For more radical Patriotes 'the battle against the English' was perceived at the same time as a battle against their kind of progress. Essentially this meant a battle against English capitalism, which had to be waged on several fronts. Patriote ambivalence about government support of a geological survey in Lower Canada is clarified by their attitudes to the ironworks of the St Maurice River near Trois-Rivières. The forges were disparaged by Patriotes as a base for a nascent local bourgeoisie which fed upon English capitalism and threatened the primacy of agriculture in Lower Canada. This negative view of the iron forges contrasted starkly with that of Upper Canadians, who saw the ironworks at Marmora and Madoc as justification for scientific searches for more valuable minerals.[10]

The geological inventory encouraged in Canada was not a reflection of the gentleman amateur tradition. Those influenced by ideas from Edinburgh, whether they practised or merely advocated geology in Canada, based their actions upon strongly utilitarian reasoning. In the Canadian as in the Scottish context, these arguments were linked to ideas of overcoming a state of relative economic backwardness. Economic progress meant the growth of industrialism, invariably the motivation for a scientific search for useful minerals such as coal and iron ore. Already during the 1830s, the use of geology to facilitate and rationalize such an inventory of Canadian resources helped to reshape the science from a systematic method of classification to an ideology forged in the image of British – and especially Scottish – industrial civilization.[11]

II

The use of science to foster the growth of an incipient industrial economy intensified during the reorganization and recovery which followed the rebellions of 1837. Its Scottish foundations were further reinforced by the personal predilections of the new English governor general of British North America, Lord Durham. John George Lambton (1792–1840) had been a pupil of Dr Thomas Beddoes, a utilitarian natural philosopher with a medical degree from Edinburgh who had been reader of chemistry at Oxford from 1787 to 1792. Sympathetic to the French Revolution, Beddoes left Oxford and set up his Pneumatic Institute at Bristol, where he became the physician and friend of Durham's father. Upon the elder Lambton's death, Durham and his brother were sent to Bristol to live with Beddoes from 1789 to about 1805. Beddoes, a firm believer in a broad and useful education, imbued his young charges with a basic knowledge and a sense of the importance of natural history, chemistry, and mechanics along with the traditional classical subjects.[12]

Durham's home environment offered unique opportunities to become acquainted with the new role of science, particularly geology. By inheriting the

Durham estate he became 'an English landed gentleman whose crop was coal.' Located in the lucrative great north-eastern coalfield of England, the estate had been built into a 'small mining empire' linked by a network of private railroads to transport coal to the coast, to be shipped in Durham's own vessels.[13]

Rather than leasing the lands to mining adventurers, the landed gentry in these northern coalfields participated in working the mineral, and Durham was no exception. A scientific and economic connection between mining and geology was recognized by the local Literary and Philosophical Society of Newcastle-upon-Tyne, founded in 1793, which concerned itself with local geology and mines safety. Durham in particular demonstrated his utilitarian and entrepreneurial interest in science by co-operating with Humphry Davy in testing a safety lamp for use in his mines. By promising 'the parallel advancement of rational knowledge and local wealth' through co-operative local scientific inquiry, the Newcastle society worked to harmonize various local interests and social classes during years of general political tension.[14]

Durham carried these valuable lessons with him to Canada. His attitude towards economic development, typical of growing numbers among the English landed gentry, was entrepreneurial and technological in its conception of science as a utilitarian activity productive of wealth. Although fallacious in its ready acceptance of such simplistic connections, this view welcomed scientific theory and experiment as conducive to the rise of industrialism. Durham and his fellows shared in a fashionable interest in science, in a belief that innovation and improvement contributed to the national good and were thus patriotic, and in a fear of other social classes who could appropriate science for their own purposes.[15]

In 1837 Durham would probably have been elected president of the 1838 Newcastle meeting of the British Association for the Advancement of Science, which was to focus upon geology and mining. Instead he accepted his political appointment in British North America and arrived at Quebec in May 1838. An important element in his intellectual baggage was the preconception that the way to achieve political and social harmony was through economic development. Nor was his widespread reputation as a political reformer all that preceded him to Canada. David Chisholme noted in the Montreal *Gazette* that Durham possessed one of the most productive landed and mining estates in England and accordingly exercised an 'influence of a preponderating stamp.' The Natural History Society of Montreal was aware of this connection and resolved unanimously that Durham become its honorary patron. The personal interest in natural history of its former patron, Lord Dalhousie, had benefited the society even after Dalhousie's departure from Canada, and it was hoped that ties with Durham could do the same. Since the Literary and Historical Society of Quebec (LHSQ) expressed similar hopes, Canadians witnessed a public rivalry for Durham's attentions.[16]

When Durham initiated a wide-ranging inventory of political and social institutions in Lower Canada, progress reports were requested from both scientific societies. The president of the LHSQ, William Sheppard, submitted details of legislative grants to the society in 1830 that had been used to purchase mineral and zoological cabinets and a library. But no public funds had been received since 1834, when the Patriotes gained a majority in the assembly.[17]

The president of the Natural History Society of Montreal, A.F. Holmes, emphasized his society's attempts to provide unique educational opportunities for Canadians. But, he added, these efforts were continually frustrated because the Legislative Assembly failed to look beyond local individual interest to the NHSM's public benefits. Financial preference was enjoyed by the LHSQ, Holmes argued, because both government and military officers resided there. In contrast, the Montreal society had 'not latterly flourished,' for it was 'placed in an uncongenial soil.' The NHSM solicited papers on the mines and geological structures of Canada, but none of those submitted had achieved a world-class calibre of original inquiry, and that very year the prize had been withheld. Holmes felt his society had little choice but to bide its time until Canadians were ready to pursue scientific careers. He cited the society's mineralogical and geological cabinet as its most valuable 'department.' Holmes regretted, however, that the Canadian part of the collection remained unclassified and incomplete.[18]

When Durham visited Montreal in July, the NHSM sent a deputation to request that he become its patron for life. Such an honour, they urged, would benefit not only the society but also the development of the intellectual and physical resources of the country. Durham assured them that he fully appreciated the advantages that must accrue from societies working to develop 'their latent resources.' Such a society, he realized, provided 'common ground for persons of different political opinions to meet together, for the advancement of scientific knowledge.' He himself came from a part of Britain where 'many valuable objects of natural history were discovered,' and he offered to write to Newcastle for specimens of local minerals. Durham was happy in particular to patronize a society directed to the development of the resources of the country. It would, he declared, give him pleasure to recommend grants to aid the society as well as to 'afford his own assistance to the praiseworthy objects for which they were associated.'[19]

Unfortunately for the NHSM, Durham aborted his mission sooner than anyone anticipated and suffered a decline in health which ended his life in 1840.

But his *Report on the Affairs of British North America* embodied assumptions which Durham had brought to his task of recommending the future path of imperial colonial policy towards Canada. Durham's legacy was the political realignment and economic reorganization which characterized the Union period in Canada, from 1841 until Confederation. Ever a man of his time, Durham visualized his task in terms which reflected the Victorian idealization of 'Baconian' fact-collecting and inductive reasoning. He regretted in his *Report* that 'obstacles have prevented me from annexing a greater amount of detail and illustration, which, under more favourable circumstances, it would have been incumbent on me to collect, for the purpose of rendering clear and familiar to every mind, every particular state of things, on which little correct, and much false information has hitherto been current in this country.'[20]

The interests involved in Canada's future, Durham emphasized, were considerable, because they involved the destinies not only of the million and a half inhabitants of Upper and Lower Canada but also of 'that vast population which those ample and fertile territories are fit and destined hereafter to support.' Durham justified his hopes by pointing out that no portion of the American continent possessed greater natural resources for 'the maintenance of large and flourishing communities.' An 'almost boundless range of the richest soil' remained unsettled and uncultivated; 'inexhaustible forests of the best timber,' not to mention 'extensive regions of the most valuable minerals,' also offered untouched sources of wealth.'[21]

Some analysts of Durham's *Report* marvel at his clear vision of the future which awaited British North America and of the role science was to play in its unfolding. But his predictions that rail and steam navigation would 'bring all the North American Colonies into constant and speedy intercourse with each other,' and before long would 'materially affect the future state of these Provinces,' were quite in keeping with the outlook of this 'protégé of the industrial revolution.' In Durham's view as an entrepreneurial landholder, natural resources acquired value only when enterprise could develop them. His main object was to restore Upper and Lower Canada on the rightful path to industrialism and economic progress, which in his view would 'do more for the present pacification of the Canadas than anything else.' Indeed, many people had objections to the *Report*, but few rejected its call to get on with the job of economic development. When Durham's successor, Lord Sydenham, took up the reins in 1839, his orders were to carry out Durham's utilitarian plan to rid the Canadas, soon to be united into one province, of their institutional inefficiency and to foster internal improvements.[22] This more stable atmosphere greatly favoured the institution of a geological survey of Canada.

III

Charles Poulett Thomson, Lord Sydenham (1799-1841), had represented Manchester in the House of Commons and served as president of the Board of Trade before becoming governor general of the Canadas in 1839. A Liberal and a trained businessman, Sydenham's task was to carry out Durham's recommendations to unite Upper and Lower Canada, to introduce municipal institutions, and to inch towards responsible government. From the larger perspective of imperial economic policy, the program aimed at the survival of Canada in its unique North American circumstances and recognized the entrepreneurial role of the state as a necessary element in this survival.[23]

Even if a geological survey of the newly united Province of Canada was not an obvious priority in Sydenham's framework of doing good business, the idea of such a survey had direct links to economic progress and to sound financial practice. One author of a popular guide to Montreal echoed his friend David Chisholme when he wrote that the geological character of the province had hardly been adequately investigated: 'The quarries of excellent stone found near the mountain, render it not less an object of interest to the citizens of Montreal for the purposes of building, than the facilities it offers for the investigation of its geology do to the naturalist for the purposes of study.'[24] For its part, the Natural History Society of Montreal continued to solicit essays on natural history, including geology, in either English or French. Then in the autumn of 1839 the society received a letter from the Literary and Historical Society of Quebec, stating its 'concurrence' in the NHSM's proposal for a geological survey of Lower Canada. Spurred on by this gesture, the NHSM appointed a committee of A.F. Holmes, J.S. McCord, Frederick Griffin (a Montreal lawyer), and a Mr Scott to carry the torch and to communicate its progress to the LHSQ.[25]

Public attention was drawn to the subject of a geological survey by the publication of Abraham Gesner's second report on the geology of New Brunswick. Gesner (1797-1864), a London-trained physician, was the first government geologist appointed in a British colony, a fact noted frequently by supporters of a similar project in Canada. He was assisted by James Robb (1815-61), who was educated at Edinburgh and Paris and had been a lecturer at King's College, Fredericton, since 1837. Gesner's report was reviewed by David Chisholme as 'unquestionably' the most interesting work he had ever perused on the physical construction of any part of British North America. The report constituted 'ample reward for the liberal and patriotic exertion in the cause of science, as in that of practical experience.' By the same token, it was regrettable that similar surveys had not been established in all the British North American provinces. Yet, he added, Newfoundland was already following suit, and Nova Scotia would have done so long before but for the monopoly of its mines and minerals held by the General Mining Association in Britain. Perhaps, Chisholme hoped, 'under the auspices of a better order of things' a geological exploration of Canada would 'speedily succeed' that of New Brunswick, since there was no longer any doubt that the effort and expenditure would prove worthwhile. Gesner's new discoveries of coal, he predicted, would elicit 'the most important consequences to the future interests – not of New Brunswick alone, but of the whole of British North America.'[26]

Other papers, such as the Quebec *Gazette*, acknowledged Gesner's report in similar tones, speculating that most of North America harboured 'valuable treasures which have not yet been discovered or explored.' Gesner's findings reached even the British Association for the Advancement of Science, meeting at Glasgow in 1840, in a paper by F.H. Baddeley on the geology of Canada. In the ensuing discussion William Buckland, the Oxford geologist, argued that boundary disputes with the United States necessitated a geological survey of British North American territories. Otherwise, undiscovered valuable coal fields might unwittingly be exchanged for fields of granite. Buckland likened 'such a proceeding' to 'the foolish men at Troy, who exchanged gold for brass.'[27]

Late in October 1840 Lord Sydenham visited Montreal and received a delegation from the Natural History Society. The recent death of Lord Durham had moved the NHSM to ask future governors general to act ex officio as patrons. Trusting again that Sydenham's sanction and influence would promote the 'usefulness and prosperity' of its goals, the society hailed Sydenham's earlier 'exertions and support of the benefit and humanizing influence of science.' Sydenham recalled his 'deep interest' in scientific pursuits earlier in his life and promised the society free access to him, 'when he would be glad to communicate with them on every subject connected with their laudable objects, and assist them by every means which his station and influence could command.'[28]

The union of Upper and Lower Canada in February 1841 combined pressure in both sections of the new province for a geological survey and established a much broader testing-ground for Sydenham's declarations. The Legislative Assembly of the Province of Canada, which convened at Kingston in the spring of 1841, was empty of 'obstructionist' radicals who had sat in both assemblies before 1838. It was cleared of all members of the select committee of 1836, because Robert Dunlop was in failing health and had been replaced as representative of Huron County by his brother William, elected in 1841. But more important was the general realignment which took place in favour of Durham's emphasis on economic progress. The cause of the survey was pushed onward by the NHSM whose resolution that 'a Geological Survey of Canada would be of the greatest service to the cause of science, and the utmost utility to the Province, by developing its resources, which may ultimately be of great extent' was conveyed to the government. An address to the governor general and petitions to the legislature were drawn up by A.F. Holmes, J.T. Brondgeest, and J.S. McCord, amid fears that Sydenham was not actively enough interested in the project. Durham's well-known support and promotion of science, it seemed to the society, had given way to Sydenham's own priorities, 'equally favourable to literature and the arts.' The NHSM committee worked feverishly to keep the issue alive.[29]

The petition reached the assembly in July via Benjamin Holmes, a Tory member for Montreal. It proceeded to a select committee composed of Holmes, general manager of the Bank of Montreal; John Neilson, editor of the Quebec *Gazette* and a moderate reformer; F.-A. Quesnel, a lawyer and a moderate reformer from Montmorency; William Hamilton Merritt, builder of the Welland Canal and now a moderate reformer from Lincoln County; and H.H. Killaly, soon to become chairman of the Board of Works, from London. In August a similar petition from the Literary and Historical Society of Quebec followed. The select committee proposed funding a geological survey, for which the Committee on Supply granted £1,500 on 9 September for one year.[30]

Related political developments help to explain the geological survey's rather tenuous beginnings. It was caught in a major transitional phase in the relationship between legislative and executive branches of government in Canada, a shift that especially affected its finances. Only two days before the granting of supply, John Neilson had moved the assembly to resolve that it not be bound to future payments and that public officers be required to report the amounts paid for their services at the opening of each session. Parallel developments, such as the establishment of the provincial Board of Works, strengthened the principle that no public work was to be commenced unless it could be completed for the sum appropriated. W.H. Merritt initiated a trial period even for the Board of Works.[31] Neilson and Merritt transposed these ideas to the select committee's deliberations on the geological survey.

Once the proposal reached the Committee on Supply, yet another snag developed. Thomas Aylwin, a radical Reform lawyer from Portneuf, declared that he would not support a geological survey unless the government was prepared to name the person who was to carry it out. Aylwin suspected 'it was a mere job – and for aught he knew, got up for the purpose of employing [two men] who had been employed as a commission to investigate the late riots in Toronto.' Upon the attorney general's indignant refusal to state the name of the person who would be employed, 'even if he knew,' Aylwin let loose a 'strain of invective and abuse of the most personal kind,' so as 'completely to disgust every member in the House.'[32]

Aylwin's efforts to focus the debate on government expenditures and away from the merits of a geological survey provoked the interesting revelation that Benjamin Holmes had been requested by 'several scientific men' to present the petition. Sydenham, he explained, had first thought of appointing Royal Engineers to the task, in order to save costs. But he concluded that 'though they might be very good Engineers, they may not be eminent Geologists.' After 'mature consideration' the present plan was adopted. 'Communications had been made,' declared Holmes, 'and negociations [sic] were now going on, for the appointment of the most eminent man that could be found to engage in it.'[33] Aylwin's response, not preserved for posterity, cleared the galleries and drove the chairman from the chair. Aylwin inadvertently acted as a foil for Holmes, who gave a 'great deal more satisfaction' to the House by his 'unassuming observations' in support of the survey. A final factor against long-term funding of the survey was the widely publicized prospectus of John Rae's 'Outline of the Natural History and Statistics of Canada,' a forthcoming publication 'based on science' and using material collected during Rae's eighteen-year sojourn in Canada.[34] There would be little point in expending public money if a scientific survey might already have been carried out. Rae, as we have seen, had influenced the survey's history before and now seemed to pre-empt its work again; but his book was never published.

The overriding fact remains that a British government loan recommended by Lord Durham to aid the completion of the St Lawrence canal system made funds available for works like the geological survey, and the NHSM had good reason to feel gratified at its success. Still, its struggle was not over. Only ten days after the funds were granted, Lord Sydenham, who would have appointed a director of the survey, died of complications after a riding accident. All the painstaking work of securing the attention, goodwill, and confidence of the new governor general would have to begin again – and quickly, lest the survey become a dead letter.

IV

Sydenham was succeeded by Sir Charles Bagot (1781–1843), a career diplomat from the Conservative landowning class. Bagot arrived at Kingston in January 1842 in failing health. Like Sir Robert Peel, the head of the government Bagot served, Bagot admired science and its practitioners, and he did not resist the NHSM's request that someone be named without delay to activate the geological survey.[35] The appointment of William Logan as provincial geologist merits closer scrutiny because his selection was not unanimous and reveals the weight of Logan's social and professional connections both in Britain and in Montreal.

Bagot met with two separate though interconnected campaigns, the one to activate the survey and the other to appoint William Logan to the task. A significant contributor to the latter campaign was David Chisholme, who worked to keep the subject in the public eye. In November 1841 the Montreal *Gazette* reprinted an address by William Buckland to the Royal Institution of South Wales. Buckland expressed his appreciation of a theory recently put forward by Logan on the *in situ* formation of coal. Logan had been the first to recognize the relationship to coal-beds of rootlike formations, or *stigmaria*, found in the underclay below them as evidence that coal was formed by the geological compression of ancient fossilized plants. The eminent Professor Buckland confessed that he had devoted many years to studying the formation of coal. He acknowledged that Logan's theory, published in the *Transactions* of the Geological Society of London in 1840, was 'founded upon most accurate observation' and had solved the problem quite convincingly.[36] Chisholme condensed those details of Buckland's speech that were unrelated to Logan and hastened to draw out the implications of the lecture's 'very distinguished reference' to Logan, 'a native of this city, and well known to many in our community for his amiable qualities in private life, and his unwearied devotion to many departments in science.'

Chisholme used the speech as a springboard for his own plan to nominate Logan to carry out the geological survey of Canada, which 'could not be entrusted to a better or more accomplished geologist.' 'Mr. Logan, during a recent visit to Canada,' he argued, 'was at much pains to make himself acquainted with the physical outlines of different sections of this country, and if officially employed to explore them more thoroughly, there is every reason to believe, that the benefits arising from such employment to the country, would be very great.' Chisholme insisted that his remarks were not prompted by either Logan or his friends. But he believed he expressed the feeling of 'every friend of science in the Province' that 'nothing could reflect greater credit on the country, than the accomplishment of the proposed geological survey by one of its natives.'[37]

William Edmond Logan (1798–1875) was born in Montreal and educated as a medical student at the University of Edinburgh. Employed in his uncle Hart Logan's mining and construction interests after 1816, Logan moved to the Forest Copper Works in Swansea, South Wales, in 1831. There an amateur interest in natural history blossomed into a professional knowledge of coal-seams, which Logan recorded on highly accurate topographical and cross-sectional maps. These maps were adopted by the Geological Survey of Great Britain, establishing both Logan's professional reputation in British scientific circles and his personal friendships with prominent British geologists. Logan began publishing his theories of the *in situ* origins of coal in 1837, and his work greatly facilitated the practical search for coal-seams. But the death of his uncle led him to resign his position in Wales in 1838. In 1840, still unemployed, he presented an important paper entitled, 'On the character of the beds of clay lying immediately below the coal-seams of South Wales, and on the occurrence of Coal-boulders in the Pennant Grit of that district' to the Geological Society of London. Logan then resolved to visit North America for three months, both to visit his brother in Montreal and to gather more evidence of the *in situ* formation of coal in Pennsylvania and Nova Scotia.[38]

Logan actually remained for over a year, from August 1840 to October 1841. He geologized in the Montreal area, 'cracking away at all the boulders' that came within his reach, and he was visiting the coalfields of Pennsylvania when the grant for the survey was made by the Canadian legislature. He was deeply impressed by the potential for 'incalculable wealth' afforded by coal and iron deposits in the north-eastern United States and toyed with the idea of seeking work in Pennsylvania because coal workers there seemed eager to learn from such a 'practical coal miner of education' as he fancied himself to be. But Logan's keen instincts drew him in another direction.

On a side trip to New York he saw the famous English geologist Charles Lyell on the street and contrived to meet him. Lyell was in America on a speaking tour and seeking out new evidence for his *Principles of Geology*, three volumes published between 1830 and 1833, which provided a new theoretical framework for the earth's formation. Lyell's *Principles* swept away the limitations of the static Wernerian approach to geology and revived its rival, the 'Plutonism' or 'Vulcanism' suggested by James Hutton (1726–97). Hutton had theorized that sedimentary processes were insufficient to explain all the earth's rock formations. He postulated instead that certain 'metamorphic' formations had solidified from magmas and that fire in the earth's core constituted as important a geological agency as water.

Vulcanism won out in the widespread acceptance of Lyell's *Principles*, because Lyell embellished Hutton's theory with enormous amounts of evidence accumulated since the eighteenth century. He transformed Vulcanism into an elaborate uniformitarian theory of the earth. This uniformitarian approach promoted a historical outlook in geology, emphasizing its investigation of the successive changes that had taken place in nature over far longer periods of time than previous theories had imagined. Lyell redefined geology as inquiry into the dynamic causes of these changes and their influence in modifying the surface and external structure of the earth. Using an actualistic method to explain forces that had operated upon the earth from the beginning of time and still continued to do so, he confirmed a consensus about the earth and its history that had been growing since the 1820s.[39]

Lyell was now spreading uniformitarian geology to North America. His immensely successful tour through the United States and British North America generated public interest as he visited Toronto, Niagara Falls, Montreal, Quebec, and Nova Scotia. The Montreal *Gazette* quoted *Silliman's Journal* on the positive social impact of the visit of 'the accomplished strangers' in Lyell's party, which promised to 'promote every good national and personal feeling.'[40] William Logan no doubt agreed, but his own aims were more personal. Logan recognized the relevance of his own geological work to Lyell's uniformitarian approach and was determined to attract Lyell's attention. He remained an extra weekend in New York and succeeded in getting an hour-long interview in Lyell's hotel rooms. Lyell, it turned out, was aware of Logan's coal theory and was also very keen to attain information about the geological structures of British North America. He encouraged Logan to continue geologizing around Montreal and promised to quote his work in the revised edition of his *Principles*. 'I am better pleased,' wrote Logan, 'that he should have quoted my name in any laudatory way than that I should have put forth the observation myself. It will be more beneficial to me for a particular object that I have in view.'[41]

Logan's object was to conduct the geological survey of Canada. Even before funding had been approved, he had considered offering himself as a candidate. He was, he had decided, 'not afraid of competition with anyone I know of.' The geological work he saw in Pennsylvania convinced him that 'I could beat it hollow.' Logan admitted his intentions to two American geologists, the Rogers brothers, whom he met at Wilkes-Barre on 4 September 1841, and he told his brother he felt assured that 'Canada would not limp far behind if I had [the survey] to do.'[42]

Logan continued his geological tour with a visit to the Pictou coalfield in Nova Scotia, where he once again encountered Charles Lyell. Lyell introduced him to John William Dawson, then a young local geologist and later principal of McGill College, whose lifelong mentor Lyell became. He also encouraged British North American geologists to co-operate in working up the extensive field which lay before them and conveyed his conviction that British North America was 'destined to surprise us yet' with important geological discoveries.[43] Both Dawson and Logan admired Lyell's interpretive work and fashioned their own researches on the model he provided; both contributed to subsequent editions of his influential books. He inspired them to use uniformitarian geology to develop new insights into the structure of the continent. Within this larger context, North America became not just a primitive, more barbaric version of

Europe, but rather a unique and valuable repository of cosmological information.

Back in England in late November 1841, Logan set about doing what he could to secure the position of provincial geologist. First he solicited testimonials from Sir George Murray, master general of the Ordnance Department, which oversaw the Geological Survey of Great Britain. Logan gained access to Murray through Anthony Murray, a relative of the master general and a business associate of Logan's brother Edmond. He directed these memorials to the Colonial Office rather than directly to Bagot, who had already been named governor general. Then in December 1841 Sir Henry De la Beche, director general of the Geological Survey of Great Britain, and William Buckland were approached for testimonials to Bagot, while Logan sought to fuel the fires by reading 'a few papers' before the Geological Society of London 'on points connected with the colony.' Indeed, he had already chosen Anthony Murray's brother Alexander as his assistant, provided De la Beche could teach him some geology.[44]

v

Logan's connections in the Montreal business community could only help as the hour of Bagot's decision approached. First, he was well known to the Molson family. Second, his brother James was a prominent Montreal merchant with high-level business and political links to the Molsons and to other prominent Montreal interests. Third, James Logan had high-level business ties with supporters of the geological survey in the Natural History Society, such as the Armour family and J.T. Brondgeest. To cite only the most prominent example, the expanding economic activities of the Molson family – from brewing into distilling, banking, and steamship and railway interests – during the 1830s had important ramifications for a geological survey of the province; the Molsons' activities implied a concomitant growing interest in the availability of coal for fuel. The very nature of their brewing and distilling enterprises allowed them to prosper even during times of general business stagnation. By the 1840s they had expanded into lucrative new fields such as mining, whose prerequisite first step was to locate workable and accessible mineral deposits.[45]

One final factor brought all these forces into interaction in Logan's favour. In 1839 Logan had joined the United Grand Lodge of Freemasons in London. Membership in the Brethren of the Craft of Freemasons during the nineteenth century meant little more than sharing in a system of philosophy and morality that emphasized mutual aid, charity, and benevolence. But from a social perspective, the tie could be most beneficial. The Provincial Grand Lodge of Montreal and William Henry and its predecessors comprised several generations of long-established Montreal families and, through its affiliation with the United Grand Lodge to which Logan belonged, Englishmen and Scots who emigrated to Montreal. During the 1830s the higher ranks of Montreal Freemasonry included several council members of the NHSM as well as John Molson Sr, and Peter McGill. Although internal and organizational problems and the death of John Molson, Sr, in 1836 considerably diminished the activities of the Provincial Grand Lodge in Montreal, a deeper affinity linked those who supported a geological survey to the rituals and beliefs of Freemasonry. For along with their ordinary religious affiliations Freemasons embraced the idea of a 'Great Architect of the Universe,' a notion that encouraged the natural urge to explore the Architect's work scientifically. These beliefs derived from the mystical underside of the Baconian rationalism so prevalent in Victorian thought.[46] They were consistent with the belief of Lord Durham, the deputy grand master of England in 1834, in the ability of Freemasonry, along with science, to submerge political differences for social and benevolent purposes. The Masonic connection could only have added weight in favour of Logan's appointment, and circumstantial evidence bears this out.

The campaign on Logan's behalf stepped up after Bagot's arrival in Canada. James Logan reached Bagot through Peter McGill, now mayor of Montreal, who presented Bagot with a memorial from the NHSM. Logan asked that his brother's name as 'likely to be a candidate' for the survey, along with Buckland's speech printed in the Montreal *Gazette*, be mentioned to Bagot. McGill added his own support of Logan, 'a Canadian by Birth,' and 'many years a Merchant in London. He is a gentleman of great attainments and has an intimate knowledge of the Geography of Canada.' The NHSM's memorial was far more vague, submitting to Bagot's consideration the 'expediency of adopting such measures as Your Excellency may deem most advisable for prosecuting a project beneficial alike to the cause of science, and to the improvement and future prosperity of the colony.' Bagot apparently combined these two requests to formulate a dispatch to the Colonial Office, adding that he would 'be anxious, if possible, to select a person who is either a native of this country or feels a strong interest in it.' Logan had been favourably recommended, he wrote, but Bagot thought it best to consult first the Geological Society of London and British professors of geology. When he visited Montreal in February 1842 a deputation from the NHSM publicly presented its memorial, to which Bagot promised a public reply.[47]

There was evidence by this time of a crack in the common front of the NHSM over the question of how the survey was to be carried out and, in particular, who was to direct it. On the one hand, Dr Michael McCulloch, a prominent physician who had long served on the council of the society, had asked in

January that Logan's name be mentioned to Bagot as 'a *native* of this Country now in England.' McCulloch had watched Logan geologizing in Montreal and felt 'convinced that another individual of his great experience and rare abilities is not likely to be met with.' He hoped the province would 'be so fortunate as to obtain his services,' and he too made a point of enclosing a copy of Buckland's speech. McCulloch advised Bagot to seek the opinions of prominent geologists in England rather than leave the final decision to a commission appointed in Canada.[48]

On the other hand, as McCulloch may well have been aware, three days earlier A.F. Holmes, still president of the NHSM, had informed Logan that he would not support him for the directorship. He judged Logan to be deficient in mineralogy, chemistry, and fossil conchology (palaeontology). Ironically, Logan had assumed Holmes would be willing to aid the survey in these areas.[49] Holmes had been aware of Logan's work since at least the summer of 1840, but he had different ideas about the priorities necessary to the selection of a provincial geologist. In a lengthy letter to Bagot, he recounted his own efforts while Sydenham was still alive. Sydenham had written to his friend the British geologist Roderick Murchison about the matter, but Murchison had just embarked on an expedition to Russia. Sydenham had then turned to geologists in the United States, who he thought would be 'much better acquainted with the peculiarities of Rock-formations and geological features of this continent, and the modes of examining them.' Holmes claimed that neither Lord Sydenham nor the Legislative Assembly had supposed £1,500 adequate to fulfil the survey's purpose, but instead had intended to renew the funding later.[50]

Holmes listed six objects of a geological survey of Canada: to ascertain the mineral riches of Canada, especially coal; to explore the geology of the country scientifically; to discover and identify new minerals; to conduct chemical tests for the value and usefulness of ores, soils, and mineral waters; to collect specimens for museums; and finally to compile the results in a major report. Of course the provincial geologist would have to possess 'a complete knowledge of the Geology of America, as well as more or less that of Europe.' He also needed 'a competent knowledge not only of the mineral nature of the Rocks but of their fossil nature.' Not only did Holmes strongly imply that Logan was insufficiently qualified, but he also claimed he had offered the position to Charles Upham Shepard of the Geological Survey of Connecticut. Shepard had allegedly proposed to carry out a joint survey of Canada with Edward Hitchcock of the Geological Survey of Massachusetts, and for a reasonable fee. Holmes was persuaded that Shepard and Hitchcock were the most highly qualified for the task, and he had recommended to Sydenham 'their selection for this important enterprize.'[54]

Bagot had previously known nothing of these negotiations and, it seems, was not swayed by Holmes' arguments. By April the Colonial Office had received testimonials from De la Beche, Murchison, Buckland, and Adam Sedgwick, professor of geology at Cambridge. On the basis of such impressive recommendations Lord Stanley, the colonial secretary, offered Logan the position of provincial geologist on 15 April. Sedgwick, it seems, had recommended Logan even though he was not familiar with Logan's work; had circumstances been otherwise, he would have recommended his own former student, J.B. Jukes, who had recently completed a geological reconnaissance of Newfoundland.[52]

While David Chisholme declared his satisfaction in the Montreal *Gazette* that 'nothing could prove more gratifying to scientific men throughout the Province' than Logan's appointment, it was probably no coincidence that as of the annual meeting in May, A.F. Holmes was no longer listed among the members of the Natural History Society of Montreal. He was replaced as president of the society and chairman of the council by William Badgley, a prominent Montreal lawyer, businessman, Freemason, and in-law and business partner of the Molsons. When Bagot visited Montreal in June and was asked to become patron of the NHSM, he seized the opportunity to draw public attention to Logan's appointments by quoting yet another 'private' testimonial from William Buckland to the bishop of Oxford, Bagot's brother.[53] As Logan himself observed, the letter 'had a wonderful effect in Canada,' especially once the NHSM gave it to the newspapers. 'I am afraid,' he mused, 'I must look as wise as an owl & practise humbug a little to keep up a character of Sir Oracle.'[54]

VI

Bagot the diplomat had understood his politics and thanked his brother for forwarding Buckland's letter. All the papers were highly pleased, he noted: 'Logan is a Canadian born.' But how well had he understood his science? The decision to ignore the proposed candidacy of two experienced and apparently willing geological surveyors, at least one of whom was well known as an ideal state geologist, spoke eloquently of his priorities. It supports the observation that 'the entire British Empire takes on a somewhat more permanent status than is always politically obvious,' even when one is considering geological approaches to the rocky sheath of Canada, in the early Victorian period.[55]

The political, social, and intellectual origins of the Geological Survey of Canada were closely tied to a growing concern with the country's destiny. Its proponents were particularly interested in alleviating Canada's relative economic backwardness. They shared certain hopes of what Canada could become and derived their expectations from a common admiration of the British experience. Yet there was a question that had to be answered first: was Canada capable of reproducing that experience? Political history showed that this was no simple act of imitation, and modifications would have to be made to account for differences between Canada's political situation and British traditions. Canada's economic and scientific adaptation was to prove equally tortuous.

Arguments for a geological survey repeatedly expressed the necessity of finding in Canada those resources that had made Britain great, especially coal and iron ore. Symbolic in the Victorian mind of progress, power, and even civilization itself, these two minerals seemed to some more desirable than gold. In order to motivate the search, supporters of a geological survey pressed into service the ideologies of Baconian and Scottish utilitarianism. The fundamental decision to adopt coal-based methods of production was partly ideological and differed from the decision in France, for example, to reject coal and the mass production it implied when the same choice was offered.[56]

Although the proposed geological survey of Canada served the interests of an incipient industrial community, its inventory component also served more abstract purposes. It promised to shed light upon great geological questions of the day. One of William Logan's major accomplishments would be to supplement and even to surpass Roderick Murchison's work on Pre-Cambrian formations, whose archetype Logan found in the Canadian Shield. These theoretical aspects were seen, at least by geologists, as but one end of a single spectrum of activities that were also practical. Logan's appointment resulted from his successful representation of both the practical and the theoretical where it counted most.

A corollary of both the material and the theoretical aspects of the geological survey was a developing assumption that there was a community of interest in Canada. In some ways the growing popular interest in geology resembled a freemasonry. It signalled a widening angle of vision to accommodate more of British North America in its field. The very act of imitating the values epitomized by the British experience carried within it the seeds of dissolution for the reciprocal relationship that mother country and colonies had hitherto enjoyed. An editorial by A.J. Christie in 1843 exemplified the growing sense of self-importance felt by British North Americans:

When we reflect on the many thousands of pounds that have been expended by the British Government in exploring the interior of barren Africa, and other regions, the acquaintance could only be a matter of speculative curiosity, and the little that is known of the cultivable portion of British North America, and the gross ignorance that prevails respecting the interior of this valuable and interesting Colony, we cannot divest our minds of the conviction, that every effort made by individuals to extend our geographical and statistical knowledge of Canada, has a strong claim to public patronage and general support.[57]

As an institutionalized act of faith in Canada, the geological survey opened a way for rudimentary forms of nationalism to find continued and even amplified expression. Whether such a soil as Canada could then nurture some of these forms to maturity would depend to some extent upon what Logan discovered beneath it.

3

Logan's Geological Inventory: 'Construction and Extension,' 1842–1850

It is the most unpoetical of all lands; there is no scope for imagination, here is all new – the very soil seems newly formed; there is no hoary ancient grandeur in these woods, no recollections of former deeds connected with the country.

Catharine Parr Traill, *The Backwoods of Canada* (1836)

The letters of Catharine Parr Traill gave eloquent expression to a deep sense of physical, cultural, and even historical isolation impressed upon immigrants to the Canadian backwoods. Pioneer settlers were struck above all by the apparently recent formation of the land and by the absence of any qualities which gave it historical significance. Yet within ten years after Traill's arrival, William Logan's annual reports of progress to the Canadian government began to foster a reinterpretation of Canadian land-forms as an important key not only to the history of the earth but to the very future of British civilization. What evoked such imaginative new interpretations was the dual nature of the provincial geologist's task: to facilitate the location and exploitation of 'economic' mineral deposits in Canada, while at the same time advancing scientific knowledge.

Logan's dual geological achievements have been ably documented in the official history of the Geological Survey of Canada (GSC) but the ideas with which they came to be associated warrant closer examination. On the one hand the work of the scientific survey apprised Canadians for the first time of the extent and quality of the country's natural resources. Since material wealth and political power appeared to be directly related, Canada's future was implicitly bound up with the results of Logan's geological inventory. On the other hand, Logan's growing international reputation, and even the survey's more dubious discovery of the controversial 'fossil' *Eozoon canadense* in 1858, elevated the Canadian imagination to new heights. Geological metaphors conveyed a sense of awe and pride in the Canadian Shield as the repository of earth's earliest lifeforms. By the late nineteenth century allusions to Canada as the cradle of 'the dawn of life' embellished the rhetoric of a Canadian nationalist mythology, which exalted Canada's prehistorical existence as a portent of the country's future greatness:

United closely, as she shall be from the Atlantic to the Pacific by a common nationality, our country will go on, increasing from age to age in wealth, in power and in glory; and it may not be too much of a stretch of the imagination to think that as it is the latest developed portion of a new world – as it was the first by millions of years, to nurse and cradle in her bosom the first spark of animal life in the eozoon, – it may be the country where at last great and fully developed humanity may find its fitting habitation and abode.[1]

The changing phases of William Logan's career as provincial geologist closely paralleled and indeed influenced several aspects of Canadian development during the early Victorian age. They fall into three main divisions: the initial reconnaissance of the 1840s and public reaction to Logan's preliminary geological findings; the flowering of Logan's reputation and authority during the 1850s, when he successfully showcased Canada's resources at several Universal Exhibitions; and the zenith of Logan's career during the politically unstable 1860s. Logan finally overcame serious financial setbacks to secure the permanence of the GSC as a public institution and to publish his major report, known as *Geology of Canada*, in 1863. To a great extent, the idea of a transcontinental Canadian nation grew from the degree and quality of self-understanding made possible through the practice of Victorian inventory science. Logan's geological inventory crystallized the abstractions with which Thomas D'Arcy McGee associated the prospect of a British North American nationality; it kindled the faith that this new nationality could actually be realized.

I

Previous studies have frequently noted William Logan's fitness to direct the Geological Survey of Canada. He is said to have combined 'a complete dedication to the pursuit of scientific truth' with 'a respect for practical endeavours, a good business head, an ability to deal with all sorts of people, and a fair capacity for managing employees' and to have applied precisely these abilities in order to 'establish, preserve, and expand the geological survey.' No one would deny the competence and dedication that Logan brought to his arduous duties. But the emphasis of his elegists upon Logan's 'altruism, his complete devotion to the interests of the Survey, and his undeviating belief in the institution and the work it was doing' should not be permitted to limit our understanding of Logan's complex character. As Morris Zaslow has pointed out, Logan's primary aim was to conduct the survey in a manner which would ensure its continued existence, and he planned a strategy accordingly.[2] He designed his survey to conform to the aims and outlooks of those who determined its (and Logan's) fate, namely the government and business communities of Canada. The extent of his success was much more a measure of this conformity than of any altruism on Logan's part.

But at the same time, Logan's survey extended the vision of many influential Canadians beyond the political horizons of their own province. Logan perceived his overall task in very broad terms which predated nascent expansionist views of his day. Publicly he predicted that the grant of £1,500 in 1842 was insufficient to 'float him over 25^0 of longitude and 10^0 of latitude'; he already feared that within the confines of the survey 'I shall never be allowed to descend into such minutiae as to ascertain whether small divisions of strata in one part of the province are contemporaneous with certain small divisions in another. I shall not be able to do more than to arrange a general skeleton of the subject.'[3] Privately he revealed anticipations which waxed more radically expansionist; in 1845 he confided to his mentor that his work in Canada actually bore far greater significance: 'Just look at Arrowsmith's little map of British North America,' he exhorted Sir Henry De la Beche.

You will see that Canada comprises but a small part of it. Then examine the great rivers and lakes which water the interior between that American Baltic, Hudson's Bay, and the Pacific Ocean – some of the rivers as great as the St. Lawrence, and some of the lakes nearly as large as our Canadian internal seas, with a climate as I am informed, gradually improving as you go westward, and becoming delightful on the Pacific. It will become a great country hereafter. But who knows anything of its geology? Well, I have a sort of presentiment that I shall yet, if I live long enough, be employed by the British Government ... to examine as much of it as I can, and that I am here in Canada only learning my lesson, as it were, in preparation.[4]

In Logan's sweeping geological overview, the extension of the territory under his purview formed part of the natural course of scientific investigation. There was no scientific reason to restrict the survey to Canada's political boundaries, which could contain neither his pride nor his geological vision. Logan utilized public disgruntlement in British North America over the Oregon boundary settlement in 1846 to urge the geopolitical importance of a geological survey of the north-western territories of British North America. 'When the British

Government gave up the Michigan territory at the end of the last American war, with as little concern as if it had been so much bare granite,' he warned, 'I dare say they were not aware that 12,000 square miles of a coal-field existed in the heart of it ... ready to supply American steamers with fuel on the lakes, while ours on the same waters, in case of war, must depend on wood, or coal expensively transported from Nova Scotia or Cape Breton Island, or across the Atlantic from the United Kingdom.'[5]

The range of Logan's personal experience combined with his geological approach to develop this expansionist outlook. First, ever since his student days Logan had shown himself eager for distinction. The achievements of family members, and later of the staff of the GSC, were to him mere extensions of this desire for recognition of his own efforts and achievements.[6] The pains he took to secure the directorship of the survey demonstrated that a strong competitive spirit was present also in Logan the geologist. He was motivated to a great extent by personal pride and he shrewdly understood how to make himself indispensable in every task he undertook.

Second, Logan's training as a businessman enabled him to shift his frame of reference easily between the practical and the theoretical poles of geological thought. Copper smelting in South Wales led him to mineralogy and geology, which in turn induced the study of coal-seams and his theory about the origins of coal. As early as 1833 Logan expressed a very practical interest in locating lucrative copper ores in Canada: 'Did you ever hear of any copper ore in Canada, or anywhere near it?' he asked his brother James in Montreal. 'If any were discovered it might become a matter of profit to us, if we could get hold of it, and it proved of good quality. Recollect this, and keep the matter before you.' From news of recent discoveries in Connecticut and Nova Scotia, Logan declared it would not be surprising to discover copper in Canada within reach of the St Lawrence.[7] This focus on the lucrative combination of copper and coal was to form the matrix of his initial approach to the GSC.

A third determinant of Logan's approach was that his main reference point was the work of the Ordnance Geological Survey of Great Britain, directed by his friend and mentor, Sir Henry De la Beche. De la Beche had assured him in 1839, 'I shall always feel gratified if you call upon me to be useful to you.' Logan in turn acknowledged the usefulness of various American state geological reports. But he also felt a particular need for the 'advice and positive assistance' of De la Beche in palaeontology, since no proper reference collection of Canadian fossils had yet been assembled. In Logan's view, comparison with the British master collection would avoid confusion over nomenclature and might even make Canada the standard of geological comparison between Europe and America.[8] However, over the years Logan's dissatisfaction mounted with De la

Beche's continuing failure to respond to his inquiries. Eventually Logan and the officers of the GSC shifted their perspective, first to American fossil collections and then to increased self-reliance in unravelling the geological strata of Canada.

For the moment, Logan attuned himself to the contemporary work of British geologists, who were busy sorting out the strata of the oldest known fossiliferous rocks. The Upper Primary or Palaeozoic systems lay below the coal-bearing or Carboniferous rocks. Great strides had been made in this branch of geology during the 1830s by Adam Sedgwick and Roderick Murchison, who by 1839 had managed to distinguish three Palaeozoic systems which they called Devonian, Silurian, and Cambrian. What obscurities remained by the 1840s could not be clarified without puzzle pieces from the formations of other continents.[9] The vast extent of unexplored Primary formations in this country boded well for the organization of the GSC at that time.

Moreover, there remained a miscellaneous group of still older and largely non-fossiliferous primitive or Pre-Cambrian formations, about which little was yet known. The absence of fossil remains was not the sole challenge of these Pre-Cambrian rocks, often so contorted and chemically altered that the proper order of their strata was almost impossible to establish. Only an extensive and contiguous territory of exposed rock could uncover the Pre-Cambrian secret. By the 1840s geologists were only beginning to grasp the chemical and mechanical processes which had forged these metamorphic rocks, and the presence of the Canadian Shield offered Logan a splendid and timely opportunity to seek answers within or, perhaps, even beyond the domain of his own survey.

As usual, Logan had both practical and theoretical reasons for doing so. On the one hand, he believed that while Secondary formations supported most of the settlement in Canada, the Primary constituted 'the metalliferous portion of the country.'[10] On the other hand, Logan's own vividly imaginative and keenly analytical mind was intrigued by the puzzle these rocks presented. Ever curious about the forces of nature, and following Charles Lyell's uniformitarian theory, Logan found on the icy St Lawrence near Montreal a geological classroom which taught him the workings of metamorphic processes:

There is no place on the St. Lawrence where all the phenomena of the taking, packing, and shoving of the ice are so grandly displayed as in the neighbourhood of Montreal. The violence of the currents is here so great, and the river in some places expands to such a width, that whether we consider the prodigious extent of the masses moved, or the force with which they are propelled, nothing can afford a more majestic spectacle, or impress the mind more thoroughly with a sense of irresistible power. Standing for hours together upon the bank overlooking St. Mary's Current, I have seen league after league of ice crushed and broken against the barrier lower down, and there submerged and crammed beneath ... From the effect of packing and piling, and the accumulation of the snows of the season, the saturation of these with water, and the freezing of the whole into a solid body, it attains the thickness of ten to twenty feet, and even more ... Proceeding onward with a truly terrific majesty, it piles up over every obstacle it encounters, and when forced into a narrow part of the channel, the lateral pressure it there exerts drives the bordage up the banks, where it sometimes accumulates to the height of forty or fifty feet.[11]

Just as the search for coal was intimately linked to Logan's own theories concerning its origins and location, so also was the search for valuable minerals and metallic ores inseparable from his desire to comprehend the Pre-Cambrian formations. In Logan's view, there was no real conflict between practical and theoretical aspects of his task, only a balance to be struck between two extremes. In his preliminary 'Remarks on the Mode of Proceeding to Make a Geological Survey of the Province' in 1842, Logan established the principle that for the foreseeable future the GSC would proceed 'but a short distance beyond the limit of settlement.'[12] But already by the first 'Report of Progress' in 1843, the unique contingencies of Canadian land-forms were conspiring to expand the geographical horizons of this narrow preliminary vision.

II

There is no better example of Logan's deft handling of both public opinion and the expectations of his sponsors than his first years on the survey. He defused public disappointment in his own scepticism concerning the presence of coal in Canada by hiding behind the mask of scientific caution. Logan always managed to find some new promise of untold mineral wealth in order to maintain public interest in his endeavours. His performance was a *tour de force* which was indispensable for the future survival of the GSC.

When Logan returned from England in 1843, he planned his survey with definite ideas about what he would (and would not) find. These preconceptions had three major sources. First, he had seen the tremendous importance of coal and copper in the industrial and commercial prosperity of the Swansea district in Wales. Second, Logan had perused the available reports and maps of neighbouring American state geological surveys, as well as the earlier geological reports on British North America by Bigsby, Baddeley, Bonnycastle, Bayfield, Green, and Wilson. Third, he had attended the opening of the Canadian legislature in September 1842. Logan treated the new session as an opportunity to collect 'the floating knowledge' bearing on his subject from 'so many persons of intelligence ... from different and distant localities' in the province.[13]

From these preliminary sources and his extensive knowledge of current geological theory, Logan confidently determined the rough limits of a 'well-marked zone of limestone,' a major structural feature of the south-western part of Canada. This information pointed out directions in which to search for metals and coal, even though the widespread limestone deposit that underlay the province was barren of such valuable minerals. 'Geological experience,' noted Logan, 'teaches that the metalliferous rocks are below [the limestone], the carboniferous above.' These geological rules of thumb were brought home with 'peculiar force' in Canada, which lay in the immediate vicinity of the productive coal-bearing formation of the north-eastern United States.[14] Rather ominously, Logan recognized that profitable coal measures existed *above* limestone,

which here spreads out so vastly; and the geographical position of this once fixed, it will be by transverse sections in the direction of its dip that we shall gradually approach to coal; but in consequence of the small removal from horizontality the limestone in so many places exhibits, the lineal superficial distance between the two formations will probably be very considerable.[15]

The implication of his purposely technical language was that coal was not likely to be found within the settled confines of the Province of Canada. Logan muted the import of this information by suggesting that the dip of the limestone formations in the Gaspé region gave more reason to begin a detailed search for coal there. Privately, though, he doubted the outcome even before travelling to the Gaspé in June 1843, owing to his earlier examination of coal formations in Nova Scotia and New Brunswick.[16]

Logan submitted his first 'Report of Progress' to Governor General Charles Metcalfe in November 1844, though he had written the bulk of it the previous April. The delay, he apologized, resulted from having to wait for supplementary statistics from Britain, information which never arrived. He had hoped to be able 'to shew the Canadians that they have something of value in their country.'[17] This had been the second of several such disappointments, as Logan complained to J.W. Dawson: fossils he had carried to Britain two years earlier were either 'too scanty or perhaps (a more likely case) they were examined too carelessly' by the secretary of the Geological Society of London to lead to a correct conclusion. 'I wish I had trusted more to myself.'[18] During the winter of 1843–4 Logan instead turned to Albany to compare the fossils he had collected with those found in the formations of New York state. The contrast with his experience with the British scientists prompted Logan to note in his next

'Report of Progress' the readiness with which James Hall (1811–98) and Ebenezer Emmons (1799–1863) of the New York State Geological Survey facilitated his investigation.

In the same report, Logan depicted the geological features of Canada as parts of more extended geological provinces of North America. His so-called 'Western Division' included the north shore of the St Lawrence Valley east to Quebec City; the 'Eastern Division', the south shore of the St Lawrence and eastward; the 'Northern Division,' a continuous line of Primary rock north of the limestone which formed a backbone across the entire province. The survey of the enormous limestone formation from Georgian Bay south to Lake Erie was carried out by Logan's assistant, Alexander Murray. Murray confirmed Logan's suspicion that it was unwarranted 'reasonably to anticipate the occurrence of any part of those true [coal] measures' in that district.

Having struck this first blow to popular hopes, Logan was not about to follow through without some cushioning. While Murray was busy with the limestone, Logan had worked to ascertain the northern limits of the coalfields of Nova Scotia and New Brunswick. He held that the peculiar conditions of the area for the time being justified a non-committal stance on the question of coal in the Gaspé. The Eastern Division, it seemed, was distinguished by 'violent contortions of the strata, the altered nature of some of the rocks, and the want of conformability in probably more than one member of the series of formations.' Pending another season's work in the Gaspé, Logan would say only that some materials of economic value had been observed there.[19]

Logan's hesitation in substantiating his suspicions that workable coal deposits would not be found in the Gaspé appeared to him to be necessary for political as well as scientific reasons. The desire for coal and for the industrial potential it implied had been a major concern of those who had originally supported the GSC. The 'great importance' to the Province of Canada 'that its Mines and Mineral wealth should be properly worked and brought into useful operation' was reiterated in the Canadian legislature with regard to 'the winning and getting of Coal' in the Gaspé district during Logan's first survey of the area. Many insisted that Canada *must* contain workable coal deposits, and they did not wish to be told otherwise. Logan was well aware of this fact, and he understood only too well one of the ironies of his task: the very predictability afforded by his scientific method could easily foster public disillusionment. A measure of this possibility appeared in the reaction of one of the survey's earliest and most ardent supporters. William Dunlop ridiculed Logan's conclusion that there was no coal in Canada as 'a statement of theoretical reasoners which he thought had no foundation.' Dunlop protested that 'he could not see because the coal vein in Pennsylvania dipped upward that therefore there should be coal up in the moon.' He insisted that 'he himself had seen coal on the banks of the Ottawa.'[20]

If Logan hoped to be able to justify continued public funding of the GSC, it was important to be seen to be accomplishing something positive. He accordingly devoted the remainder of his first report to the non-fossiliferous formations of the Northern Division, including the northern shores of the upper Great Lakes. These 'Inferior Rocks,' he suggested, contained iron ores in an 'extraordinary abundance,' which 'may render them of great importance in an economic point of view.' Arguing from rather questionable analogy, Logan pointed to similar valuable ores in New York State:

Such extraordinary masses of iron ore, one would suppose, cannot fail to become of national importance, and when we consider that valuable deposits of the same mineral quality are already known in Canada, in the townships of Marmora, Madoc, Bedford, Bastard, Hull and other places, and reflect upon the great extent of the primary regions in so many parts of which the magnet is deflected from its meridian, most probably by the proximity of the magnetic oxide, it is not unreasonable to hope that a diligent search may disclose provincial beds of equal consequence.[21]

Logan stretched the bounds of caution and moderation still further when he volunteered:

It is at the summit of the rocks under description, in the peninsula lying between Lake Superior and Lake Michigan, in a great range of trap interposed between the transition series and a metamorphic group, which rests upon the granite, that Mr. Douglas [sic] Houghton, the State geologist of Michigan, has made the discovery of an important collection of copper ore veins, which are likely to become of considerable economic value, and it yet remains to be ascertained whether an analogous condition of circumstances may not extend to Canada.[22]

For had not copper ores already been found, he added, in several localities there?

Logan had in fact never visited these Primary rocks of the Northern Division, and he drew most of his bold inferences from little more than hearsay. Only later did he justify his conjectures by referring to Douglass Houghton's Michigan report of 1841 and his understanding that Houghton 'had visited the British shores of Lake Superior, and considered their mineral character much the same as that of his own side of the water, though I believe he has made no published statement to that effect.' A truer picture emerged from Houghton's own report on the south shore of Lake Superior during the summer of 1840. Houghton did not set foot on the British shore for very long, if at all. He had only surmised from the geological character of that general district that mineral veins might be found 'either directly upon or not far from the coast of the lake.' Houghton was later credited with creating a positive image of Michigan's 'hitherto despised Upper Peninsula'; but he admitted that because of the wild terrain even on the south shore of the enormous lake, he had 'traced no one vein for a further distance than one mile, and usually for a distance considerably less.'[23]

Although Logan undoubtedly did not intend it, his allegations of valuable ores in the Primary formations of Canada unleashed an important chain reaction of events. The report of 1843 appeared just when Canadians were beginning to question Canada's role in the future of British North America. But increasing scales of economy and technology also conspired to hurry these intellectual changes along. Inspired by Logan's report, some influential Canadians turned their interest to territories far beyond the limits of Canadian settlement.

III

The link in Logan's outlook between the unsuccessful search for coal and the more promising search for copper and iron ore deposits increased in importance as the time to renew the GSC's grant drew near. Logan and Murray spent the summer of 1844 as promised, back in the Gaspé district, attempting to sort out the contorted and inverted strata at least well enough to decide whether the region contained coal. The 'Report of Progress' was written with special care 'to produce an effect' on members of the legislature. Logan intended to curry favour by proposing the establishment of a museum of economic geology like the one that so impressed the public in South Wales. Experience had convinced him that the larger the specimens he displayed, the more valuable the deposits would appear to be, particularly 'in the minds of the unlearned.' Logan therefore asked Murray to procure 'a thundering piece of gypsum ... as white as possible,' along with a huge slab of lithographic stone.[24]

In the meantime, Logan's disappointment in the British geological survey drove the Canadian survey to closer co-operation with the New York State survey. Not only had De la Beche continued to neglect Logan's letters, but for over a year he had not even acknowledged receipt of Logan's Nova Scotia specimens, let alone classified them. Logan was further disillusioned when the Geological Society of London refused to publish his detailed report on the coal section at Joggins. Once again he undertook the two days' journey to Albany, where he was well received and permitted to copy all the collected and classified fossils for comparison with the growing Canadian collection. 'Until I can contrive to get a comparison instituted with British fossils,' he concluded, 'I fancy I shall be obliged to content myself with the American nomenclature.' But by the end of 1844 Logan's conception of Canadian geology had shifted towards a more independent stance. Not only did he now concede that 'it will only be when the detail is ascertained that generalizations here will be usefully compared with generalizations in Britain'; he furthermore decided that despite the availability of American work he would 'refer to nobody's system in coming to conclusions.' Logan resolved to refrain from making 'one rock contemporaneous with another until I can run them to a junction or something like it.'[25]

The 'Report of Progress' for 1844 was submitted in May 1845. Of necessity, it presented as much a topographical as a geological survey of the Gaspé, a 'herculean task' aided somewhat by the introduction of Rochon's micrometer to measure rivers in unexplored country. The rocks were violently contorted and their strata often inverted. Frequent reports of bituminous minerals resembling coal forced Logan time and again to state that 'none of the material where it has come before me *in situ*, bears any analogy in the mode of its occurrence to workable coal.' Fishermen who referred to deposits of jet black shale as coal were mistaken, he explained, because the Carboniferous series in Canada was too low down to be associated with profitable seams of coal. The discovery of such coal-seams in the Gaspé district would, Logan finally admitted, be 'contrary to present geological experience.'[26] The bitter conclusions were clear: Logan considered a narrow margin on the north shore of the Bay of Chaleur to be the northern limit of the great eastern coalfield of North America. There was no hope of finding workable coal-seams in the Gaspé, hence in all that was known of the province of Canada.

Once again Logan attempted to soften this blow with positive news. He reported topographical details which, he explained privately, could be used 'in case the economic facts of Canadian geology should turn out a negative quantity.' The coastal margins contained fertilizers to improve the agricultural productivity of these rocky soils. In addition, Logan presented fossil evidence of a new species of trilobite; he wished to honour Canadians by calling it *Brontes canadensis*. Finally, he alluded vaguely to the nearby presence at Bathurst, New Brunswick, of copper ores. Even though earlier mining efforts there had been abandoned, Logan insisted that 'mines of this character have, however, occasionally been successful, and a locality, which I had occasion to visit in 1834, in the Spanish Pyrennes [sic]' contained similar deposits in even shallower beds and still 'promised a profitable return.' These suggestive comments bordered on the misleading and fell outside the realm of Logan's usual claim that 'facts are the great desiderata at present in Geology, & he deserves well of geologists who takes pains to collect them.'[27] Once again political motivations joined Logan's personal interest in and experience with copper ores and coal; for early in 1845 the GSC faced its first crucial political test.

Well aware that the original grant was insufficient to carry out a complete geological survey of Canada, Logan had quite consciously used the first two years of the survey to 'make friends, show the utility of the undertaking, and excite some interest in the subject among the legislators.' Despite negative results in the search for coal, he succeeded in all of these social aims and perhaps most of all in the latter one. Not wanting to limit support for the survey to any political party or faction, Logan adopted a cross-bench strategy: 'Avoiding politics as I would poison,' he explained, 'I made friends on both sides of the question.'[28]

This strategy attained a special importance because of the mixed reaction of the Canadian public to Logan's work. Many Canadians remained unwilling to accept his denial of the existence of coal in the province. In one incident at Weston, near Toronto, a group of adventurers who discovered bituminous shales took matters into their own hands because they feared the government would confiscate their mineral rights if they did manage to strike a potential coal mine. They employed an old miner to bore a shaft ever deeper below the level of the Humber River, halting the operation only when hope outlasted their finances.[29]

Certainly the eccentricity of Logan's personal habits contributed to the suspicion with which both he and his science were sometimes received. Whether in the city or the field, Logan displayed a singular disregard for appearances, and he was often mistaken for the caretaker of the survey's offices in Montreal instead of the wealthy and cultured man that he was. During the weekend in New York when he met Charles Lyell, Logan seemed equally concerned about a pet turtle that refused to catch flies; following a chat with Lyell he ran out to hunt for worms. And after three months in the field, he described himself as a scarecrow,

hair matted with spruce gum, a beard red, with two patches of white on one side, a pair of cracked spectacles, a red flannel shirt, a waistcoat with patches on the left pocket, where some sulphuric acid, which I carry in a small vial to try for the presence of lime in the rocks, had leaked through – a jacket of moleskin, shining with grease and trousers patched on one leg and with a burnt hole in the other leg; with beef boots – Canada boots as they are called – torn and roughened all over with scraping on the stumps and branches of trees, and patched on the legs with sundry pieces of leather and divers colors, a broad rimmed and round-topped hat, once white but now no color and battered into all shapes.[30]

On more than one occasion in the field, observers became convinced that Logan had escaped from a nearby lunatic asylum. Given the rather eccentric behaviour typical during field studies, and the rarity of such a scene in Canada, this assumption may not have been as absurd as it first appears. While geologizing in the Gaspé, Logan measured sections of exposed strata at Percé by pacing diagonally across the beach in repeated patterns, mumbling numbers and stopping only to jot these in his notebook. He hammered at ordinary-looking rocks, wrapped pieces of broken stone in paper, and hoarded them in his room at the local boarding-house. Quite oblivious to his surroundings, except where local topographical features appeared relevant to his recorded measurements, Logan reacted to anyone who crossed his path politely, but nevertheless as an interruption of something more important. He often neglected to explain the purpose of his activities to local citizens unless they dared to ask. In a bizarre episode near Quebec, the driver of a calèche hired to take Logan to geologize at Beauport mistook the leather covers of his prismatic compass and clinometer, both hanging on his belt, for a holster and a sheath. When Logan halted at intervals to hammer nearby rocks, the imaginative driver panicked completely, believing the stones would be used as additional weapons against him. He drove without stopping to the nearest asylum, calming down only after Logan overpaid him so generously that other drivers offered to risk their lives by taking Logan anywhere he wished to go.[31]

When the issue of continued financial support for the GSC came before the House of Assembly in January 1845, it took the form of a government bill that Logan himself had written. Logan claimed 'there was not a dissentient voice' in discussion and that the survey was 'the only subject of the session in which all agreed.' The 'Act to Make Provision for a Geological Survey of This Province (8 Vict., c. 16) was in fact discussed on at least three occasions as a 'bill to grant a sum of money for the *completion* of the Geological Survey' (italics added). Prior to the meeting of the committee of the whole House, Attorney General James Smith declared himself 'anxious to meet the suggestions' of several members. Robert Baldwin (Reform, York) and George Moffatt (Tory, Montreal) had asked to see collections of specimens preserved and exhibited. William Dunlop, still disappointed by the apparent lack of coal, supported the survey as an instalment in a larger statistical inventory of the resources of the country, which he hoped would eventually be conducted in 'the fullest possible manner.' Mines were desirable, he argued, as a means to build up a home market and so that 'the farmer might be enabled to convert his corn into some more durable commodities.' Such mines, he hoped, would soon be located by the survey.[32]

Logan succeeded in achieving legislative recognition that it was 'expedient that the said Survey should be continued to a completion.' Since the act passed on 17 March specified only that funding would be granted at the rate of not more than £2,000 annually for the next five years, the terms of 1845 did not set a time limit for the actual completion of the survey. The grant could conceivably be renewed or even increased after five years, if Logan could show the amount 'inadequate for the effectual investigation of so extensive a Territory as is comprised within the limits of the Province.'[33]

In exchange, the provincial geologist's duties were now specified. He was to provide an 'accurate and complete Geological Survey' of the province and 'a full and scientific description of its Rocks, Soils and Minerals,' with 'proper Maps, Diagrams, and Drawings, together with a collection of Specimens'; and to distribute duplicates of this provincial collection to literary and educational institutions. Logan also reluctantly agreed to submit annual reports of progress to the governor general, privately determining to report 'facts and no theory.' Logan was concerned that even a merely factual report might 'enable others to anticipate any results that might be beneficial to me.' However, he decided that his 'final, full, and scientific description' of Canada's geology would be submitted only after the completion of the survey. Meanwhile, he would concentrate on 'such parts of the subject as may have an immediate bearing on matter[s] of economic importance, whether of a positive or negative character.'[34]

A final result of public debate over the continuation of the GSC was public access to Logan's earlier reports to Governor General Metcalfe. These reports drew attention to Canada's geographical extremities, both east and west, and to Canada's changing relationships to these regions. The new interest was reflected in a rash of newspaper articles on the political and economic implications of Logan's remarks on Canada's coal and copper resources. The hope that coal could still be discovered as an outcrop of the great eastern coalfield did not die easily. In 1845 the incorporation of the British-backed Gaspé Fishing and Coal Mining Company was hailed as proof of increased public confidence in the future prospects of Canada. The success of the undertaking, it was felt, would 'accelerate the filling up of the colony's waste lands with the surplus capital and population of the Mother Country,' lands that had been left idle too long or 'abandoned to the intrusion of alien Americans.' The result would be to 'add materially to the strength of the loyal British population of Canada.'[35]

The horizons of this eastward vision extended beyond the dubious prospects of the Gaspé region to the certain and well-known coal deposits of Nova Scotia. Business in Canada had been lively for the previous twelve months, noted the *British Colonist*, and less firewood was being produced for fuel. It was therefore unlikely that firewood prices would ever be lowered again. 'What was to be done for fuel?' the paper asked pointedly. Nova Scotia coal mines, suggested the editor, could more profitably supply coal to the St Lawrence region than to the United States. Soon Canadian towns and cities would convert to coal altogether – 'We hope the Nova Scotians will think of it.' For the time being, large-scale British North American trade was hampered by imperial duties, which made the colonial economies incompatible with one another.[36] This situation would soon change with the British move towards free trade.

IV

At the same time, public attention was drawn to the westernmost boundaries of the province, to the shores of Lake Superior. Several years earlier Douglass Houghton's geological reports had precipitated a prospecting boom for copper in Michigan's Upper Peninsula. As a result of Logan's reports, editors in both sections of Canada entertained no doubts that the Canadian side of the lake was 'as rich as that to the south-west' and that 'the natural characters of the country are precisely the same.' Developments on the American side were followed as 'highly interesting here, because the eastern shores of Lake Superior and the Sault Ste. Marie belong to this Province, and offer, beyond doubt, an equally promising field for mining operations.'[37]

If these mines fulfilled their promise, they would be of great importance to British North America, not only by increasing the mining population and encouraging fishing, smelting, and lumbering industries but also by revitalizing the flow of trade from the excellent harbours of Lake Superior through the St Lawrence system. As a result, one editor predicted, 'Ten years from now we shall be as independent of Great Britain or any of her colonies, for the article of copper, as we are now of the world in that of lead.' Another projected that 'the time is not far distant' when the Lake Superior region 'will be regarded, from a better knowledge of its treasures, as the most valuable portion of the British possessions on this continent.'[38]

By 1846 Logan's reports had excited more than abstract interest in the geology of Canada among its legislators. It was precisely the businessmen and speculators among them who took up financial interests in the copper ores of Lake Superior. In this way, the expanding perceptual horizons of certain influential Canadians and their reliance upon science to support their interests were fast becoming an institutionalized feature of Canadian development.

Several factors help to explain the rising interest in Lake Superior copper. In addition to its age-old uses in construction industries, shipbuilding, coinage, and the military, copper ranked with iron and zinc as the three indispensable metals of the Victorian industrial age. The only metal that occurs abundantly both in native form and in various ores, copper was known as a conductor of heat and electricity. It promoted the burgeoning electrical industry, and by the 1840s it was used in the wiring of electric telegraphs.[39]

William Logan's experience at the Forest Copper Works near Swansea had crystallized his predisposition to view copper as a symbol of the ineluctable progress of British civilization. Changing technologies in copper smelting elevated Swansea to the copper capital of the world, mainly because coal was introduced as the smelting fuel. The proximity of coal deposits brought the smelting industries to Swansea rather than to Cornwall, where the copper was mined, since it was cheaper to transport copper ores than coal. Swansea's decline began only after Logan had left for Canada with memories of the peak of its productiveness.[40]

Logan instinctively considered commercial factors when deciding which territories to survey. He did not believe that the time was ripe for financial investment in the mineral resources of Lake Superior, since he was well aware of the need to consider location and related transportation costs when assessing the commercial value of mineral deposits. But he held out the hope that although there was no coal in Canada anywhere near the Lake Superior copper deposits, some might eventually be discovered in British territories farther to the west.[41]

By 1846 the Canadian government had received more than 160 applications for licences to explore the mineral lands on the Canadian shores of Lakes Huron and Superior. Most of the applicants were not themselves prospectors but prominent businessmen and politicians from major cities and towns. The sessions of 1846 and 1847 were bombarded with petitions to incorporate mining companies involving many of these same people.[42] The resulting discussions raised several crucial questions: who was responsible for the colony's economic development; how was control to be maintained over these regions; and what path of development was to be pursued, and how quickly?

The first ad hoc decisions were made through a committee chaired by Étienne Parent in 1845: companies of foreigners would not be incorporated; non-residents would not be licensed; and non-British subjects would not be permitted to work the mines. Nor would transfers of interest be recognized unless the parties were at least British subjects, so as to avoid 'the perversion of the intentions of the Government in granting these Licences.' For along with the 'necessity of developing the mineral and other resources of these regions' went the fear that 'if Canada did not do so, the Americans would.'[43]

This fear was played upon by licensees lobbying for more mining privileges. When the Reform watchdog Thomas Aylwin protested against the apparent advantages enjoyed by those with easy access to the government, John Prince

(Ind., Essex) pointed to his own explorations on the shores of Lake Superior as an indication that 'every possible means had been taken to make known the resources of the country.' Instead of complaints, he felt he deserved thanks for removing 'the reproach so frequently thrown out against the Canadians by the people of the other side' of the border, 'that their sluggishness was so great that it would be necessary to root them out, to make room for more active men.'[44]

Representatives of business and government united in the assumption that science could arbitrate disputes over developmental policy in Canada. Parent's committee designated 'a rough Geological Inquiry by a scientific man' as the prerequisite for future decisions and suspended licensing grants 'until the Provincial Geologist, or some other scientific Agent of the Government' established official boundaries of mining lots and filed a general report. John Prince, in response, went to considerable expense to obtain reports and specimens gathered by 'scientific men' in order to impress his case upon the Crown Lands Department.[45]

The blatant rhetoric which inevitably accompanied such appeals was carried a step farther in 1846 with the petition of the British American Mining Company for incorporation to explore and work mines not only on the shores of Lake Superior but anywhere in the province. The founders of the company were Montreal businessmen, including Peter McGill, George Moffatt, and James Logan. The trio petitioned both the assembly and the governor general for special privileges. In a personal appeal to the Earl of Cathcart (Lord Greenock), James Logan was not the silent partner, as he usually was in his business dealings. Cathcart, the interim governor general between Metcalfe and Elgin, was an accomplished geologist who had written several papers for the Royal Society of Edinburgh and had discovered a new sulphate of cadmium known as Greenockite. Cathcart promised Logan his support, contingent upon that of the assembly. The sponsor of the bill was John A. Macdonald, who withdrew it when it was attacked as a 'dangerous assumption' of power by a joint-stock company with only limited liabilities.[46]

McGill, Moffatt, and Logan were more successful a year later with the incorporation of a far more ambitious enterprise known as the Montreal Mining Company. Moffatt described the company as 'a voluntary association of persons to whom, individually, the government had granted the right to explore' lands around Lakes Superior and Huron. The original list of grant-holders included prominent Montreal businessmen as well as W.B. Jarvis of Toronto; Dr James Wilson, the amateur geologist of Perth; S.B. Harrison (Reform, Kent); and Sir George Simpson, governor of the Hudson's Bay Company. The new company claimed rights over most of the land granted for exploration in that area, but its incorporation was nevertheless agreed to on the ground that it was 'of great importance to this Province that its mines and mineral wealth should be properly worked and brought into use.'[47]

William Logan suggested in 1846 that this principle should indeed be carried out, since he adhered to the view of the government as the landholder and therefore the owner of the minerals it contained:

If a system of leases should be adopted by the Provincial Government the term granted should in my opinion be a long one. I should not feel disposed to place confidence in the bona fide *mining* intention of any company of adventurers who would take a short one. A mine ... can scarcely be properly worked without considerable outlay to put it into productive condition, particularly in a new locality, at a distance from a well settled country, and from a market with which to establish a traffic; and it is but reasonable that the adventurers should have ample time to receive it back, with a large profit to reward their enterprise.[48]

Logan's advice was followed quite closely, and he was appointed the final arbiter of 'the general interest.' It appeared to be in the general interest to spend money to develop mines; since government had none to spend, it was relying upon private enterprise to bridge this gap.[49]

As provincial geologist, William Logan wielded considerable power to interpret the natural resources of the country, and to determine which regions and which minerals were to be developed.[50] He played the role of an authoritative middleman between government and business interests. Business, to a great extent, meant the Montreal business community, of which Logan had never ceased to be both a financial and a spiritual member. Logan's dual allegiance to science and the Montreal business world, and George Simpson's to both the Montreal Mining Company and the Hudson's Bay Company, resonated in tones which heralded the path of Canada's future development.

v

Simpson had been an active founder of the Montreal Mining Company. In 1846 he retained the services of an American promoter named Forrest Shepherd to report on mineral lands of Lake Superior and to select the locations of the company's lots. In November Shepherd issued a glowing report on the margin of Lake Superior as 'one vast laboratory for the fabrication of metalliferous veins' even more promising than the American side. Never shy of hyperbole, Shepherd claimed that even William Logan was 'wearied in marking down the noble veins he discovered in every quarter'; Logan, on the other hand, described Shepherd's lots in his private journal as 'not so striking as I anticipated.' Shepherd reported that he had joined Logan in his survey and therefore had to postpone the extraction of copper ores until the following season. After all, he reminded shareholders, the company needed a favourable report from Logan if they were to obtain title to their lands. Logan's labours, Shepherd shrewdly noted, not only enhanced the value of the lands, but also 'secure for him in the scientific world a well deserved reputation as enduring as the rocks and waters' which he studied.[51]

At a general meeting, the shareholders of the Montreal Mining Company were assured by the firm's first president, George Moffatt, that they held 'a strong interest in a national point of view' and that the company would 'appear before Parliament as a benefactor, having done much for the advantage of the country.' Several Montreal papers took up this cry, adding that the illustrious names linked to the company were 'sufficient guarantee that nothing will be wanting to ensure success' and 'command the confidence and cooperation of the community at large.' Both the *Herald* and the *Gazette* envisioned the Lake Superior region as the 'Cornwall of Canada, occupied with a thriving busy population, and increased employment given to our shipping in transportation of ores and supplies.'[52]

Upper Canadian papers were generally less sanguine regarding the clandestine government activities of which they felt Montreal insiders had been permitted to take advantage. The entire tract of land above Lake Superior, they argued, 'ought to have been carefully examined by competent men' – not, in other words, by interested men like Forrest Shepherd. But even the Toronto *Globe* agreed that 'none will rejoice more sincerely than we will if the Lake Superior mines turn out well; all we want is justice to the public in the disposal of the lands – and caution in entering on the undertaking.'[53]

Such caution was well justified, as the company's perilous beginnings proved. The formidable difficulties involved in setting up a mining industry in such isolated territory required long-term financial commitments. In this light, Shepherd's advice to the shareholders 'to set off at least a portion of your property for the capital of some enterprising and experienced adventurers,' so as to 'secure a fair return to the pioneers and proprietors of the soil,' seemed reasonable. But the conservative policy of tying up the mineral lands north of Superior without expending capital for their development has often been blamed for the relatively slow economic growth of the region. Indeed, it was not long before William Logan showed his disgust with the management of the company, first by advising his brother James to reduce his stock holdings and later by refusing to comply with the company's request that he boost its stock values by verifying his previous surveys and 'reporting to the stockholders if the promise of it still held good.'[54]

George Simpson had engaged the services of Forrest Shepherd at a time when, as governor of the Hudson's Bay Company (HBC), he was hardly looking to end the isolation of the Lake Superior region. Since its merger with the North West Company in 1821, the HBC's trade rights excluded all the area that drained into Lake Superior and lay within the province of Canada. But new sources of opposition to the company's monopoly emanated from this region, which had thus become for Simpson a frontier to defend his own interests and provide 'a cover and protection to our own proper Country.' To the end of the 1830s, these new sources of opposition included not only the rival American Fur Company but also lumbering and fishing companies which threatened to detach Indian allies from the HBC's control. The latter developments appeared all the more threatening to the HBC because its truce with the American Fur Company expired in 1844. Now the discovery of copper and other minerals heralded the advent of a mining population, which Simpson feared would further destabilize the delicate traditional balance. The advance of settlements did not in itself necessarily contravene the long-term interests of the HBC, but the implied loss of control did. For the purpose of maintaining company control over its territories, Simpson used techniques similar to Logan's; he cultivated friends in the House of Assembly.[55]

But Simpson's personal interest in the Montreal Mining Company also made formidable enemies for the HBC. Among them was Allan Macdonell (1808–88), a Toronto lawyer and promoter who helped organize the rival Quebec and Lake Superior Mining Association in 1847. Macdonell's interest in the north-west and his distrust of the HBC derived from family connections and were intensified by his mining experiences. Macdonell's grandfather had been a fur trader whom Simpson disliked, and Macdonell's resentment increased when the HBC denied his association the aid of its guides and voyageurs over the rough terrain north of Superior. The HBC instead gave every assistance to the Montreal Mining Company's expedition, led by Forrest Shepherd. In return, Macdonell refused to report to HBC forts along Lake Superior, thus arousing even deeper suspicions in the company, which maintained a vital post at Michipicoten.[56]

The Quebec and Lake Superior Mining Association began operations on Michipicoten Island after 'formidable preparations' that included the erection of smelting furnaces. But the expense of extracting ores rich enough to bear the cost of smelting and transportation proved nearly prohibitive. After selling his shares in the association in 1849, Macdonell led an '*émeute*' of Indians and Métis to drive out the miners. He claimed their aim was to draw government attention to the land claims of native peoples against the mining companies. The loss of this 'Michipicoten War' led to a treaty signed in 1850 between W.B. Robinson, on behalf of the Canadian government, and Macdonell, among others, on behalf of the Ojibway people inhabiting the shores of Lakes Huron and Superior. The Ojibway ceded their claims to the land in exchange for reservations and an annual payment.

While Macdonell thus apparently lost out to Simpson and his allies in the Reform government, at least insofar as the government granted control of this annuity to the HBC, this victory would in the long run prove costly.[57] Most of the mining population left the north shore of Lake Superior by 1848. But for the ambitious Macdonell and his cohorts, the mining potential of the Lake Superior region described in Logan's reports was only the thin edge of the wedge; next he helped form the ideological vanguard of a movement to drive the HBC out of its territorial domains in British North America and to replace its hegemony by that of a greatly expanded Canada.

Logan's reports had cast the western fringes of Canada into the limelight and inspired Macdonell and others to strengthen links between the periphery and the settled core of the province. One of these links was a proposed ship canal at Sault Ste Marie. The first mention of such a canal appeared in Maj. George Phillpotts's second 'Report on the Inland Navigation of the Canadas,' submitted to the British government in 1840. Phillpotts, a Royal Engineer, had been appointed by Lord Durham to conduct a survey with the aim of opening water communications between Lake Erie and the Atlantic Ocean. At the time, Phillpotts was alone in his belief that his task would be incomplete if notice were not taken of Lake Superior as well. By 1847 the situation had altered considerably. First, in 1846 a Canadian government inquiry into the province's expenditures on public works, especially canals, had vindicated the principle that such works promoted the general interests of the country. Second, the economic blow delivered by Britain's repeal of the Corn Laws in 1846 motivated commercial interests to snatch victory from the jaws of defeat by taking command of trade on the Great Lakes.[58]

Directors of the major mining companies on the lakes – the Montreal Mining Company, the Quebec and Lake Superior Mining Association, and the British North American Company – favoured a canal at Sault Ste Marie for obvious reasons but did not form a cohesive lobby. The *British American Journal of Medical and Physical Science* went so far as to suggest that such an improvement would enable Canadian products to compete with English mines, even if they had to be transported to Swansea for smelting. In 1847 the commissioner of public works, Hamilton H. Killaly, was sent to survey the St Mary's River for possible canal construction sites and he reported that geography actually favoured the Canadian side of the river for such purposes. Killaly pointed out that copper mines had already laid the future foundation of a considerable trade in the region; he did not see 'the slightest probability of any improvements ever taking place there, unless they should be induced either by the construction of the proposed Canal, or by the improvement of the Portage Road, with a good Pier at each end.'[59]

Yet even Killaly had to admit that a dilemma existed. He recognized that financial returns for the foreseeable future could not yet justify the expenditure for such a canal. Nevertheless, the House of Assembly continued to receive petitions for a canal at Sault Ste Marie, usually from representatives of mining companies. In 1851 Allan Macdonell and his brother petitioned for incorporation as the Sault Ste Marie Canal Company, but after several divisions in the House the question received a six months' hoist. This process was repeated two years later, and added to it were several other attempts, in particular by W.B. Robinson, former chairman of the Board of Works, to force Francis Hincks's Reform government to sanction the construction of such a canal, whether by public or by private enterprise. All efforts were thwarted personally by Hincks, who seemed so extraordinarily determined to prevent this measure that some suspected Hincks owned shares in the American Sault Ste Marie Company. Yet no connection between Hincks and any interests opposed to Allan Macdonell's project were ever proven.[60]

As was no doubt feared by the HBC, the attention of Macdonell and his associates rapidly progressed beyond the headwaters of Lake Superior and the boundaries of the province. One of the most prominent of these associates was George Brown, owner (with his brother Gordon) and editor of *The Globe* of Toronto. Brown had been critical of the copper mining speculation on the shores of Lake Superior and had carefully differentiated for his readers the two crucial parts of its paradox, namely the undeniable enormous mineral wealth of the Lake Superior region, on the one hand, and the entirely questionable probability that mines could be worked there in the near future, on the other hand.[61]

In 1850 Brown published a lengthy article on the value of 'Lake Superior and the Northern Country.' He was convinced that the 'injurious and demoralizing sway of the Company' would 'ere long be brought to an end, and that the destinies of this immense country will be united with our own.' Two key arguments, as Brown saw it, were, first, that the Lake Superior copper mines and the vast resources of the lands farther west would provide Canada with an important source of trade; and second, that a canal at Sault Ste Marie would secure for Upper Canada all the trade flowing through Superior and Huron, and from there by rail to Toronto. Brown concluded: 'Independent of the hope that the high road for the Pacific may yet take this direction – there is a field of enterprize presented sufficient to satiate the warmest imagination.' But if the work were much longer deferred, he added, 'our [American] neighbours will no doubt be at it.'[62]

Allan Macdonell's contagious enthusiasm had expanded his own views as well. In 1851 he and his associates petitioned the assembly for a charter to build a railway from Canada to the Pacific Ocean, knowing full well that they did not have the consent of either the imperial government or the HBC for such a project. The petition received consideration in the Standing Committee on Railways and Telegraph Lines, chaired by Allan MacNab, who also had investments in Lake Superior copper mines and in whose law office Macdonell had been employed. Although the committee issued a pro forma negative response to the petition, it reprinted for distribution a pamphlet by Macdonell on the subject.[63]

Macdonell argued in Newtonian language that in world commerce Britain was 'what the principle of gravitation was to the material world – that which regulates and upholds it all.' With a canal at Sault Ste Marie and a railway to the Pacific, this commercial lead could be safeguarded against parallel American advances. 'Civilization, with all its influences, would march, step by step, with the road,' according to Macdonell's plan, across a territory that 'abounds in mineral.' He even speculated that, 'above all the blessings and advantages that can be conferred upon a country like this,' coal was 'abundant and easily obtained; it crops out in various parts of the [Saskatchewan] valley.' While admitting that there was 'something startling' in his proposition for a transcontinental railway, Macdonell urged his readers not to 'insult the enterprize of this enlightened age by denouncing as visionary and impracticable the plan of a simple line of rails over a surface of no greater extent without one half of the natural obstacles to overcome' that the pyramids of Egypt or the Great Wall of China had overcome. 'To do so would evince a forgetfulness of the vast achievement of this age.' Macdonell urged with an air of scientific detachment that 'nature invites the enterprize,' which would unite all the British territories 'by means of her Canadian Empire, bound together with sinews of iron.' In addition, Canada was offered all the opulence and power represented by the Far Eastern trade: 'Like the Genii in the fable,' the prospect offered 'the casket and the sceptre to those who, unintimidated by the terrors that surround it, are bold enough to adventure to its embrace. In turn Phoenicia, Carthage, Greece, Rome, Venice, Pisa, Genoa, Portugal, Holland, and lastly England, has won and worn this ocean diadem; Destiny now offers it to us.'[64]

For the next decade Macdonell's pamphlet served as propaganda for promoters of Canadian expansion to the north-west, catalysing the ultimate demise of the HBC in those territories.[65] While intertwining the grand themes of the mid-

Victorian age (the immense profits of colonization and development, with minimal outlay; material progress; and imperialism) Macdonell's appeal was also a striking example of central Canadian expansionist rhetoric. The idea of a transcontinental nation in this case was an outgrowth of the expansionism McGee thought to be its progeny, a grandchild of the geological inventory which had helped inspire Macdonell's sentiments.

VI

The first decade of the Geological Survey of Canada witnessed changing options and priorities for the province of Canada. Debate over the survey's future revealed the aspirations and uncertainties involved in this process, making some goals appear more viable and more attractive than others. This was especially true during the period of widespread commercial insecurity precipitated by the repeal of the Corn Laws in 1846. Insecurities ran considerably deeper than mere trade problems and gave cause to question Canadians' relationship with the rest of the British empire.[66]

A related question was whether Canada needed manufacturing industries in order to guarantee its separate existence in North America. 'We are aware,' wrote the Toronto *Globe*, 'that the want of that important, but yet undiscovered article, coal, must greatly retard the progress of manufactures, and regard their establishment to any great extent as unlikely, and perhaps not desirable, for a long period, unless absolute necessity arises.' Others bemoaned the fact that Canada had been deprived of 'one important element of national prosperity' in the absence of coal. Logan himself pointed out that since it was more economical to carry copper to coal than vice versa, Saginaw Bay appeared to be 'naturally destined' for the reduction of copper ores from Lake Superior; predating the National Policy by a generation, he warned that without technological change, such as applications of electricity he had read about, 'it can only be the operation of fiscal laws that will prevent the Canadian ores from finally reaching the same destination.'[67]

The absence of coal in the colony raised the spectre that within its political boundaries the province of Canada could never be economically self-supporting. The *British American Journal* thought it a 'mortifying historical fact' that after the British conquest the mineral riches of Lake Superior, long known and worked by Indians and Frenchmen, had been ignored and forgotten. The 'unaccountable and disgraceful apathy and ignorance of British Ministers' had lost all the territory between the forty-second and the forty-ninth parallels of latitude. A mass meeting of 'more than two thousand of the most respectable and influential' citizens of Montreal in January 1846 called for the annexation to

British North America of most of the land north of the forty-second parallel. The reasoning behind this dramatic gesture was that as things stood, even the British North American colonies together did not seem to constitute 'a territory of convenient geographical shape to form a nation – that is, anything approaching a square, where each touched upon the other, so that the trade of all tended to one common centre.' Instead, they seemed to resemble more closely 'the selvage along a piece of cloth.'[68]

The futility of the Annexation Manifesto in 1849 convinced many of these people that life on their own selvage was perhaps preferable to a better position on someone else's cloth. One development that drove Montreal businessmen to sign the manifesto was the near-collapse of the international copper trade, a consequence of the European revolutions of 1847–8. The resulting loss of confidence in Canadian mines was mitigated somewhat when the government ordered Logan to resurvey some of the mineral lands in 1849. A difference between those who signed the manifesto and those who instead joined the opposing British American League formed at Kingston was not interest in mineral lands but rather the direction of their orientation. Annexationists, who included the westward-looking Allan Macdonell, saw no encouragement for Canadian manufacturing in the eastern colonies. Yet the British American League, presided over by George Moffatt, believed instead that a union of British North America would foster domestic manufacturing by extending the field of operations for intercolonial trade. Halifax, it was noted by the league, was at the centre of the most extensive coalfields in the world. One member was indignant that, as the eastern colonies produced only coal, timber, and fish, Britain allowed American ships 'to come into *our* three or four ports on the same terms as our vessels; they convey *our* coal, *our* timber and *our* fish to their ports' (italics added).[69] Another delegate believed that

these provinces might form the nucleus of a great and mighty nation. When he looked to the vast extent of territory, and natural resources that the connection would give us, the inexhaustible treasures of coal, the fisheries and timber of the other provinces, he was convinced that nothing but a wise system and policy was necessary to make us a great and prosperous people, and if we could only draw closely around us the bands of union, we should soon be enabled to stand on our own feet, and maintain our position and rights among the nations of the world.[70]

The two groups were divided more by the crisis atmosphere than by ideology. Increased economic stability after 1849 dissipated the crisis, but public awareness of both the potential and the limitations of Canada's natural resources continued to be reflected in the attention paid to the colony's territorial extremities. In 1852 the House of Assembly charged a select committee of 'the Magdalen Islands, and the Western Part of this Province, above Lake Huron,' with investigating under what tenure the lands were occupied; the condition of agriculture, fishing, and other branches of industry, 'whether mines, minerals, or otherwise,' and possible improvements; and the commercial value of the region to the province. With members ex officio interested in these regions, it was a foregone conclusion that both areas would be judged of 'great importance to the future prosperity of the Province.' The committee urged that Canada extend a more visible presence into both regions by establishing court-houses and jails to control lawless American fishermen. The Magdalen Islands constituted Canada's first line of defence on the Gulf of St Lawrence, and the committee expressed concern that Nova Scotia would annex the islands for similar reasons. Such an event, it declared, would be a 'public misfortune' for Canada. One resident described plentiful supplies of gypsum and possibly coal on the islands and requested that the provincial geologist be sent to ascertain the true value of these deposits.

The committee was astonished to learn from the commissioner of Crown lands that his department possessed no reliable data with which to pinpoint the exact northern and western boundaries of Canada. Most of what was known of the lands north of Lake Huron was culled from Logan's reports; so it was assumed that the area was so rocky that its only sources of wealth were fish and metalliferous ores.[71] Logan and his science had already come to be regarded as 'something of an oracle,' a font of knowledge indispensable to the future progress of the province.

The first decade of Logan's directorship witnessed an extensive survey rather than an intensive geological inventory of the province's rock formations and mineral resources. The renewed financial support of the Canadian public purchased a continuation of this trend, with its concomitant extension of the idea of what Canada was and could become. The growing desire for public acquaintance with the work of the GSC resulted in 1849 in a motion in the House to reprint all Logan's reports of progress. Debate centred not on their presumed value but rather on the issue of government expense for public use. Francis Hincks, a leader of the retrenchment-minded Reform government, argued that if the country were truly to profit from these reports they should be reprinted by a private publisher independent of the House. Supporters of the survey agreed with the decision to refer the question to the Standing Committee on Printing, but the issue provided an opportunity to ridicule publicly those who still doubted Logan's authority as a man of science. These sceptics committed the folly, 'still to some extent prevalent, of *expecting to find* COAL in Canada, notwithstanding the decided *adverse* opinion of our gifted Provincial Geologist on that point.' Since the Act of 1845 was due to expire in 1850, the *British American Journal* emphatically cited 'the puny and undignified, *un-British* scale of the staff of our Provincial Survey, as a national work, compared with the magnificent arrangements of several of the neighbouring States.' Thomas Sterry Hunt had recently joined the survey staff as a chemical analyst, but the serious understaffing of 'so important a work' had, according to the *Journal*, nevertheless 'indefinitely protracted its completion.'[72]

This criticism appeared to represent the general consensus, as little political opposition was raised when the GSC's term came up for renewal in 1850. Logan and his staff indeed trod a fine line between the two main factions in the assembly, striving to appear neutral. Even Alexander Murray earned a stiff private warning from Logan against 'political meddling' shortly before renewal was discussed in the House. Murray apologized that he had 'only been provoked once or twice to express myself publicly either in print or otherwise, by the shameful attacks which have been frequently made upon one of my oldest and kindest friends.' He wished, however, 'to be entirely controlled' and pledged neither to 'write nor talk upon such matters again, until circumstances occur to justify my doing so, in *your* opinion as well as my own.' Logan in the meantime lobbied various members of the legislature – 'out of sight, out of mind' was what he feared most for the survey.[73]

Logan had one other concern, namely the issue of retrenchment, which divided the Reform-dominated executive council. The leader of a movement to curtail expenditures was William Hamilton Merritt, chief commissioner of public works. In 1850 Merritt unveiled a sweeping plan to streamline the public service, to introduce efficiency into the system, and to generate funds for public works on the St Lawrence canal system. Logan feared that Merritt viewed the GSC as added competition for an already limited budget. If Merritt could not abolish the survey, he might reduce its allowance. But the majority of the executive council, led by Francis Hincks, favoured its continuation, and Logan obtained assurances from opposition leaders like Allan MacNab (Hamilton) and even Clear Grit members like Malcolm Cameron (Kent), who sided with Merritt on retrenchment, that the government would not abandon the survey. Logan's reports had proven the survey a desirable means of developing the resources of the province, and the House agreed to revive and continue the Act of 1845 for five more years.[74] Logan could well be satisfied with this first phase of his career as provincial geologist.

4

'Grandeur and Historical Renown,'
1851–1856

The renewed guarantee of its existence, albeit only for a limited period, heralded a new phase in the development of the Geological Survey of Canada as a Canadian institution. Logan and his assistants were drawn into provincial preparations for participation in a series of Universal Exhibitions at which, thanks to Logan's conscientious efforts, Canada received more international recognition for its scientific accomplishments than anyone had dreamed possible.

The first of the series was the Great Exhibition of the Works of Industry of All Nations, held at the Crystal Palace in London during the summer of 1851. Logan's deep involvement, first in preparing and then in presenting Canada's exhibit, initially detracted from the normal work of the GSC. The time-consuming task pre-empted his work in the field for an entire season. Yet his painstakingly collected and systematically organized exhibit of Canadian minerals and his service as a juror at the exhibition repaid this loss with exponential gains.[1] The recognition earned by the Canadian exhibit consolidated Logan's international reputation as both a theoretical scientist and a practical explorer for economic materials. At home it precipitated a collective sense of pride, self-confidence, and self-respect associated with the land and a growing appreciation of its potential value as revealed by the methods of science. The pattern was repeated during this second phase of Logan's career.

I

London's Great Exhibition has been lauded as 'the apotheosis of the lofty [Victorian] ideal of "rational entertainment," ' the climax of a long series of local exhibitions and fairs held in British towns. But as an international spectacle, it departed from British tradition and instead emulated a series of French *expositions* beginning in 1806. The Crystal Palace exhibition was supposed to herald a new era in world diplomacy, to provide both a tangible measure of the progress of mankind and an arena for peaceful competition among nations. These ideas of progress and competition were in fact closely connected. 'We had,' wrote William Whewell (1794-1866), the Cambridge philosopher and historian of science,

> offered to our review, the choicest productions of human art in all nations; or, at least, collections which might be considered as representing all nations. Now in nations compared with nations there is a difference; in a nation compared with itself at an earlier time, there is a progress ... By annihilating the space which separates different nations, we produce a spectacle in which is also annihilated the time which separates one stage of a nation's progress from another.[2]

The organization of the Great Exhibition posed a characteristic problem for the Victorian mind: to find the proper system of classification for the articles exhibited. Organizers originally planned to juxtapose similar items regardless of place of origin, but logistical difficulties instead necessitated a geographical arrangement of foreign and colonial articles. As a result, the Canadian exhibit was allotted space to be presented intact, in effect to stand on its own merits. The Crystal Palace became, according to Whewell, 'the cabinet in which were contained a vast multitude of compositions – not of words, but of things, which we who wandered along its corridors and galleries might con, day by day, so as to possess ourselves, in some measure and according to our ability, of their meaning, power, and spirit.'[3] William Logan, with his experience in establishing the Museum of Economic Geology at Swansea, acted as Canada's curator. In his view, the exhibition was, after all, 'nothing more than a grand and instructive display of the same kind' as the Museum of Economic Geology in London.[4]

Canadian interest in the Great Exhibition was aroused early in 1850, at first in Montreal. Public meetings there nominated a committee to encourage more widespread co-operation. The Conservative Christopher Dunkin foresaw several practical results of a successful exhibit, including increased trade and immigration. More important, he pointed out, Canada would become 'respectable and respected': 'We should reward ourselves by developing the resources of our country, and by bringing together the people in more kindly unison.' George Étienne Cartier agreed that 'Canada ought to take pride in doing her share, or she would be unworthy' not only of 'the rank she holds as a British colony' but of 'her geographical position.'[5] The committee would invite co-operation and assistance from all parts of the province, and it included many prominent Montreal businessmen along with William Logan.

At the same time, a select committee of the House of Assembly took the British invitation 'to compete with us in a spirit of generous and friendly emulation' quite literally. The committee recommended that partial efforts be merged to embrace the whole of united Canada. It feared that without preliminary selections and eliminations, multiple articles of the same kind might be exhibited; this would 'in fact bring the different sections of the Province into competition in England.'[6] In order to present a united front to the world for the first time, a provincial exhibition was held in Montreal in October 1850, subsidized by a government grant for awards and expenses. Commissioners, including Logan, were appointed to select the Canadian exhibit from prize-winning submissions.

The classification of objects at the 1850 Provincial Exhibition followed the pattern determined for the Great Exhibition. Of four main sections, the first, Raw Materials and Produce, illustrated 'the natural productions on which human industry is employed,' and was subdivided into mineral, vegetable, and animal kingdoms. Minerals were classed as metallic, chemical, glass and pottery, or stone. Eligible specimens included 'only those remarkable for their *excellence*, for *novelty* in their occurrence or application, or *economy* of their extraction or preparation; or those remarkable as *illustrations* of some further processes of manufacture.' Logan's own catalogue of Canadian minerals was reprinted in the Montreal *Gazette* to guide and inspire prospective local contributors. A united effort, it was hoped, could overcome cultural and even class conflicts in Canadian society. The prize-winning articles which formed the core of the Canadian exhibit at London were seen to represent 'a detailed account of the geological survey and its fruits' in Canada.[7]

A major theme of Logan's Canadian mineral exhibit at London was that the minerals would become important commercially if only their locality and means of transport could be made known and then developed. Although not all the organizers agreed that mineral specimens should be linked conceptually to the means of extraction and the processes by which they were made useful, Logan certainly used this criterion to determine the specimens he chose as the focal point of the Canadian collection.[8] The key to the exhibit was not copper, since only the lodes at Bruce Mines had yet been worked. The absence of bituminous coal might also be perceived as detrimental to Canada's industrial future, so that its availability at Cleveland or Nova Scotia had to be pointed out. Logan highlighted 'the ores of *iron*,' in which he declared the country 'abounded' from Labrador to Lake Huron. Water-power and wood, he added, offered 'fuel in abundance near all the localities.' Bog-iron ore processed with wood charcoal in Canada was less liable to crack than even high-quality iron imported from Sweden and could, he believed, be produced competitively. Logan utilized the very absence of coal as a selling point to British smelters, who therefore had no reason to apprehend competition with their coal-smelted iron exported to colonies. Logan reported that he had convinced the British public 'to regard the great beds of magnetic oxide as national magazines' which 'it is always satisfactory to the inhabitants of a country to know is within their reach and control.'[9]

Logan knew how to derive advantage from a growing popular concern in Britain that a coal shortage loomed in the not-too-distant future. By the 1850s there was already some fear of resource depletion brought on by decades of reckless exploitation. Coal and iron had made Britain a world power, and unlike other staples, neither was renewable. Unless new mineral fields could be brought into operation in the foreseeable future, Britain would have to 'retrograde' by modifying its manufacture of iron.[10]

Appropriately, the marvellous ferrovitreous 'green-house' construction of the Crystal Palace itself epitomized the 'age of iron' in which it was built. Indeed, some saw iron consumption as a 'social barometer' by which to compare 'the relative height of civilization among nations.' By this measure, the British North American colonies were at last becoming 'civilized.' They grew increasingly capable during the 1850s of producing iron themselves as technology shifted the keystone of the iron industry from the location of coal to the location of the ores.[11] Whether or not all these long-range changes were very widely perceived in British North America, the attention bestowed upon the Canadian exhibit in London created a ripple effect at home.

II

When the Crystal Palace opened in the spring of 1851, Canada was compared to a débutante entering the world stage, and no Canadian, not the governor general and least of all W.E. Logan, entertained expectations of success in competition with the rest of the world. The most that could be hoped for was that Britons would learn something about the province, and that Canada 'should take this opportunity of speaking for itself, in a language pertinent and practical.' Comparisons with other exhibits were inevitable, and it soon became apparent that among the British colonies, Canada's exhibit stood in the first rank for extent and variety. Much meaning was read into the location of the Canadian exhibit on the main floor of the palace, 'at the right hand of Britain.' Canadians took special delight when the Americans appeared to have underestimated the competition and were 'annoyed' that Canada was, for once, 'superior to them.' The *Quebec Morning Chronicle* surmised that the exhibition had even killed off any remaining annexationist sentiment. Traces of embarrassment or inferiority quickly vanished from the Canadian press reports.[12]

Observers who felt that at the outset 'Canada was sneered at, she was nothing,' reported their astonishment at the recognition earned by the Canadian exhibit and drew lessons of increased collective energy, self-reliance, and pride from the experience. 'Our object,' wrote one editor, 'is simply to exalt ourselves ... so that we may be led to increased self-reliance.' The exhibition 'will induce Canadians to think more of themselves and less of their neighbours than they have hitherto been in the habit of doing.'[13] Another was persuaded that the sooner Canada managed 'to do everything within herself the better off she will be.' One had only to think of the 'thousands and thousands of pounds which have been sent out of Canada, and are still going, for machinery, which could be made there, and which would help to enrich the country by keeping our mechanics in it'; it was high time for the country to wake up.[14] The high praise reserved by the committee of jurors for the Canadian exhibit was proof, wrote the *Globe*, that in both skill and raw materials 'Canada stands very high in the scale of nations.'[15]

A particular aspect of the exhibit that was singled out for praise was the mineral collection. Logan himself never tired of citing the impartiality of the jury reports, and of quoting the French inspector general of mines, a juror at the exhibition: 'Of all the British Colonies,' declared M. Dufrenoy, 'Canada is that whose exhibition is the most interesting and the most complete, and one may even say it is superior, so far as the mineral kingdom is concerned, to all the countries that have forwarded their products to the Exhibition.' This superiority arose from the collection's arrangement, for 'the study of it furnishe[d] the means of appreciating at once the geological structure and mineral resources of Canada.'[16]

Awards were carried off by the Montreal Mining Company for smelted cake copper from Bruce Mines, by the Hon. James Ferrier for wrought iron from bog ore smelted at the St Maurice forges, and by the Marmora Iron Company for cast iron extracted from local magnetic oxide. Among the honourable mentions was Dr James Wilson of Perth, for minerals discovered through his own researches. Logan felt sure that the whole collection would have received a higher award if he had not been appointed a juror in the same category.[17]

A permanent result of the distinction earned by Canada at the Great Exhibition was a widespread belief that the enormous potential inherent in the rich natural resources exemplified by Logan's mineral collection now demanded a wider scope for comparison than provincial limitations allowed. Science, and especially geology, grew in status and popularity as it served to extend to a world scale the very measure of Canadian achievement. A more immediate result was Canada's participation in the Industrial Exhibition in New York in 1853, where Logan's mineral collection awakened Americans to Canada's mineral wealth. Indeed, Canadians too were becoming convinced that geological inventory, while not likely to discover gold, nevertheless could reveal 'the geological formation of the country, and expose the more useful minerals which now lie hid to us.' In the long run this rational procedure was thought to afford more general prosperity than mere gold digging could ever do.[18]

Increasing public faith in the Geological Survey was reflected in a growing appreciation of Logan himself. When Logan lectured on Canada's geological structure to a literary soirée held by the Natural History Society of Montreal, he explained yet again that the strata of rocks in Canada, while situated below the Carboniferous, were above those in which deposits of metallic ores were found. His reminder that Canada 'had coal upon all sides of it' was repeatedly and 'enthusiastically cheered.' Logan's refusal of the presidency of the NHSM was much regretted in the press, and soon thereafter he was nominated to the senate of the University of Toronto.[19]

Popular interest in geology no longer flourished mainly along canal excavation routes. The enormous expansion of railway construction in the 1850s offered access to more remote regions of Canada that were seen as wide and interesting fields open for investigation. This new attitude was exemplified in Sandford Fleming's survey of the Toronto-Collingwood route and culminated two decades later in the active co-operation of the GSC in Fleming's survey of a railway route to the Pacific. Increasing donations of geological specimens to the public through scientific societies attracted attention to geology as a 'magnificent science' which ranked second only to astronomy in 'the grandeur of its speculations and the imposing aspect of its present developments' and 'in its bearings upon the progress of the arts and the happiness of mankind.'[20]

The process of geological inventory and the widespread acclaim earned by the GSC enhanced public appreciation of the value of the land within Canada's boundaries. At the same time it exposed inadequacies represented by those same boundaries. Although the sense of optimism represented by the former was thought to encourage the growth, even among immigrants, of an '*amor patriae* which will knit them to the soil, and that vital energy which will secure their social assimilation' and unite them 'to form the nucleus of a nationality,' the latter fostered a sense of incompleteness that afforded 'a solid basis for organic expansion' within North America.[21]

The intellectual links between inventory science and the developing consciousness of a Canadian 'national' interest were reflected in the changing orientations of the GSC under Logan's direction. Logan never forgot the 'mortification' he suffered with De la Beche's failure to identify two crates of fossils he had sent to London a decade earlier. His sojourn there during the Great Exhibition allowed him to pursue the matter further, only to discover the two crates still unopened in the basement of the premises of the British Geological Survey. Logan brought the specimens back to Canada, still unexamined, and determined to appoint a palaeontologist to his own staff.[22] Funding for this desideratum and other votes of public confidence rewarded Logan's successful appeal to Canadian self-awareness and pride before a select committee appointed in 1854 to decide, once again, the survey's fate.

III

In September 1854 increasing instability in the Canadian political system gave rise to a Liberal-Conservative coalition, led by A.N. Morin and Allan MacNab. A resolution on 20 September empowered a select committee to determine the best means of publicizing the information already obtained by the GSC and of completing it within a reasonable time. The chairman was John Langton, of the moderate wing of Conservatives led by John A. Macdonald. Langton was born in England and educated at Cambridge. He emigrated to Peterborough in 1831 and earned respect both for his efficiency in guiding administrative reforms which elevated the Canadian government to a more rational working basis and for his personal integrity as a politician. He became Canada's first auditor general in 1855. Langton was a close friend of John Young (Reform, Montreal), a prominent merchant and a close friend of William Logan.[23]

Only a few months before his appointment to the committee, Langton had addressed the Peterborough Library Association on 'The Importance of Scientific Studies to Practical Men.' The age, Langton told his audience, was characterized by 'the practical application of our knowledge to purposes of immediate and obvious utility.' As a result, 'we are apt to undervalue everything that does not at once come up to our standard of utility.' Yet in science, he argued, 'a truth once ascertained is an accession to our knowledge, the importance of which can never be known till you view it in connection with all around it.' Langton advocated scientific training for mechanics 'for the sake of science itself,' because therein lay the 'seeds of great discoveries,' whether 'new and useful in practice, or something important in principle.'[24] He was favourably predisposed to the case of the GSC, and most of the committee's conclusions bore the mark of Langton's opinions.

Langton was joined on the committee by other members similarly disposed. Robert Bell, a Conservative (later Reform) merchant from Carleton Place, was a son of Rev. William Bell and an uncle of the younger Robert Bell, who would join the staff of the Geological Survey in 1857. Bell came from a family of amateur naturalists and geologists, including his father and his brother, Rev. Andrew Bell, who testified before the committee and was the father of Robert

Bell of the Survey. J.C. Taché (1820–94), a physician and surgeon educated at the Séminaire du Québec (from which another witness was called), was a member of the Board of Agriculture of Lower Canada and interested in scientific agriculture. A.N. Morin had been commissioner of Crown Lands, responsible for mineral lands, since 1851.

Also among the witnesses called before the committee was James Hall, the state geologist from Albany, who praised Logan's efforts in showing the country to be 'rich in all those mineral products (with the exception probably of coal) which lie at the foundation of modern progress and civilization.'[25] Rev. Andrew Bell stated that he was heartened by support for Logan, a 'growing taste for Geological studies, and an appreciation of the advantages to be gained from them' which he noticed among Canadians:

Nothing is plainer to me from my own experience, than the fact that there is a gradual breaking down of the prejudices which have been entertained in regard to Geology; and amongst the whole circle of my friends and acquaintances throughout the Province, I have marked a growing desire for information in regard to it, as well as a growing conviction, that there is a definite and orderly arrangement of the rocks, and that it is only in certain rocks that certain useful minerals are to be obtained, – in short that it is science, that points the way.[26]

With this observation Logan agreed, adding, 'An excellent vein of geological knowledge seems to run up the Ottawa.' Logan regretted the widespread misapprehension 'that all our researches have the precious metals for their object,' which too often had the effect of 'immediately freez[ing] up the fountains of communication' in settled areas. In contrast, settlers along the Ottawa had 'got beyond the chance of such an epidemic,' through the influence of local amateur geologists and of the *Ottawa Citizen*, which 'occasionally gives them a good sound geological leader.'[27]

The *Ottawa Citizen* was owned and edited part-time by yet another Robert Bell (1821–73), a surveyor and civil engineer of Irish descent, unrelated to the Scottish Bells mentioned above. Bell represented Russell County as a Reformer in the assembly after 1861. Judged by one of his contemporaries to be 'among the intellectual giants of the age,' Bell promoted railway and settlement schemes in the Ottawa Valley. In 1850 he wrote a lengthy and thoughtful editorial on the future federation of British North America as a necessary counterweight to the threat of American annexation. 'We must not forget its importance,' Bell urged his readers,

as a foundation for a future national existence, and that, besides the increased confidence in our credit that would be created by our consolidation, it would produce a powerful national sentiment, which, with the accompanying feeling of certainty as to our future career, would induce a more united and energetic interest in all that concerned the welfare and the honour of our common country; – for then we would feel that we had one worthy of our best aspirations – presenting ample scope for the development of such a degree of true patriotism and righteous ambition as might enable it, in future times, to rank among the great Nations of the earth.[28]

In his view these ambitions were contingent upon an assumption that 'Canada and Acadia would be acknowledged as the ultimate heir' to all the dominions of Great Britain in North America, which would be linked from Atlantic to Pacific by a rail and water system of inland navigation. 'Good government and national progress, like good farming,' he insisted, required 'certainty of tenure.'

Bell's Canadian expansionism was inspired by knowledge of the rich mineral deposits that gave the territories both west and east of Canada 'decided advantages.' William Logan's achievements at the Great Exhibition had encouraged his views, for whereas Canada had once been seen as 'a sterile field' for businessmen, its image had been transformed into 'an extensive field for future industry and enterprise,' especially when one considered the iron ore along the Ottawa.[29]

The focus of the *Ottawa Citizen* upon geology was sharpened in the autumn of 1852 when Elkanah Billings (1820–76), future palaeontologist of the GSC, began contributing articles to the paper and sometimes editing it. In October Billings began a nine-part series reviewing Logan's 'Report of Progress' for 1851–2, using local experiences to amplify and simplify the more abstruse theoretical passages. The *Citizen* reiterated in 1854 that just as accelerated railway construction in Canada propelled the iron trade, so also would follow the growth of a 'manufacturing business out of our own ores, in our own country, and by our own labour.'[30]

An implicit motivation for the appointment of the select committee of 1854 had been political reaction to criticism of the GSC reprinted by newspapers throughout the province. Appended to these critiques was usually a refutation of the charges, by editors like Bell and Billings. One charge levied by critics was that in ten years' time Logan had failed to produce a geological map of Canada. Logan's many defenders suspected a smear campaign launched by Count E.S. de Rottermund, an embittered former employee of the survey with political and business connections in high places. In an appearance before the committee, de Rottermund conceded that Logan, 'that scientific gentleman[,] could only have described what he was enabled to see or know.' Canada's territory was vast, and de Rottermund admitted that the stretches which Logan had been able to 'pass in review' had indeed been surprising. But the necessary corollary to this feat, he continued, was detrimental to potential investors: 'vague information, published under the sanction of Government; for this kind of description by no means gives information, which may aid in opening the works of the Mines, but creates sufficient excitement, to entail ruin by the great confidence which it inspires or by the permanent discouragement consequent upon failure.' The problem was particularly acute, he added, in reports on auriferous soils, platinum, copper, and iron ores. His proposed solution was to create a department of mining separate from a department of geology, both directed by a council rather than by a single individual. To a limited extent, this suggestion was heeded when de Rottermund became inspector of mines in the Crown Lands Department some months later.[31]

A second charge stemmed from regions where it seemed that not enough was being done to develop local resources. Written testimony from Louis Agassiz of Harvard University hinted at the Canadian government's concern with dissatisfaction among French Canadians who wanted more practical results from the survey in Canada East. One French-Canadian newspaper argued, along similar lines, that Logan's style was too officious and arid to do justice to a science capable of bringing immediate practical benefits to much of society.[32]

It was not surprising that the select committee called as a witness Rev. Prof. Édouard Horan, who taught geology and mineralogy at the Séminaire du Québec. Horan proved a most co-operative witness who extolled both the scientific and the utilitarian benefits of the survey. He urged that Logan's reports be republished in condensed form as aids to the teaching of Canadian geology using Canadian examples. Republication should begin immediately, Horan advised, 'so that the credit may be given where it is due, and the Province may derive honor for its efforts in favor of science and industry.' For it was 'the universally received opinion' in Europe that 'science and practice are mutual aids: and practice, when unaccompanied by science, is mere empiricism, and is a fruitful source of error. It is only by science with practice that a country can improve its resources.'[33]

It was left for William Logan to state his own case, and he used a number of well-placed appeals, both emotional and scientific. Logan did not hesitate to quote once again Dufrenoy's praise of the Canadian mineral collection at the Great Exhibition and to recall his own consequent induction into the Royal Society of London. 'I esteem the honour more,' he declared, 'as I believe I am the first native Canadian who has been elected a member, or at any rate the first native Canadian who has been elected for work done in Canada.' Logan added that he had incurred personal financial losses for the benefit of the survey. Moreover, he revealed, he had rejected a lucrative offer from the East India Company in 1844 to investigate its territories for coal. Despite the certainty that 'the investigation would lead to a very extended reputation' for himself, Logan said, he had declined because he was 'influenced by a rooted attachment to this country, and feeling that perhaps some favor had been extended to me by the community in the present investigation because I am Canadian.' Nor could he forget his well-known map of the coal district of South Wales: 'It cannot be disgraceful for Canadians to know,' he noted, that a Canadian geologist had 'executed a small portion of the best work of the British Survey.' Logan came equally prepared to quote favourable opinions of his survey by well-informed 'foreigners,' who included Charles Lyell and J.W. Dawson.[34]

When asked about the importance of his scientific achievements, Logan cited his recent identification of the 'Laurentian' series of the Canadian Shield as metamorphic formations. His claim was corroborated by the survey's chemist, Thomas Sterry Hunt, who explained that the Laurentides, an ancient mountain chain extending from Lake Huron east to the Gulf of St Lawrence, were 'composed of the oldest known rocks, not only of North America, but of the Globe.' 'On this continent,' Sterry Hunt continued, 'they are so far as yet known confined to British America,' with possible small exceptions south and west in the United States. This unique feature of British North American geology substantiated Lyell's theories of metamorphism by revealing 'the existence at the remote epoch of their formation, of Physical and Chemical conditions similar to those, which have accompanied all the succeeding Geological periods.' These formations also illustrated the connection between the geology and the agricultural capabilities of a district, affording the basis of a rationale for planning future settlements.'[35]

As for the practical advantages arising from the survey, Logan told the committee that these were of necessity 'only beginning to be felt.' There was a direct correlation between geology and the distribution of manufactures, and a map of the former would define both the potential and the limitations of an area's natural resources. The work of the GSC was essentially 'economic researches carried on in a scientific way,' because the two kinds of activities were necessarily symbiotic. But, Logan cautioned, 'the area of Canada is so large and the explorers so few, that we could not satisfy public expectation if we dwelt a very long time on one district.' This statement was Logan's only public attempt to explain his organization of the GSC as an extensive rather than an intensive process. Logan's own political astuteness certainly weighed heavily as a contributing factor, but the vivid example of Sir Henry De la Beche, 'the master operator of the nineteenth-century scene,' accounts for the rest. De la Beche's anxiety to demonstrate the territorial progress of his own geological survey was well known, and it served as Logan's model.

This expansive approach seemed to contradict Logan's earlier experience as a 'miner and metallurgist' who concentrated meticulously on the details of one small section of coal strata. Logan was himself well aware of this contrast and cautioned that he 'would not have the Committee for a moment suppose that such work can be done in this country for a century to come.' Logan had acted upon his conviction that 'coal and iron are two of the materials, the presence or absence of which should be immediately ascertained,' since they could usually be relied upon to constitute 'a safe foundation for Manufactures.' But his overall task was still to determine the basic physical structure of the province, to divide the surface into parts, and thereby to define the limits of 'what is known or what is to be known.' His priority was to 'facilitate discoveries, and make available to a multitude of his co-inhabitants, whatever mineral product the intelligence or good fortune of any and every individual may enable him to bring before the world.'[36]

The report of the select committee on 29 March 1855 pronounced 'with confidence, that in no part of the world has there been a more valuable contribution to geological science for such a small outlay.' At the same time it was 'mortifying' that 'results of so much value' were inaccessible to the public, who were therefore still unacquainted with what the survey had achieved. While strongly urging that the unity of the survey be maintained under Logan's direction, the committee recommended additional funding for a palaeontologist and a mining officer. Deputy provincial surveyors and railway surveyors were also ordered to co-operate with the GSC. Most important, publication of at least twenty thousand copies of Logan's consolidated reports were guaranteed funding. The survey's annual allotment was to triple.[37]

By the time John Langton had leave to introduce a bill 'to make further provision for the Geological Survey of this Province' in April 1855, the session was almost over, and Logan was attending to the Canadian contribution to the Paris exhibition.[38] Official renewal of the GSC's mandate would have to wait, but in the meantime the course of events continued to favour Logan.

IV

In December 1854 Edmund Walker Head replaced Lord Elgin as governor general of the British North American colonies. Head (1805–68) was Oxford-educated, an accomplished geologist and crystallographer, and a personal friend of Charles Lyell. As lieutenant-governor of New Brunswick since 1848, Head had established a reputation for his utilitarian attempts to rationalize the administrative and educational systems there and for his encouragement of railway construction and scientific agriculture. The appointment of Head to Canada furthermore placed in a highly influential role one who had at least since 1851 shared the ideas expressed in Robert Bell's editorials in the *Ottawa Citizen* on the federation of British North America. Head set the tone for an imperial policy that favoured among the British North American colonies both a continued allegiance to the British Crown and 'the feeling that a national destiny of their own' was before them.[39] It was a situation in which Logan's expansive survey could thrive.

A further step in the growing appreciation of the province as the basis for a new nation resulted from participation in the Universal Exposition held in Paris in 1855. The executive committee for the Canadian exhibit, chaired by Francis Hincks and including J.C. Taché and William Logan, 'earnestly press[ed] on the public the importance of systematic and, when practicable, scientific arrangements' and 'insure[d] unity of action as well as efficiency.'[40]

While Logan was busy collecting 'economic specimens' from the still-frozen and snowy Canadian terrain for the mineral exhibit, he was also preparing a relatively detailed geological map which embraced British North America as a single territorial unit. Logan assumed that for the sake of continuity such a map would have to include at least those geological formations contiguous to Canada, and he arranged to credit the work of James Hall in Albany and James Robb in New Brunswick. He was not satisfied with these limitations, as he pointed out to J.W. Dawson, by then principal of McGill College in Montreal: 'Our geographical map of Canada generally includes Nova Scotia, as we can scarcely shew the eastern extremity of the province & its boundaries without bringing in the latitudes & longitudes of your part of the world. This will give an opportunity of shewing your geology on the map, as well as that of New Brunswick, both of which have important relations to Canada.' Dawson replied that his own geological mapping of Nova Scotia, Prince Edward Island, and part of New Brunswick for his *Acadian Geology* came 'much nearer the truth than anything that has yet appeared,' and ought therefore to be used, 'if you introduce our province onto your map.' 'It would,' he perceived, 'be a great advantage to us to be attached to your report, and the report would be rendered more satisfactory to geologists abroad.'[41]

Once at the Universal Exposition, Logan continued to rely on his own ingenuity for the benefit of Canada, as circumstances arose. The geographical arrangement of exhibits at the Crystal Palace in 1851 was not preserved at Paris in 1855, where national exhibits were broken up so that visitors would concentrate upon specialties. But Logan managed to extricate his exhibit from these restrictions by pleading that Canada's was a special case. He argued successfully that Canada's manufactured articles were 'not of that showy description which will so much figure in the *palais* or main building,' and requested instead to be permitted to display them with the primary products in the *annexe*. 'By keeping her materials as much as possible together,' he reasoned, 'Canada will display a more creditable front than would result from a classification and distribution of objects in the different buildings.' Once again he reached into his own pocket to afford expensive fittings for the exhibit, 'in order that we may not be behind the rest of the world.'[42]

Logan's extra efforts paid off more handsomely even than in 1851. Both he and Taché were named Chevaliers of the Legion of Honour by Emperor Napoleon III as a reward for their accomplishments. A reviewer for the London *Times* found the most striking feature at Paris exemplified by the Canadian exhibit:

What can be more delightful than to be able to watch the first stages of progress in infant communities – how hardily and industriously, applying every resource of modern science and still, they hew out wealth and independence for themselves from primaeval wilds – how they subjugate nature with a rapidity and completeness unknown in any past age of the world, and, self-governed and self-relying, tread with confidence, in the face of the nations, the path of greatness to which their destiny manifestly calls them!

The country was rich in copper and iron ores, and if it had no coal,

who shall say for what wise purpose? Perhaps to stimulate their industry in clearing away those interminable forests, interposed between western civilization and the Rocky Mountains. Certainly we may hope to enable Canada to compete with Sweden in supplying our iron trade with an abundance of the finest quality of iron smelted with wood charcoal.[43]

The explicit assumption that Canada would supply raw material for British industry was noted by Taché. For the time being this might have been true for iron for steel production but surely, he countered, Canada was capable of more. Had not the GSC shown that Canada possessed 'untouched riches' – a wealth of the same materials which accounted for Britain's industrial and commercial greatness? The preconditions of that greatness were 'the light of science, and extensive enterprise.' It was important to think big. If Canadian samples of cast iron seemed inferior to others exhibited, Taché suggested, the problem was not so much the quality of the iron, but rather the simplicity of its design.[44]

Arguments about the appropriate degree of industrialization for Canada were lost on much of the Canadian public, who understood more simply that the task of the GSC was 'to precede civilization, and, penetrating into unknown regions, to point out sources of mineral wealth hitherto unknown, preparing thus the way for the industry of civilized men who shall replace the savages.' A stirring editorial appeared, for example, in *The Canadian Statesman* of Bowmanville in September 1855. Entitled 'Canada Against the World!' the article recognized that 'though we are living in the Country itself, we would not have known half its worth, had we been without our Geologist,' who had 'presented things in a new light.' 'We are so much enraptured with the love and pride of country,' gushed the editor, 'and we feel so much indebted to Mr. Logan for what he has done for it, that we are at a loss to express how deeply we feel on the one hand, or how much we are elated with patriotic pride on the other.' But at the same time the editorial exposed a darker side of the transfer of emotional allegiance away from Britain, in a growing intolerance of cultural heterogeneity which blamed French Canadians for failing to pull their weight at the exhibition.[45]

A suggestion by *The Canadian Statesman* that a knighthood for Logan would bring great honour to all Canadians was soon gratified. The matter was already set into motion by November 1855, and if he did not instigate the process, Logan participated in the backroom work necessary to secure the title. Ever circumspect in his behaviour, Logan was concerned during the Paris exhibition to avoid adverse publicity in Canada where, he felt, any errors 'will be made matters of state' and, still worse, 'a foundation for a newspaper war.'[46]

Logan's knighthood unleashed a public pride compounded early in 1856 by his receipt of the Wollaston Medal from the Geological Society of London. George Simpson wrote to Logan that he could 'scarcely recollect any proceeding of Her Majesty that gave equal satisfaction to the Province at large; – the compliment gracefully paid to your own merits, is looked upon here as paid to the Province which you represented.' Logan's habitually well-placed replies to such comments graciously drew his audiences into sharing his achievements and, incidentally, made superb newspaper copy:

I am proud to think that it was, perhaps, more because I was a Canadian in whom the inhabitants of the Province had reposed some trust, that the honour which has been conferred upon me by Her Majesty was so easily obtained. That I am proud of the honours which have been bestowed upon me by the Emperor of France, in respect to my geological labours, and also by my brother geologists in England, there can be no doubt. But I have striven for these honours because I considered that they would tend to promote the confidence which the inhabitants of the Province have reposed in me, in my endeavours to developе the truth in regard to the mineral resources of the Province.[47]

He was widely fêted as an *enfant du sol* with whom all Canadians shared distinction.

The practical upshot of these emotional responses was to quell any opposition to the GSC which remained at home and to urge upon the assembly the immediate implementation of the Langton report of 1855. In March 1856 an article in the *Ottawa Citizen* asked rhetorically, 'The Geological Survey – Of What Use Is It?' The author depicted the survey as 'a national process' of taking stock: 'Every merchant knows the value of this operation, and so should every nation.' Delineating the geopolitical importance of this inventory process to the coal question, the writer continued: 'The national wealth of the people inhabiting any particular region must be drawn from so much of the earth's surface as is contained within the boundaries of the country which belongs to that people.' Since it was a maxim of political economy that a nation should confine its industry to those things for which it is best adapted, it followed that 'without a Geological Survey, no Political Economist can direct the industry of Canada, or say what should or should not be done.' And since Sandford Fleming had recently addressed the Canadian Institute on 'The Geological Survey and Its Director,' copies were to be circulated with some flattering remarks by Sir Roderick Murchison appended. On 31 March the House of Assembly heard a petition of the institute 'that further provision may be made for carrying the Geological Survey of this Province on a more extended scale.' Attorney General John A. Macdonald and Governor General Head steered it through; in May £5,000 annually for five years was voted.[48] It was not quite what Langton had recommended, but close enough. Now there was no more mention of the GSC's projected completion date.

By the end of this second phase of his career as provincial geologist, Logan's international reputation seemed further proof that he and his survey were indispensible to Canada's future development. At the pinnacle of his success, he used his authority in the rough-and-tumble of Canadian politics to secure an impregnable position for himself and for the survey as a Canadian institution. Logan's vitality and energy reflected that of the mid-Victorian British Empire. Like entrepreneurs and scientists throughout the empire and indeed elsewhere in the world, he was quite willing to promote an ideology advocating the potential benefits of an informal Canadian empire beyond the confines of the province proper. It was this expansion of intellectual boundaries which gives him his full historical significance.

5

'Permanence,' 1857–1869

> The Captain was very experienced in this sort of surveying. He had brought with him the necessary instruments and he started on it at once. He instructed Eduard and some of the local trappers and peasants who were to assist him. The days went very well. He spent the evenings and early mornings on his map, drawing the contours and hatching the heights. Soon everything was shaded and coloured and Eduard saw his possessions taking shape on the paper like a new creation. It seemed to him that only now was he coming to know them, only now did they really belong to him.
>
> Goethe, *Elective Affinities* (1809)

Logan's triumph at Paris and the Act of Continuance of 1856 heralded a third phase in his geological inventory of Canada. Although the question of the Geological Survey of Canada's existence had been successfully placed beyond the reach of party politics, three problems troubled the final years of Logan's career. First, it became more and more difficult to function within Canada's increasingly unstable political system. Second, the survey was still vulnerable to the financial adjustments which inevitably accompanied swings in the economic cycle. In 1859 a policy of retrenchment halved the GSC grant, necessitating the cancellation of field explorations for an entire season. Third, Logan's health had been broken by the strain of his endeavours at London and Paris. He felt oppressed by the 'confinement' that sedentary administrative duties forced upon him.[1] This last phase of his directorship was marked not so much by more intensive explorations of the province as by a consolidation of earlier extensive work. It culminated with the publication in 1863 of *Geology of Canada*, with preliminary measures to extend the purview of the survey's investigations, and with the establishment of the GSC as a permanent institution.

'Permanence'

I

By the late 1850s geology formed an integral part of many people's understanding of Canada. One of the major links of the province to its wider geographical context was the perennial 'coal problem.' The push to build railways during the 1850s intensified the hope that Logan's scientific verdict about the unlikelihood of workable coal deposits in the province might be mistaken and that Canada would be able to fuel locomotives as well as its own iron and steel industry.

By this time, however, Logan's judgment found support among a growing phalanx of British North American geologists, who reviewed his reports in scientific publications. They interpreted the 'forbidding documents' as a new branch of Canadian literature which revealed the uniqueness of the Canadian experience. The work of the GSC, argued J.W. Dawson, permitted the first clear conceptualization of Canada as it really was. He speculated that the geological map of Canada displayed by Logan at Paris in 1855 had influenced the selection of Ottawa as the seat of government in 1858. Bytown was a 'last outpost to the northward, of the Great Silurian plains of Canada.' It might therefore be regarded as 'a favourable point for bringing the wealth and population of the more valuable parts of the province to bear on the improvement of the rocky and intractable Laurentian country.'[2]

But long-term speculations were irrelevant to those who harboured a financial interest in the discovery of coal in their vicinity. One influential promoter of such possibilities was Count de Rottermund. As inspector of mines for the Crown Lands Department, de Rottermund conducted surveys of the Lake Superior, Lake Huron, and Quebec City regions. In 1855 and 1856 he reported finding specimens of a 'carbonaceous combustible' that showed 'much analogy to real coal' in a geological formation 'coeval' with that of coal. Corroborated by several members of the Muséum d'histoire naturelle in Paris, de Rottermund, whose arrogance detracted from what valid ideas he did express, echoed the concern of many Canadians that 'we shall be called at a future period, perhaps not very far distant, to extract from our own soil the combustible which we now obtain at great expense from foreign countries.'[3]

De Rottermund's 'coal' was actually a large bed of shale that underlay an area around Quebec City. Those who, like Joseph Cauchon (Bleu, Montmorency), were involved in financing the North Shore Railway believed that even if what de Rottermund had found was 'not scientific coal, it was coal nevertheless.' A more insidious game than the misguided count's was played by speculators who planted coal specimens in order to enhance property values and to discredit Logan. The most notorious of these scandals occurred at Bowmanville in 1858, when a cheese sandwich accompanied the last bucket of coal brought out from such a 'mine'; J.W. Dawson defended Logan by chiding the Canadian public for its gullibility:

The thing that we cannot have, is always that which we most desire, and the more richly we are endowed otherwise, the more earnestly do we long for the one object that may have been withheld. So it would seem to be with the Canadian public in the matter of coal. All the riches of the earth and of the hills and of the deep beneath, have been thrown into its lap, except this; and like the child whose toys are all valueless because mamma cannot give it the moon to play with in its own hand, it turns its eyes away from all its other treasures, and cries for coal.[4]

The restless impatience and frustration evinced in the desire for coal was in part a reaction against the idea that emulation of British industrialism was an unnatural rather than an inevitable expectation. This reaction was reflected in the revival during the late 1850s of the ideas espoused by the British American League a decade earlier, during another economic recession. 'Why,' asked Alexander Morris, in his enormously popular *Nova Britannia* in 1858, 'with all that modern civilization is doing for us, should not British America follow in the footsteps of her parent?' In 'the natural course of events, such changes will come, as surely as the child becomes the man, or the feeble sapling becomes the sturdy monarch of the forest.' To Morris and others like him, the dream of an industrial future for Canada seemed feasible when linked to the idea of a federation of the British North American provinces in a 'union of interests' in which Canada could play a major role. This was so because the eastern provinces, in particular Nova Scotia and New Brunswick, possessed enormous workable coal-beds, and the possibilities inherent in 'the magnitude of their combined resources' impressed these advocates of federation immensely. In contrast, opponents of such a federation like Joseph Cauchon, who continued to believe that coal would be found near Quebec, argued that the Atlantic region had nothing of value to offer Canada and would instead become a drain on the public revenue.[5] But eastward was not the only direction in which to turn for the elusive fuel.

II

Although the two were by no means mutually exclusive, another perspective was suggested to those who looked westward. In 1855 George Brown editorialized in the *Globe* that there was no reason any longer to doubt William Logan's statement that coal did not exist within the boundaries of Canada. There was, however, reason to believe that coal did exist in the Hudson's Bay Company's territories. Brown argued further that those territories 'must ultimately be incorporated into the province.' When a Pacific railway was constructed through British North America, he predicted, 'the locomotives which will wake the echoes of the shores of the Eastern ocean at one time, and at another, the banks of Lake Superior, may be supplied with British coal'; but for the present Canadians had to content themselves with seeking the precious mineral in the territory of their neighbours.[6]

When the issue of renewing the charter of the HBC arose in 1856, Logan and the GSC were drawn into the controversy. George Simpson, governor of the HBC, knew well that Logan took a personal interest in 'any matter touching the mineral wealth of the Hudson's Bay Territory.' Simpson had aided the survey's expeditions and had even sent Logan boxes of specimens. Logan knew Alfred Roche, an ardent expansionist in the Crown Lands Department who accompanied Chief Justice William Draper to London in 1857 to represent Canada before the House of Commons committee on the HBC's claims. Roche was convinced that Logan shared his expansionist interests and that the British exploring expedition led by Capt. John Palliser 'should see or should communicate with' Logan as a counterweight to the influence of the company upon the expedition's findings. Roche hoped that Logan would advise the geologist of the party, Dr James Hector from the University of Edinburgh, whom Roche considered too young for the enormous responsibilities of his appointment. However Palliser, as Alexander Murray reported to Logan later, 'has given us the slip and gone by the N. States.'[7]

When the Canadian government decided to send its own expedition into the north-west, Logan selected the scientific staff to be attached to the party. Since Murray was physically unfit for the task and the other members of the GSC had already left for the season's field operations, Logan nominated H.Y. Hind, who was eager for the position. Hind, who taught geology at Trinity College in Toronto, was known to Logan as a fellow member of the Canadian Institute, editor of the *Canadian Journal of Industry, Science and Art*, and an advocate of Canadian expansion into the north-west. Although Hind claimed no more than 'to know what to observe of general features, and to describe what I see,' Logan deemed his abilities sufficient to collect the 'facts and materials available for the proper understanding of some of the main geological features' of the territory. He was, he wrote, 'fully persuaded' that Hind 'would make an excellent explorer and produce a valuable report,' a report whose expansionist leanings were thus a foregone conclusion.[8]

Although both the Canadian and the British governments appointed parliamentary committees to consider the future of the HBC territories, testimony before the British committee reflected more clearly the diversity of opinion on the mineral wealth of these territories. While scientists like J.H. Lefroy and John Rae agreed that they had seen no signs of workable coal deposits, others such as Sir John Richardson, the explorer, and David Anderson, the bishop of Rupert's Land, countered that nothing was certain because the company had never commissioned a geological survey of its lands. When asked about the failure of copper mining companies north of Lake Superior, Richardson was led by Edward Ellice's line of questioning to agree that this failure was due mainly to Canadian mismanagement and might not occur if British companies were to take over. Alfred Roche testified that although he had never visited the HBC territories, he believed them to be much more valuable that they had been represented to be, especially in minerals. One of the strongest arguments was offered by Alexander Kennedy Isbister, who described the Mackenzie Valley as a 'mass of minerals' and stated his conviction that the mineral wealth of much of the region exceeded the fur trade in value.[9]

Isbister (1822–83), a Métis, was a nephew of William Kennedy, the Arctic explorer employed in 1854 by George Brown and his associates to assess the north-west as an economic hinterland for Toronto; he was born in the company's territories at Cumberland House. After some years in the employ of the HBC, he studied at Aberdeen and Edinburgh from 1842 to 1844. Isbister developed Canadian expansionist ideas in reaction against the HBC's policies. In 1855 he read a paper entitled 'On the Geology of the Hudson's Bay Territories, and of portions of the Arctic and North-Western Regions of America' before the Geological Society of London; the paper was published, accompanied by a coloured geological map, in the society's *Quarterly Journal*. The essay recapitulated observations of geologists and travellers and concluded that their evidence pointed 'to the existence of a vast coalfield, skirting the base of the Rocky Mountains for a great extent, and continued probably far into the Arctic Sea.' The work collated geological knowledge about north-western and Arctic British North America and gave Canadian expansionists a splendid reference source during the second half of the nineteenth century.[10]

Isbister's approach harmonized with the outlook and hopes of a growing number of Canadians. In addition to the group around George Brown and Allan Macdonell in Toronto, another area of expansionist interest was the Ottawa Valley, especially around Perth. In March 1857 the House of Assembly received three petitions from Lanark, Renfrew, and Argenteuil counties for the HBC territory to be annexed to Canada, and its boundary with the United States clearly established. Except for the suggestion that newspapers in these communities emulated the Toronto *Globe*, this apparent anomaly has never been explained.[11]

A number of factors fostered this regional pocket of expansionist nationalism. The Perth area was, of course, originally the site of defensive British military settlement after the War of 1812. J.W. Dawson might well have had this fact in mind when he predicted: 'Gradually there will grow up in the glens of the Laurentian territory, a race of hardy Canadian hill-men, who, if sufficiently leavened by the elevating influences of Christianity and education, will be of inestimable value to the country, both in peace and war.' Many of the settlers were Scots who had acquired and disseminated a knowledge of geology and an appreciation of the potential value of the rocky terrain, the more so with the excavation of the Rideau Canal. By the 1850s William Logan's geological 'Reports of Progress' confirmed the presence of valuable mineral ores and economic materials in the Laurentian rocks which characterized the northern and western edges of the region. If only a source of coal could be made available, it was believed, local iron ores could be worked to great advantage.[12]

The region also lay on the historic canoe route of fur traders who travelled from Lachine over the Ottawa to Grand Portage, and by tradition it looked westwards to Lake Superior and beyond. This tendency to look westwards was hardened by the realization after mid-century that the best lands in the southern parts of the province had been settled. Concern mounted among those backed against the Shield to open up new and better lands for settlement. 'Experience shows,' wrote J.W. Dawson, 'that the energy and force of character of the population of such districts rise to meet the difficulties that surround them.' Thus 'these regions become nurseries of the patriotic feeling and of the mental and bodily energy, that are too apt to die out on the more fertile plains.'[13]

It was for similar reasons that the growth of a 'national sentiment,' self-respect and self-reliance, seemed 'unmistakably evident' to Alexander Morris, a native of Perth. Morris urged upon Canadians their 'plain and manifest responsibilities': 'They must,' he insisted, 'evince an adequate appreciation of their duties, and must possess a thorough knowledge of the advantages which they possess, and of the vast resources which Providence has placed at their disposal, in order that they may advance steadily toward that high position among nations which they may yet attain.' The means for conducting this necessary inventory, Morris recognized, were at last available through 'the trustworthy testimony of scientific exploring expeditions.'[14] Canadians were, however, already emboldened by the GSC to reconsider their place in the world.

III

As early as the hearings of the Langton committee in 1854, E.J. Chapman, professor of geology and mineralogy at University College in Toronto, had agreed with Andrew Bell that public interest in geology was indeed on the rise in

Canada. Chapman suggested that in a few years, as geology became more widely cultivated in the province, an annual scientific gathering might be organized at Montreal. The idea, he conceded, was for the time being premature. Yet two years later H.Y. Hind lamented in the *Canadian Journal* that 'the science of Canada has, as yet, no recognized or independent existence, and its students, if they would place themselves in *rapport* with those other lands, can only do so by a sacrifice analogous to the naturalization by which a foreign emigrant attains to the privileges of American citizenship.'[15]

Feelings of inferiority were assuaged when the American Association for the Advancement of Science (AAAS) accepted an invitation from the city of Montreal and its Natural History Society to meet there in 1857. A peripatetic society modelled on the British Association for the Advancement of Science (BAAS), the AAAS was seen to differ 'in magnitude and universality from all the lesser and more local societies,' and the event was touted by Montrealers as one of the most important and interesting that had ever occurred in Canada. The meeting was also expected to attract a number of British scientific luminaries, although some declined in fear of Montreal's reputed sweltering summer weather. As might be expected, Logan and his staff were active as both organizers and participants. Logan chaired the geological section and presented several papers analysing perhaps his most important discovery in Canada, the Laurentian and Huronian formations. The discovery in turn had enabled James Hall of Albany to introduce at the same meeting a major theoretical concept in geology, that of the geosyncline. Geologists applied the idea of geosynclines, great downward flexures in the earth's crust, to explain the formation of mountain chains.[16]

The success of the meeting released another wave of public pride in what one reviewer termed 'a virtual Scientific annexation of American to British American minds, in their action in the wide field of physical and natural science.' 'May we not,' asked the Natural History Society of Montreal, 'indulge the hope that a Canadian Scientific Association may soon be organized, and take an honourable place alongside of similar institutions in Europe and America?' J.W. Dawson outlined Canada's role as a link between Great Britain and the United States: 'A practical union of the American Association with its older and greater sister of Britain is much to be desired,' he declared. 'Why may not Canada, as a middle ground, some day secure a joint meeting of both of these bodies?' Nor was this proposed relationship limited to the realm of science, which offered a springboard into culture and politics precisely because the society believed that science knew no political limits. 'The great physical laws of the Universe are the same in all lands. Geological structure and animal and vegetable life are everywhere framed on one uniform type. We cannot attempt to nationalize science without losing its greatest results.' If the role of link was acceptable for Canada on the international level, then how much more appropriate it seemed for Montreal, situated 'at the meeting point of the two races that divide Canada,' to disseminate the 'high moral and intellectual influence' of science to the British North American regions beyond. There was, after all, a 'bond of brotherhood which should unite all the cultivators of science in every country.'[17]

When the excitement in Montreal had died down and the staff of the GSC returned to its annual field explorations, Logan at last took up the task of compiling and condensing his 'Reports of Progress' and completing the geological map provided for in the act of 1856. He had bided his time in producing these eagerly awaited signs of the survey's progress, he explained, until the palaeontological branch came into its own. Otherwise students of Canadian geology would have to depend upon what had been accomplished elsewhere. Elkanah Billings had made promising contributions to the study of Canadian fossils since 1856, but the GSC faced new financial woes when the act of 1856 expired in 1861, ironically because neither the report nor the map had yet materialized.[18]

The fate of the GSC soon became entangled in a wider political crisis. The minister of finance since 1858 had been Alexander Tilloch Galt, a son of John Galt, formerly of the Canada Company. Galt had joined the Conservative cabinet just when the end of the first railway era had combined with an international financial depression to increase public debt while decreasing revenue. Galt's financial statement for 1860 noted an excess expenditure by more than $7,000 in a category including 'the Geological Survey, the militia, and arts, etc.' His supplementary estimates in 1861 again listed the GSC and the militia in tandem. The Macdonald-Cartier government was caught in a double bind. On the one hand, it viewed the American Civil War as an emergency requiring greater militia expenditures, but on the other hand it was being hounded into defensiveness by the Opposition's insistence that the deficit be reduced. Since debts incurred by the Grand Trunk Railway and the new parliament buildings under construction at Ottawa continued to grow, the budget had to be trimmed elsewhere. As much as Galt supported the GSC in principle, the fact that Logan's map and report had not been submitted left him without a compelling justification to renew the five-year commitment to the survey. It reverted instead to the list of annual contingencies, its grant for 1861 reduced from $20,000 to $9,000 without any objection from the assembly. The voting of supply in 1862 increased the amount to $25,000, but still only for one year, with an additional $8,000 for a Canadian contribution to the London exhibition that year. With that the government fell, victim to an even more expensive militia bill. The situation appeared grave to Logan, overworked overseas once again at the London exhibition. Financial unpredictability caused him severe anxieties, which filtered down to arouse confusion and insecurity among the survey staff.[19]

Nevertheless his *Geology of Canada* was finally published in 1863, and the accompanying map three years later. Containing the substance of all of the survey's previous reports and 'much original matter,' the work nonetheless made no pretence of answering all of the difficult questions confronting students of Canadian geology. It was, in Logan's words, 'not intended to be a history of what everyone has done' in the subject. The volume of more than nine hundred pages was widely noted and very favourably received. J.W. Dawson exaggerated more than a little when he remarked that 'every useful rock or mineral occurring in the country' was noticed in the work, but he was correct in pointing out that it thus placed Canada 'on a level with those older countries whose structure has been explored, and the knowledge of it made the common property of the world ... In some departments of geology,' he recognized, 'it even makes Canadian rock-forms rank as types to which those of other countries will be referred.' The term 'Laurentian,' for example, applied as the general designation for the most ancient formations of Europe as well as of America. Reviewers agreed that Logan's *magnum opus* marked an era in the history of Canadian development and represented a 'complete treatise on the geology and mineral wealth of the Province.'[20] In some ways, it might be argued, with the publication of *Geology of Canada* and the map that followed, the GSC had carried its original mandate to completion.

Instead of breaking up, the survey was about to become a permanent institution. 'No one,' declared J.W. Dawson, 'need henceforth have any excuse for professing ignorance of the labors of the Geological Survey, or for representing it as a useless expenditure of the public money.' When John Sandfield Macdonald, as premier in 1863, dared to suggest that the GSC might be a failure because the mines of the country 'had been discovered quite independently,' he nearly precipitated a political crisis by incurring a rebuttal from his own minister of finance, Luther Holton. Macdonald was lampooned even years afterwards for his display of 'gross and ludicrous ignorance.' Opponents delighted in chuckling that Macdonald objected to Logan's studies of Canadian fossils because that was 'carrying personalities too far.'[21]

When A.T. Galt became minister of finance once again in 1864, the act continuing the GSC at $20,000 annually was renewed for five years, with the survey to be placed under a government department. This commitment had indeed verified J.W. Dawson's observation that 'times may change, and editorials and acts of Parliament may become waste paper; but rocks and fossils are permanent things, and work once well done in reference to them is sure to retain its value.'[22] In another more ironic sense, some of those 'rocks and fossils' were not quite as permanent as Dawson may have wished.

IV

A focal point of the changing perceptions marked by the publication of *Geology of Canada* was the discovery in 1858 of curious markings in the Laurentian limestones near Ottawa. The fragments were suspected fossils of primitive organisms representing the earliest forms of life on earth. While conclusive evidence remained elusive to 1863, several factors encouraged belief in the organic origins of this 'supposed fossil.' First, researches by Thomas Sterry Hunt into metamorphic processes had predicted that such evidence of life would be found in the Laurentian rocks, long assumed to precede the appearance of life on earth. Hence, such a discovery would be a remarkable Canadian contribution to science and to the world. Second, more conservative geologists like J.W. Dawson, who attempted to legitimize his geological inquiries by linking them to biblical studies, were predisposed to accept evidence of the early existence of relatively sophisticated life forms such as the giant Foraminifer which the fossil was thought to represent. The discovery appeared to Dawson to provide an effective counter-example to the theory of evolution by natural selection put forward by Charles Darwin in 1858-9, because it implied that more complex forms had existed before simpler ones. Third, even Logan, who maintained a relatively detached attitude in comparison with Dawson's enthusiasm for the fragment, mentioned it as a fossil in *Geology of Canada* and put forward the evidence for his suppositions. But when further specimens were analysed late in 1863, the certainty especially of the palaeontological section of the GSC staff began to wane.[23]

By April 1864 Dawson had named the 'fossil' *Eozoon canadense* (dawn animal of Canada) and had even set out a scale of priorities for the co-discoverers of what Charles Lyell hailed as 'the greatest discovery of his time.' Logan displayed the *Eozoon* to the BAAS in 1864, where professional opinion was divided between experts in fossil Foraminifera and experts in mineral crystallography, each group using microscopes to see what they had been trained to see. The latter group eventually won out, but those in favour included Thomas Huxley. As late as December 1864 even Logan described the British reception of *Eozoon* as 'altogether ... [a] great success.' The *Canadian Naturalist and Geologist* published a special issue devoted to articles by Logan, Dawson, Hunt, and W.B. Carpenter (a prominent British expert on Foraminifera) on *Eozoon* as a fossil. Logan planned to include illustrations of *Eozoon* in a palaeontological 'decade' published periodically by the GSC. However, additional discoveries of *Eozoon* in rocks of too many ages and localities, including Ireland and Bavaria, had aroused the suspicions of both Logan and Lyell by late 1865. Dawson remained unshakeable in his belief long after the issue had ceased to exist for most geologists.[24]

The controversy over *Eozoon* also died hard among Canadian romantics who found a nationalistic inspiration in the image of Canada as the cradle of life on earth. In a collection of *Portraits of British Americans* (1865-8), Fennings Taylor and William Notman pondered the enormity of 'Logan's' discovery:

May we not add the homage of involuntary sympathy to the mysterious sense of awe which must have possessed the mind of Sir William Logan, as he stood before those Laurentide Mountains, face to face with one of the great mysteries of nature – the chosen repository of one of her amazing secrets. We can imagine the learned skill with which Sir William Logan gauged the depths and measured the heights of those rocks, but we cannot imagine what his sensations must have been as he diligently anatomized their structure and discovered that those gigantic hills which stretch from the sterile coast of Labrador to the fertile regions of the far West ... were neither more nor less than accumulated fossils, the petrified forms of what was once organic life. Thus do the stones cry out, and in their sublime majesty preach strange sermons ... [Logan's name] will be cherished by Canadians [for his labors in] a land whose antiquities have been explored by his genius, illustrated by his pencil, described by his pen, and reproduced in a manner [that has] awakened admiration in the old world as well as in the new, and has taught [people] to think of the attractions of a land on whose surface are some of the newest settlements of the human family.[25]

The Canada First movement also drew from the legacy of the *Eozoon* during the 1870s. They felt it had done more to draw the attention of the scientific world to Canada than anything else of late years. Members expressed disappointment 'that this important point in the Palaeontological history of Canada is to be brought in question, and doubt thrown upon what was considered one of the most important discoveries of Canadian geologists.' Indeed, *The Nation* chose to suspend its judgment until J.W. Dawson, 'considered the champion of this ancient problematical fossil,' could throw more light upon its true character. When Dawson reiterated all his old arguments in *The Dawn of Life* in 1875, editors of the magazine felt he had answered objections to *Eozoon* fairly, and concluded: 'It is but fitting that the oldest rock, the Laurentian, being best known in Canada, the discovery in it of the oldest known fossil remains should be made by a Canadian, who, long since, in winning laurels for himself, contributed something to the fame of his country.'[26]

Besides providing a literary vehicle for romantic nationalists to anticipate Canada's manifest destiny, the geological inventory performed by the survey worked, in Dawson's words, to 'inoculate our statesmen with a healthy belief in the geological future of Canada.'[27] The appropriate scientific antibodies began to do their work during the 1860s, the final years of Logan's career.

V

We have already seen evidence that as early as the 1840s William Logan had anticipated that one day he would conduct a geological survey of the entire area of British North America. Much of his work, moreover, had directed public attention to the regions at the northern, western, and eastern extremities of the province of Canada. In 1860 A.T. Galt established free ports at Sault Ste Marie and Gaspé to encourage the rapid development of the province's outer fringes. Public interest in the north-west made the government aware that as late as 1864 Canada's geographical boundaries in that direction were not even precisely defined. Logan took great interest in the plans for the confederation of the British North American provinces largely because the operations of the GSC would be greatly extended under the control of a federal government. In London during the preliminary negotiations of 1866 Logan often breakfasted with Galt, Cartier, Macdonald, and Brown.[28] But expansionist tendencies in Logan's geological inventory had several foundations, not all of them based upon his own self-interest.

One of these foundations was, of course, the 'coal problem,' the ramifications of which were beginning to play themselves out in earnest. Logan's *Geology of Canada* was generally considered the final word on Canada's predicament in this regard. It stated that the only representatives in Canada of the true coal-bearing series afforded nothing but a few carbonized plants. The obvious alternative of turning elsewhere for viable sources of mineral fuel took on new meaning during the American Civil War, when the American government placed an embargo on the export of coal to Canada. At the same time, Nova Scotia exporters of coal lost out when the United States abrogated the Treaty of Reciprocity, which had allowed this coal to enter the American market duty free.[29]

To compound the problem, concern grew that Britain's enormous coal reserves were finite after all; even if they could not be exhausted during the coming century, the price of extracting coal from deeper mines would rise considerably in the foreseeable future. The assignment of a geologist, James Hector, to the Palliser expedition in 1858 reflected the imperial government's concern with future fuel supplies. Hector had predicted that the existence of a supply of fuel 'principally within the British territory, and in the plains among the Saskatchewan, will exercise a most important influence in considering the practicability of a route to our Eastern possessions through the Canadas, the Prairies, and British Columbia.'[30]

But an increasing uneasiness was epitomized by the publication in 1865 of William Stanley Jevons's influential study *The Coal Question: An Inquiry Concerning the Progress of the Nation, and the Probable Exhaustion of Our Coal Mines*. Jevons warned Britons that a 'momentous choice between brief greatness and longer continued mediocrity' was about to force itself upon them, if their consumption of British coal supplies continued at the same rate. The Colonial Office reflected its concern in a circular requesting that British North American colonies report on the extent and status of their coal reserves in 1866. The question might have been cause for some alarm in Canada, since British coal had been an important trading commodity in exchange for Canadian timber and was preferred for its quality over Nova Scotia coal by regular importers like the Molsons, who used it in both their foundry and their brewery. At the same time it was hoped by coal-owners in Nova Scotia, as well as by some idealistic Canadians, that a growing 'community of interests' based on an interchange of commodities between east and west might 'form a bond of union among British North Americans who otherwise constituted the mere "selvage of an empire". '[31]

No doubt such a 'community of interests' included the anticipation of considerable profits for some of its advocates. In 1864 Robert Bell of the GSC was inspired by a lecture on Newfoundland coal regions around St George's Bay, 'as with all its other treasures, yet unexplored.' One day, the lecturer promised 'the black smoke from the mouth of many a coal mine will darken the air of this region, and the heavy sobbing of the steam-engine, dragging up the precious treasures, will re-echo amid the forest solitudes.' The coastal areas of Newfoundland had been surveyed by the British geologist J.B. Jukes during the 1830s, but when he failed to make important discoveries of useful minerals the survey was discontinued. Undaunted by this knowledge, within two years of his original inspiration Bell schemed to apply his acquaintance with geological reports of the island to take advantage of 'unlimited' coal markets in Canada and perhaps even in New England.

Bell, financed by Donald Ross of Montreal, was to explore and map tracts of Newfoundland territory, and titles would be secured by a silent partner, Ambrose Shea, a member of F.B. Carter's pro-Confederationist government in St John's. Shea reasoned that his government was 'always ready to encourage such projects' to develop the resources of Newfoundland. But he had already touched off public suspicion by hiring Alexander Murray to survey the island under the authority of the GSC in 1864. A subsequent wave of mining speculation by outside capitalists met with a burst of anti-Confederationist feeling in the colony, one of several factors in the resounding defeat suffered by most of the union's political advocates at the polls.[32]

Bell was similarly involved in a speculative venture to develop petroleum on Canada's western fringes, specifically on Manitoulin Island, during the same years. He saw inviting prospects in the contemplated expansion of Canada into the north-west and in the island's position on the only route into that territory. Bell was only one of a number of engineers and businessmen interested in promoting the possibilities of petroleum as an alternative mineral fuel to coal; petroleum had to be processed in Canada instead of shipped to Britain because of its unpleasant odour. Bell obtained Logan's endorsement of a series of public lectures on petroleum resources in Canada, intended 'to convince the public how big a thing we have to offer them.' Logan's name and reputation were, in essence, used to promote the interests of the Grand Manitoulin Island Improvement Company, in which Alex Campbell, commissioner of Crown Lands, was also a silent partner. Other arrangements were made by Bell and his associates to develop petroleum in the Gaspé. Bell was urged by a partner to delay submission of his official report on the area to Logan 'until we see what our prospects are.' Bell's private projects were closely related to, and had a negative effect upon, the quality of his geological assignments, but they reaped few financial benefits because the market for Canadian oil shares had shrunk dramatically by the summer of 1866. As an alternative mineral fuel, Canadian petroleum was not sufficiently understood and remained financially too speculative for the time being.[33]

Expansionist tendencies in the GSC were deeply rooted in the scientific ideology shared by Logan and J.W. Dawson. A common influence upon both their careers and their approaches to geology, as has been seen, was their relationship with Charles Lyell. Lyell was recognized by his admirers as 'the reducer to order of the homogeneous,' because he encouraged the study of geology by historical period rather than by geographical region. The new approach exerted a centripetal force which ignored political boundaries as 'arbitrary and artificial' in comparison to more tangible geological units. Upon Lyell's death in 1875, *The Nation* reprinted a striking interpretation of the contribution of science to the Victorian outlook:

Marvellously as our acquaintance with the facts of science has advanced, still increase of knowledge has been less characteristic of the age than unification of knowledge. Instead of enlarging our domain, we have brought its scattered provinces under a single rule. Classes of phenomena previously held to be unconnected with each other, and studied separately, have been shown to result from the action of common cause, and to obey the same fundamental laws.[34]

Logan's accomplishment, best exemplified in his geological maps, was analogous. Robert Bell of the GSC, ever aware of the angles, correctly grasped in 1864 that Logan's 'great map would just come in at the right nick of time for the benefit of the hon. members & others of the new Confederacy as the old detached maps of the province did not tend to instruct the provincialists in one section on the geography of the others.' The younger Bell saw more clearly than Logan, who was constantly harassed by more mundane financial and technical difficulties and who fretted that the Fenian raids in 1866 had driven his newly published atlas out of the minds of the government.[35]

Whether Logan's decisions were instinctive or coldly rationalized, his actions constantly pursued geological knowledge from this expansive uniformitarian perspective. In 1864 he had sent Alexander Murray to conduct a geological survey of Newfoundland under the authority of the GSC. This meant, in effect, the authority of Logan himself, who confessed that he took 'a most intensive interest in the investigation,' which could not only reveal the natural resources of the island but also shed some light on the Laurentian formations Logan was obsessed with. As such, the geology of Newfoundland was exhibited with Logan's Canadian collection at the Paris exhibition in 1867. When the dominion government after 1867 forbade any official involvement of the GSC in the survey of Newfoundland, Logan asked one of the Canadian staff to request a leave of absence and proceed to Newfoundland anyway.[36]

The Geological Survey of Canada had been extended, at least unofficially, to Nova Scotia even earlier. Logan and Dawson, a native of the colony, had long shared a special interest in the coal regions of Nova Scotia. Yet a systematic geological survey had never been carried out there, mainly because of the control held by the General Mining Association over the coal regions. But the gradual demise of the association and the discovery of gold by the late 1850s revived public interest in the possibility of a survey. Dawson, who did not himself enjoy geological surveying, was anxious as early as 1858 to subordinate such a survey to Logan's authority, as he believed 'it would be a great thing to have the whole brought into one great work at the close.' In 1861 he formally advised the Nova Scotia government to pursue just such a course, and Logan offered gratuitous services to the provincial secretary, Charles Tupper. Although local problems, including the resentment of local geologists like David Honeyman (1817–89) and Henry How (1828–79), professor of natural history and chemistry at King's College, Windsor, prevented the immediate organization of a geological survey of Nova Scotia, such a survey was said to be one of the stipulations agreed to when Nova Scotia entered Confederation.[37]

In much the same way, when H.Y. Hind was commissioned in 1864 to carry out a geological survey of New Brunswick, whose rock formations had been used by Logan as a basis for his earlier study of the Gaspé, Hind immediately arranged a liaison for consultation with the GSC. Hind argued that he had 'every reason to believe that many of the formations in New Brunswick are repetitions of rocks which occur in Canada.'[38]

Well before 1867, then, the GSC under William Logan's direction had mapped the geological formations of Canada south of 52 degrees north latitude and east of 100 degrees west longitude, and it was informally guiding the organization of geological surveys in all the other British North American colonies except Prince Edward Island. By the time Confederation was realized in July 1867, there was no question but that the jurisdiction of the GSC would be extended to correspond to that of the new federal government, even though mines were now under provincial jurisdiction. But Logan fretted once again whether a corresponding financial increase would be just as forthcoming. Characteristically, his fears lessened somewhat when the Royal Society of London awarded him one of its Royal Medals after the Paris exhibition of 1867. It was an honour which, he believed, 'came in good time to have an influence on what may be done by the Government of the Confederated colonies in respect to our Survey.' Logan would be seventy years old in 1868, and he planned to retire as soon as a satisfactory replacement could be appointed to carry the heavy burden of such rapid territorial expansion. In the meantime, he kept his staff at work among the Laurentian rocks, where he confessed himself 'very anxious to make progress,' at least partly because the 'presence of Gold in them will interest the public.'[39]

Logan's search for a 'worthy successor' took him to the other side of the Atlantic, where consultations with A.C. Ramsay, a friend on the British survey, led him to A.R.C. Selwyn. Selwyn had been trained with the British survey in the days of De la Beche and directed the geological survey of Victoria, Australia, until its termination just before he was offered the Canadian position.[40] The concern to appoint a Canadian, which had been utilized to Logan's advantage in 1842, and to which Logan had so often referred in his public addresses, was after all secondary to his concern to appoint someone inclined to continue the expansionist trends he himself had established.

Sir William Logan spent his remaining few years after retirement in December 1869 still exerting a strong influence upon the Geological Survey of Canada. He served as Selwyn's adviser and sometimes even as acting director. His health was deteriorating, and he was certainly in no condition to take seriously a recommendation to Lord Clarendon by one of his connections in London that he be named governor general of Red River. He travelled often between his sister's home in South Wales and the rocks of the Eastern Townships in Quebec, obsessed with preserving his reputation against criticism of his interpretation of their origins by Thomas Sterry Hunt. Although later investigations were able to improve some of his conclusions, as a whole his contribution to Canadian geology was enormous and stands the test of time remarkably well.[41]

Yet the place of the GSC in the mainstream of Canadian history has seldom been recognized. That place was carved out by the conformity of Logan's uniformitarian geology to some of the major trends of the early Victorian age. Its foundation harked back even earlier, to the political and military influence of the British imperial connection to the utilitarian outlook defined by the Common Sense philosophy of the Scottish Enlightenment, and to the social conservatism of the educated professional and business classes who defined their own best interests within this context. They hoped to make Canada a viable economic entity by searching for workable deposits of coal and other useful industrial and agricultural minerals. Geology offered the obvious method for such an inventory, and the information it provided then dictated the shape and extent which such a viable Canadian entity would have to take. William Logan recognized these needs and, driven by a number of motivations, allowed his survey to become a motor of Canadian expansion in the name of 'progress.'

Logan's geological maps increased in scope beyond the political boundaries of the province of Canada throughout his career, and this progression encouraged Canadians and other British North Americans to pursue larger interests. His descriptions of 'economic minerals' certainly did so; the best example was coal, which Logan concluded would never be found within the province. In this case his survey turned attention to the enormous coalfields of the eastern provinces, and to the possibility of coal in the great north-west. Another example was copper, which Logan encouraged Canadians to seek on the northern shores of the upper Great Lakes. The new understanding offered by Logan's geology of the limitations and possibilities of the country, given the technology of the time, seemed to provide a solid foundation upon which an expanded Canada would have to be built if it were to prosper through industrialization in future.

This is not to dismiss the important political, economic, and military factors which played a decisive role in the formation of the Dominion of Canada; without political deadlock in 1864, the American Civil War and the end of reciprocity in 1865, the Fenian raids in 1866, and the growth of larger imperial concerns in Britain, Confederation would not have taken place in 1867. But the intellectual foundations reinforced by the GSC gave the idea its aura of feasibility.

For the idea that Canadians in 1867 were capable of creating a transcontinen-

III 'Permanence'

tal nation called for an imaginative idealization of reality. The perceived powers of science lay at the very root of this leap of faith and lent to it an aura of inevitability, even in the eyes of its most bitter Rouge and regional opponents. At its core lay a growing popular interest in the study and control of nature, and a concomitant faith in recognized 'experts' in this field of knowledge. 'Nature,' observed J.W. Dawson in 1868, 'has already taken hold of the mind of Young Canada, and is moulding it in its own image.'[42]

Sir William Logan's geological inventory did not solve Canada's economic problems; the most it could do in this regard was identify them. The 'coal problem' was eliminated neither by federation with the Atlantic provinces nor by expansion into the north-west. It persisted well into the twentieth century.[43] But the survey did help to change attitudes about the Province of Canada and its need for the other British North American territories. As the GSC showed so clearly, ideas that D'Arcy McGee had seen as kindred to that of a British North American nationality ('extension, construction, permanence, grandeur and historical renown'), and to which the survey gave substance, were not always its progeny. In conjunction with the economic and political developments which altered Canada's place within the British Empire and on the North American continent during the early and mid-Victorian period, they nourished the idea of a transcontinental Canadian nation.

Any concept of a Canadian or a British North American nation in the early Victorian age was still more closely identified with the state than with its people, a relationship which implied a focus more upon land than upon blood. 'Here we are,' wrote *The Nation* as late as 1875,

rooted to the soil, clinging to it by the every fibres of our being, and drawing from it sustenance for body and mind. Should we not have, then, a different feeling towards this land from any that we cherish towards others? Certainly we should, and this feeling constitutes *nationality*. We recognize ourselves as belonging to the land, as sustaining peculiar relations towards it, and as borrowing from it a name and a position in the world.[44]

But a community of material interests failed to follow as a corollary of federation during these early years. The concept of nationality also changed its emphasis, becoming increasingly racially oriented by the late nineteenth century. The nation-state replaced the state as the object of loyalty, while the nation was increasingly personified and idealized as 'a huge collective self.'[45] The new dominion could never meet such terms, because it was not designed to do so.

Still, the optimism underlying the promise of material progress, and the role played by science in it, held fast after 1880. Thanks to 'the observations of travellers and the researches of men of science,' Canada was 'known' to possess 'what our neighbours lack, coal almost the entire way from Manitoba to Victoria. The mineral wealth of the North-West has only been vaguely guessed at; but it is known that ... even beyond the Arctic Circle, gold, iron, copper, lead, and coal have been found in exhaustless abundance.'[46] Even the difficult problem of access to northern mineral resources constituted, it was argued after gold was struck in the Klondike in 1897, 'mere details to be summarily swept aside before The Great Purpose.' For such was thought to be the 'colonizing power' of mineral discovery; countless numbers followed on the track and 'the history of a new country, possibly a nation, begins.'[47]

The GSC under William Logan's leadership provided practical motivation and intellectual context for the confederation of the British North American territories. To some extent it also shaped the idea of the nationality that was to be nurtured within the extended political confines. It offered hope for future prosperity, a rationale for co-operation, and above all a mission for Canada to fulfil. Uniformitarian geology had identified in its subject a heap of building stone, and Logan's geological inventory imposed conceptual order and uniformity upon those stones. But it was not within his power to cement them in place.

PART II

TERRESTRIAL MAGNETISM AND METEOROLOGY

6

The Spirit of the Method

Such is the spirit of the method by which I persuade myself it will someday be possible to connect together, by empirical and numerically expressed laws, vast series of apparently isolated facts, and to manifest their mutual dependence.

Alexander von Humboldt, *Kosmos* (1845)

After geology, the two inventory sciences that most altered the way Victorian Canadians saw themselves and their place in the world were terrestrial magnetism and its close ally meteorology. For besides the distinctive natural history of their immense land, its most evident characteristics were its northerly latitude and harsh climate. It was natural that an interest would develop particularly in the climate with which British North Americans contended daily. During the nineteenth century, theories explaining this climate, and how it might still be evolving, directly influenced what Canadians thought their country could become, while searches for the north magnetic pole on British North American territory generated awareness of the land's uniqueness. The scientific study of the environment confirmed Canadians' pride in the features which brought them international attention and which they strove to maintain by promoting these sciences themselves.

Unlike interest in the natural sciences, which flourished quite naturally around Montreal and Quebec and in the Ottawa Valley, interest in these physical sciences was centred in Toronto. Several attempts during the 1830s to emulate the Natural History Society of Montreal and the Literary and Historical Society of Quebec had failed at Toronto. Only in 1849 did the Canadian Institute, which focused on the physical and engineering sciences, finally succeed in establishing itself. Earlier failures were attributed to a general perception that the vast unchanging flatness of the settled southern peninsula of Upper Canada did not capture the interest of prospective natural historians:

> The simplicity and sameness, over great areas, of the geological formations of this peninsula, – their comparative poverty in fossils, the absence of mountain ranges, – the limited catalogue of its mineral productions; all undoubtedly combine to deprive that delightful study of many of its attractions, and to deprive societies ... of an allurement and stimulus to individual exertions. The same physical peculiarity limits to a certain extent, ... as compared with other geological provinces of this continent, – the field of the naturalist and botanist.[1]

The organized study of physical science in the form of terrestrial magnetism and meteorology was introduced to Canada in 1839, the result of imperial participation in a world-wide network of scientific researches. Canada's political membership in the British Empire, in addition to its geography, drew the colony inexorably into the wider magnetic and meteorological inventory. The practice of these two physical sciences differed from geology in that the organization of the former was externally imposed by formal imperial policy. Yet the ideological currents underpinning geomagnetism and meteorology resembled those of geology, and some of them ran much deeper in Canadian society. One of the most important contributions of these sciences was to encourage Canadians to reevaluate their position and even their character as a northern people.

These sciences were evolving in a specific context. During the early nineteenth century geomagnetism and meteorology were recast with 'Humboldtian' aims and emphases. The new approach sought accurate and comparable measurements of phenomena observed on a world-wide scale. Its origin can be traced to the years after the Napoleonic Wars, when the European scientists Alexander von Humboldt and Carl Friedrich Gauss sought international co-operation to explain and quantify the laws of geomagnetism. By the 1840s American observers such as Joseph Henry, Elias Loomis, and James Pollard Espy were borrowing the model of Humboldt's magnetic observatories to establish laws governing great continental storms. Great Britain's co-operation in these scientific enterprises made all but inevitable the involvement of British North America a vast British and northern territory believed to ensconce not only the north magnetic pole but also the source of North America's winter weather systems.[2]

I

The magnetic qualities of the earth had for centuries held a practical international interest because of their importance to navigation. Even before the seventeenth century, it was known that the declination, or deviation, of the compass needle from true north varied irregularly with the meridian. This variation confounded navigators, who required a reliable corrective in order to stay on course. They recognized that the angle of inclination of the needle to the horizon – its dip – also differed with location. To complicate matters further, both variation and dip changed over time in three cycles whose amplitudes could be measured in centuries and decades (secular), seasons (periodic), and days (diurnal).[3] The publication in 1600 of William Gilbert's *De Magnete* added to practical knowledge the theory of an invisible force of attraction that maintained an orderly universe.

Gilbert's theory of attraction found its epitome in the work of Isaac Newton (1642–1727), whose *De philosophiae naturalis principia mathematica* (1685–6) demonstrated how the force of gravity could maintain the universe and defined universal laws of motion in the language of mathematics. By the eighteenth century the obvious remaining task within this Newtonian framework was to comprehend both electricity and magnetism with the same mathematical precision that defined gravity.

Interest in electricity and magnetism was legendary and had scarcely shed its aura of magic and mystery by the nineteenth century. Investigations had revealed the dual characters of both of these forces – each had a positive and a negative aspect – but a direct connection between them was not substantiated until Hans Christian Oersted (1757–1851) noticed the deflection of a compass needle by an electric current in 1820. Finally in 1831 Michael Faraday (1791–1867) formulated the theory of electromagnetism, in which magnetism was shown to be electricity in motion, and vice versa.

The means of applying knowledge about electricity and magnetism to the earth were less obvious than results achieved in the laboratory, and the problem called for new information. The idea of using instruments to observe and record changes in the atmosphere had emanated from *New Atlantis* by Francis Bacon (1561–1626) and had interested the Royal Society of London since its foundation. Co-ordinated networks of meteorological observers were organized soon thereafter: a Florentine system extended to Warsaw and Paris during the seventeenth century; the Royal Society made similar attempts in England during the eighteenth century; and the Meteorological Society of Mannheim organized thirty-nine weather stations, including four in the United States, all with comparable barometers, thermometers, and hygrometers, in 1780. The American government recorded meteorological observations on the North American continent from the days of Jefferson and Madison. For health reasons the American army Medical Department began to keep meteorological records in 1814, and for agricultural reasons local land offices followed suit after 1817.[4]

Similarly, by the late eighteenth century developments in chemistry and static electricity outlined a body of knowledge upon which a meteorological theory could be structured. By 1800 basic meteorological instruments such as the barometer, thermometer, and hygrometer made a science of the whole world seem possible. Although the relationships between heat and motion were only vaguely understood, the science of meteorology had advanced far enough to study mechanical, though not yet thermodynamic, aspects of atmospheric change. It would be over forty years before synoptic weather maps clarified the movements of storm systems over large areas, but these earlier advances facilitated the precise measurements upon which all later developments rested.[5]

By the early nineteenth century the pulse of scientific interest in both terrestrial magnetism and meteorology had quickened. Polar expeditions searching for a north-west passage recorded magnetic variations in high northern latitudes. The scientific search for a north magnetic pole intensified after the publication in 1819 of Christopher Hansteen's *Magnetismus der Erde* (The earth's magnetism).

An ardent follower of Hansteen's magnetic theories was Edward Sabine (1788–1883), astronomer to several British polar expeditions in 1818, 1819, and 1820. Sabine, a captain in the Royal Artillery who had served at Quebec and Niagara during the War of 1812, used science to advance his career after the peace of 1815. He supported Hansteen's hypothesis that magnetism was essentially part of meteorology; he had learned firsthand the impact of meteorological phenomena upon magnetic variation and had built his scientific reputation upon the attempt to define this relationship more precisely. In 1825 he published evidence that the locus of maximum magnetic force was not the same as the place where the needle dipped 90 degrees. There was, then, a certain ambivalence in the concept of the earth as a magnet, and hence in the meaning of 'north magnetic pole.' The discovery, Sabine wrote later, enhanced interest in geomagnetic researches because even the broad outlines of the science had to be recast.[6]

A new era in magnetic and meteorological observation dawned with the 'cosmical' outlook suggested by the Prussian Alexander von Humboldt (1769–1859). Widely recognized by the end of his long career as 'the Prince of Modern Science,' Humboldt had been inspired by the voyages of James Cook to undertake extensive scientific travels observing the complex interrelations of natural phenomena on a world-wide scale. By the 1830s the foremost authority on terrestrial magnetism, Humboldt encouraged Sabine to take up standardized instruments and methods and to pursue accurate and comparable measurements on as grand a scale as possible. Humboldt and Sabine hoped ultimately to derive from accumulated data a mathematically expressed universal magnetic law based on the Newtonian model. The earth and its atmosphere became a scientific laboratory for these Humboldtian studies, their results to be reduced and made intelligible by new methods of standardizing observations.[7]

Networks of observatories were extended ever farther afield by the mathematical curiosity of C.F. Gauss (1777–1855), as well as by Humboldt's speculations. In 1835 Gauss's German Magnetic Association tested the mathematics of terrestrial magnetism, and Humboldt attained the co-operation of the massive Russian Empire. More directly responsible for the science as it developed in Canada were the steps by which the private ambitions of a few individuals in Britain translated into a 'magnetic crusade.' By 1838 Sabine had convinced both the Royal Society and the BAAS to extend a chain of magnetic and meteorological observatories across the entire length and breadth of the British Empire. The British government was induced to fund the project under the army's Ordnance Department, with Royal Artillery officers to act as observers under Sabine's supervision. Within a year the American Philosophical Society at Philadelphia expressed its willingness to co-operate with the British network. The observatories were toasted as 'by far the greatest scientific undertaking which the world had ever seen.'[8]

The British Association for the Advancement of Science, founded in 1831 to promote co-operation among cultivators of science in different parts of the British Empire and elsewhere, appreciated the obvious importance of British North America as a quarry of information for both terrestrial magnetism and meteorology. Canada lay due south of the magnetic pole of verticity, where the needle dips 90 degrees, discovered by James Clark Ross (1800–62) in 1831; waiting to be discovered somewhere in the HBC territories lay the pole of intensity, where the magnetic force was greatest. Quebec and Montreal were designated imperial meteorological stations in 1837, and in 1839 Montreal joined a list of magnetic observatories that included stations at Van Diemen's Land (now Tasmania), the Cape of Good Hope, and Bombay. The president of the BAAS in 1839 looked forward to magnetic observations 'from Montreal to Madras.' The main goal, he explained, was threefold: to detect and quantify changes in the three magnetic elements (intensity, variation, and dip); to draw the lines of force and direction; and to discover the true cause of terrestrial magnetism. In so doing the association hoped to complete what Isaac Newton had begun: 'a revelation of new cosmical laws – a discovery of the nature and connexion of imponderable forces – all these the possible results of approaching the heights of theory on what may prove to be their most accessible and measurable side.'[9]

Because of its proximity to the north magnetic pole, British North America – Canada in particular – was accorded a special role in solving the problem of terrestrial magnetism on all three levels of investigation. And because of its vastness and northerly location, Canada was destined to become equally prominent in the study of North American meteorology. This unsolicited new calling generated interest not only among devotees of the science but eventually in Canadian society in general.

II

While it was traditional to keep daily notes on the weather, prior to 1840 few inhabitants of British North America expressed any interest in observing changes in the earth's magnetism, probably because such specialized observations required sensitive instruments. Indeed, the Royal Artillery officer who arrived at Toronto in 1839 to commence observations, Lt Charles Riddell, informed Sabine that he had not found anyone who appeared to take the least interest in the work or on whom he could rely for assistance. 'There is a College and three or four Professors,' he added, 'but I have only seen one of them[,] the shallowest little wretch I ever met. My only chance is in the arrival of some officers who may have a turn to science.'[10]

Riddell's first impressions did not show that Upper Canada was devoid of people interested in physical science. In 1833 Mahlon Burwell presented to the House of Assembly a petition from John Harris, of Woodhouse Township in the London District. Harris, a cousin by marriage of Egerton Ryerson and a former scientific officer in the Royal Navy, had served on the Navy's hydrographic surveys of the Great Lakes under Henry Wolsey Bayfield (1795–1885) and William Fitz-William Owen (1774–1857) from 1815 to 1817. He was asking for public support for an observatory at York. The petition was sent to Burwell's select committee on education, only to be subsumed by much larger issues. The committee recommended the immediate establishment of the University of King's College, with a professorship of practical astronomy, an observatory, and 'all the instruments and apparatus necessary for the study of that sublime science.' Harris, the committee's report continued, inspired 'other scientific men' in the province to request the literary and scientific education to which the youth of Upper Canada, as much as in any other 'enlightened country,' were entitled. For although astronomy required 'the exercise of the sublimest powers of the human mind to comprehend the proofs of the truths which it exhibits,' the committee insisted that its results, none the less, were 'within the grasp of every thinking man.'[11]

By linking the observatory to the more controversial issue of a provincial university, Burwell's committee effectively shelved it for a time. But the establishment of an astronomical observatory was reconsidered in 1835, when the British Admiralty complained to the colonial secretary that while 'half of the globe' was connected by a chain of observatories, the other half was 'destitute of the means of furthering the purposes of astronomy.' It was 'equally wanting in the material objects of geography. The subject,' the Admiralty concluded, had been 'strangely overlooked.' Accordingly the Earl of Gosford, governor-in-chief of British North America (1835–37), was ordered to select a proper site. Sir John Colborne, the lieutenant-governor of Upper Canada (1829–36), was asked whether the Upper Canadian legislature would underwrite at least part of the expense of such an observatory. Colborne's term in Canada ended shortly thereafter, and his successor, Francis Bond Head, met with more pressing political concerns. So the issue lay dormant through the rebellions of 1837. In 1838 the clerk of the assembly replied to a dispatch from the Colonial Office that the question had never reached the House and that no member knew anything of the matter.[12]

Besides the deep political problems that overshadowed the establishment of a scientific observatory in Canada, one other factor helps to explain the relative lack of popular interest in terrestrial magnetism before 1840. In the British popular imagination, magnetism was associated with the esoteric 'animal magnetism' practised by Franz Anton Mesmer in revolutionary France. Mesmeric seances found their way to Lower Canada through Edward Gibbon Wakefield, who accompanied Lord Durham in 1838. Public lectures on magnetism attracted Lower Canadian audiences to the halls of the Literary and Historical Society of Quebec steadily after 1840. But Durham's entourage spent only a day in Toronto, hardly long enough to organize such a sideline, and mesmeric seances did not catch on there until a decade later. When Charles Dickens visited the city in 1842, he drew attention to the magnetic observatory as one of its few positive features. With the introduction of the magnetic telegraph in 1846, Canadians witnessed even more directly the beneficial powers of the magnet as a useful tool warranting their more serious attention.[13]

In contrast to the sparse interest in terrestrial magnetism, interest in meteorology existed virtually from the earliest settlement. Meteorological records of British North America had been kept since the first arrival of Europeans. Most of these records were descriptive weather observations, and few were systematic. Such haphazard records had no value for Humboldtian purposes, since they could not withstand tests for verifiable accuracy and comparability. Meteorology's relationship to practical astronomy was exemplified in New Brunswick. In 1840 King's College at Fredericton acquired a new professor of mathematics and natural philosophy, William Brydone Jack (1819–86). Jack had been trained at St Andrew's in Scotland as a favourite pupil of David Brewster, the first observer to obtain an hourly meteorological register for an unbroken series of years. Jack's personal interests, like Brewster's, leaned strongly towards astronomy and its practical permutations, including meteorology. They saw observatories as necessary preliminaries to networks of private observers. After observatories completed the 'great network of triangulation' by furnishing standards of comparison, establishing laws governing physical phenomena, and fixing secular or normal data, private individuals could carry out the 'detailed surveying.' Jack accordingly concentrated on establishing at Fredericton a high-quality observatory supplemented by university courses in practical astronomy.[14]

Lord Dalhousie lent his prestige to amateur meteorology during his term as governor of Nova Scotia. He kept daily weather journals at Halifax and continued his hobby when he was transferred to Quebec as governor of Canada in 1820. Dalhousie's efforts, and those of Lady Dalhousie, an amateur botanist, stimulated the scientific interests of others and attracted a circle of like-minded residents of Quebec to form its Literary and Historical Society in 1824.[15]

Many scientific contributors to the LHSQ's *Transactions* were military officers stationed at Quebec, several of whom expressed an interest in meteorology. Capt. Richard Bonnycastle, Royal Engineer and amateur geologist, contributed observations of the aurora borealis in 1827. William Kelly, a naval surgeon, kept a meteorological journal at Cape Diamond, Quebec, from 1824 to 1831 and offered the society his conclusions that the climate of Canada over the previous two centuries had remained fairly constant. The implication – that little amelioration could be expected in the near future – flew in the face of the common notion that the climate would improve with the progressive clearing and cultivation of the land. Kelly shared in Humboldtian ideas that mean temperature, not the occurrence of local frosts, measured long-range climatic change. 'It is easy to understand,' he chided his audience, 'how philosophers may be occasionally led astray by pursuing a favourite theory, or philanthropists by the hope of a favourable change in anything which influences the comfort or prosperity of mankind.' But he could not as readily see why 'the opinion of a gradual improvement in the climate should be held by the colonists themselves, contrary to what, from a general view, would appear to be the real state of the matter.'[16]

Like the belief that Canada had workable coal, the myth that its climate would be moderated by clearing and cultivating the virgin forest died hard. It paralleled other aspects of early Victorian optimism and served propagandistic purposes unique to Canada. Legislators and physicians saw a need to overcome 'the prejudice, founded in ignorance' that British North America's climate was too extreme for agriculture and unhealthy for potential immigrants. The mathematical master of Toronto's Upper Canada College, Rev. Charles Dade, kept meteorological tables during the 1830s in an attempt to link changes in the weather to cholera epidemics in 1832 and 1834. In 1834 Dr William Rees proposed to research the 'medical topography and climate of North America' for just such a purpose. Although the work was never completed, future legislative committees on immigration and colonization attached similar interpretations to information gathered by growing numbers of meteorological observers.[17]

Another locus of meteorological observers during the 1830s was the Natural History Society of Montreal. At the instigation of Judge J.S. McCord, the chairman of the NHSM council and a man with a personal interest in the meteorology of British North America, the society solicited the co-operation of the Hudson's Bay Company. The HBC's governor, George Simpson, promised to have meteorological journals kept at company posts in the 'Indian territories.' The society sent instructions for observers to Simpson and anticipated valuable additions to observations already donated by officers of the company. In 1837 the NHSM's annual report remarked that meteorology was making rapid advances on the continent, and circulars were sent to corresponding members in the other provinces, calling attention to the subject and inviting co-operation. The NHSM's annual essay competition also invited papers on the medical statistics of Montreal, as well as on changes observed in the climate of Lower Canada and their possible causes.[18]

An imperial magnetic and meteorological observatory would certainly provide focus for the geophysical inventory of British North America. Neither of its official pursuits was new to the population; what no one expected, however, was that the site for the observatory would be not Montreal, as originally planned, but Toronto.

III

When Lt Charles James Buchanan Riddell arrived at New York in September 1839, his first move was to visit Prof. Alexander Dallas Bache, at Girard College in Philadelphia, and Prof. Joseph Henry at Princeton University. Bache (1806–67), an influential member of the American Philosophical Society, was an expert in astronomy, meteorology, and terrestrial magnetism. He had attended the BAAS meeting of 1838 and supported American participation in the proposed British chain of observatories. These interests and Bache's influence, Riddell believed, would make him 'a most valuable acquaintance.' Henry (1797–1878) had done considerable research in electromagnetism and, independently of Michael Faraday, had discovered electromagnetic induction in 1830, paving the way for Morse's invention of the electromagnetic telegraph. The three men arranged exchanges of information and simultaneous observations. A gradual shift in attitude developed, comparable to William Logan's experience in geology during roughly the same period, when American data provided more practical, relevant, and immediate standards of comparison than the British did. Shortly after Riddell's arrival his brother in England remarked to Sabine that surprisingly many Americans seemed to take great interest in the establishment of the first observatory on the North American continent.[19]

In selecting a site for the proposed observatory, Riddell preferred Montreal. Compared with Quebec, Montreal had a better climate; compared with any town farther inland, it was located closer to military headquarters and was more accessible by sea. This last point was paramount because of dangers to fragile instruments in a voyage up the St Lawrence to Kingston or Toronto. But Montreal's unsuitable local geology outweighed the advantages of its geography. Riddell was advised by Captain Bayfield, commander of the naval surveying service stationed at Quebec, that basalt and trap rock underlying the area were probably magnetic and would inevitably disturb his variation readings. While Riddell searched for alluvial soils deep enough to circumvent this problem, Bayfield overruled him and named Toronto as more suitable for a fixed magnetic observatory. With its base of solid limestone Toronto appeared, as Riddell wrote to Sabine, 'perfectly safe from all local attraction.'[20]

Riddell reluctantly complied with Bayfield's preference, and his arrival in Toronto brought the further disappointment that the military reserve west of Fort York, where he had hoped to locate the observatory, was largely uncleared and swampy. The situation was saved when a member of the King's College council hinted that a formal request from Riddell might well coax a grant of land from the college. In the ensuing agreement, two acres were granted on condition that the buildings would be used for scientific purposes only and would revert to the college upon abandonment by the British government.[21] Ironically, the future university now acquired its Magnetic and Meteorological Observatory a mere six years after the Burwell committee first formulated its recommendations.

An auxiliary function of the fixed observatory at Toronto was to serve as a base station for a magnetic survey of British North American territory in the area of maximum magnetic intensity. Sabine had intended to conduct such a survey himself in 1839, before he was named to superintend the colonial observatories. The survey was postponed but Sabine still intended to come to Canada, possibly in 1840 or 1841. Riddell had not been in Toronto long before he noted its convenient location, with steam communication on both sides. Two canoe trips of a fortnight or less each, he estimated, would obtain magnetic observations more than a thousand miles apart on entirely new ground. Riddell pointed out to Sabine that such a survey would be of great practical use to the country, since local surveyors were anxious to obtain observations free from interference by local beds of magnetic iron ore near the Bay of Quinte. Eager to conduct the surveys himself, Riddell made plans to assign a young officer, Lt Charles Wright Younghusband, then only twenty years old, to the observatory in his stead.[22]

Advice from the HBC resulted in three possible arc-shaped routes to Hudson Bay and James Bay: westward across Lake Superior to Lake Winnipeg and York Factory; northward across Lake Superior to the Michipicoten and Moose Rivers to Moose Factory; and eastward along the Saguenay River to Lake St John and Lake Mistassini to Rupert River and Fort Rupert. Riddell hoped to make a name for himself as a scientific explorer, and thought it 'of course very natural that I should feel great pleasure at the idea of making such a journey with such an object as this would be.' But his hopes were dashed even while his survey plans were being laid. The recurrence of chronic symptoms of dysentery, which two years earlier had caused his transfer from tropical climates, now required his removal from the cold. Riddell was invalided home to England in February 1841, with Younghusband left to superintend the daily observations.[23]

In order to guarantee the continuance of these observations after 1842, Edward Sabine needed something to justify past expenditures. The most convenient way to make results available was to publish those of each observatory separately, not in integrated form as had been planned. The first volume to be prepared happened to be the Toronto observations for 1840–1.[24] Thus, while the publication of countless observations did not in itself contribute in any obvious way to the theory of terrestrial magnetism, Sabine obtained funding for the observatories. But moreover, the Toronto observatory achieved an importance for its own location and contribution as much as for its function as a haphazardly chosen link in a larger chain. With the arrival of Riddell's successor, a second stage in the history of the Toronto observatory brought these nascent trends to fruition.

IV

The advantages of a wider magnetic survey while a magnetic observatory could serve as a primary station for reference and comparison were at least as clear to Sabine as they had been to Riddell. Even though Lieutenant Younghusband assisted Riddell gratis in hope of qualifying for such a survey, Sabine decided that the young officer would remain instead acting director of the Toronto observatory. The magnetic survey was to be conducted by Lt John Henry Lefroy (1817–90), who with Riddell had been among three Royal Artillery officers chosen to direct colonial observatories. Lefroy's aptitude for science had distinguished him at the Royal Military Academy at Woolwich and again as a student of practical astronomy at the Royal Engineers' Establishment at Chatham during the 1830s. He had served at St Helena since 1840. He exuded an air of competence and serious dedication to duty, impressing his superiors as an outstanding candidate for the British magnetic inventory.[25]

Lefroy believed that the tedium of the daily and sometimes hourly observations he knew awaited him in British North America would be compensated by significant results. Favourable results were inevitable, he wrote to Sabine before leaving England in July 1842, 'for the work is not one whose completion turns on any single point.' Each day's observation would add an integer to the sum already gathered, and 'stop when it may, what has gone before will have its full value.' Still expecting Sabine to visit Canada in 1844, Lefroy hoped to submit, 'when we meet at Toronto, as large a body of results as will in some degree answer the question that must grow out of those Ross is obtaining at the opposite pole.' Two months later, though, some of this idealism had eroded when Lefroy, still en route to Toronto, found himself 'in the middle of stumps, logfences, corduroy roads, and framed houses, therefore in some danger of acquiring the antipathy to trees which every settler seems to possess.'[26]

Lefroy had landed at Quebec in August, but he interrupted his voyage upriver to take magnetic readings and consult American scientists with similar interests. They included Bache and Henry, as well as Renwick at New York, Nicollet and Ducatel at Baltimore, and Bartlett at West Point, 'the most in earnest' of the magneticians he met. Bache, with his 'simplicity and bonhommie and high talents,' impressed upon Lefroy the importance of the connection between the Toronto and Philadelphia observatories. Indeed, Lefroy's proposed northern survey around the focus of maximum magnetic intensity was intended by the BAAS to be complemented by surveys in progress by American magneticians. Such a reinforced network, it was contemplated, would yield marvellous results for the whole North American continent, especially when extended to include observations in progress on the southern coasts of the United States and in the Gulf of Mexico.[27]

Upon arrival in Upper Canada, Lefroy was apprised by Rawson W. Rawson, the civil secretary at Kingston, of plans by J.J. Audubon (1785–1851) to lead an American exploring party. The famous French-American bird-illustrator and naturalist intended to travel to the Columbia River by a circular route whose northern arc would cut across British North America. Lefroy determined to join the party as a trial run for his own magnetic survey. Although he feared the drawbacks, as he explained to Sabine, 'of so long an association with such a crew of outlaws as the trappers are reputed to be,' he had heard that Audubon greatly resembled the well-known British author of *The Natural History of Selborne*: he was 'a pleasing and benevolent person, a sort of Gilbert White.'

Audubon's unfavourable reply quickly disabused Lefroy of his initial notions and convinced him that while magnetism might be compatible with 'the leisurely movements of a party having a particular object in view,' this was not likely to be the case with an exploring party who sought adventure. He heeded Audubon's advice that a summer's journey through HBC territory might better fulfil the objects he had in mind.[28]

Nor was the conflict of purposes which Audubon had foreseen eliminated by the *carte blanche* co-operation of the HBC guaranteed by George Simpson. On Simpson's advice Lefroy and one assistant, Bdr William Henry of the Royal Artillery, left Lachine in May 1843 with the first brigade of HBC canoes of the season, with ten boxes of instruments weighing several hundred pounds in tow. They travelled along the Ottawa route to Lake Huron and Sault Ste Marie, the ordinary limit of western travel beyond which he believed 'all was wilderness and prairie,' and from there to Fort William, the last point whose longitude had been ascertained. All along the route Lefroy and Henry performed three classes of observations of magnetic variation, dip, and intensity: two hours of daily observations; three or four hours of occasional observations at principal posts; and twelve or more hours of observations on predesignated 'term days,' simultaneously with other observatories. The rest of the brigade proceeded on its own schedule, and Lefroy often travelled until night before reaching camp.[29]

At Fort William Lefroy and Henry transferred to a 'half canoe' with a guide and a small crew. This decision was never fully explained, but Sir John Richardson, the geographer and Arctic explorer, had earlier sent Lefroy a list of scientific desiderata and had argued that the rigid schedule of the company brigade would frustrate Lefroy's primary duties. He assured Lefroy, moreover, that he could more easily overtake the brigade in a smaller vessel. George Simpson had countered that Lefroy would instead be easily separated from the brigade in a half canoe, as indeed he was. His Iroquois guide lost his way several times after leaving Fort William for the complex maze of rivers, lakes, and portages to Lake Winnipeg. Lefroy paid an unscheduled visit to Fort Garry, then moved on to Norway House, the base station for all his readings north of Toronto.[30]

As he travelled north-east to York Factory on Hudson Bay, Lefroy found the magnetic intensity decreasing. This discovery helped establish the major axis of the concentric ellipses, which Sabine called 'isodynamic ovals,' used to depict decreasing intensity as one moved away from the magnetic pole of intensity. Lefroy and Sabine hoped that co-operation with American scientists such as Dr John Locke of Cincinatti, who had measured intensity from the eastern seaboard to the Mississippi, would determine the corresponding minor axis and the point of maximum intensity where the two lines intersected.[31]

Lefroy then proceeded westwards to Cumberland House, to the Athabasca

River, and to Fort Chipewyan on Lake Athabasca. While wintering at Chipewyan from October 1843 to February 1844, Lefroy and Henry decided to take hourly observations day and night and during magnetic disturbances, at intervals of about two minutes, for hours on end. This incredibly tedious task, enlivened by long evening chess games and such soldierly amusements as casting bullets out of frozen mercury, they stoically repeated at Fort Simpson on the Mackenzie River not long after. Still farther north, observations at Fort Norman and Fort Good Hope took the party almost, but not quite, within the Arctic Circle, much to Lefroy's disappointment. Lefroy justified his decision to proceed much farther north than had originally been planned with the argument that the expense and difficulty of obtaining magnetic information from more remote regions of the continent increased in so high a ratio that he felt himself in a 'now or never' situation.[32]

Returning southwards to Fort Simpson, Lefroy and Henry rejoined the HBC brigade heading back to Fort Chipewyan. From there he followed the Peace River to Fort Dunvegan, since 'no scientific traveller' had descended the river since Alexander Mackenzie in 1789. Geographers were anxious to know where the river issued from the Rocky Mountains. From there he travelled to Fort Edmonton, which he found lay on the same isodynamic line as Toronto; to Carlton House, Cumberland House, and Norway House. Lefroy arrived at Toronto via Penetanguishene in November 1844 and continued almost immediately to Montreal to complete his business there.[33]

By late 1844, Lefroy had survived an extraordinary five-thousand-mile trek and surveyed, Sabine recognized, 'a much greater extent of country towards the Northwest than had been contemplated.' He established three principal lines of magnetic intensity north of the fiftieth parallel of latitude, roughly along the Saskatchewan, Peace, and Mackenzie rivers. Together with a fourth line determined by John Locke below the fiftieth parallel, these lines defined the isodynamic ovals of the northern part of the American continent. Lefroy's observations corrected an earlier belief that the axes ran east-west along the meridian, showing rather that these isodynamic lines tilted north-westwards.[34] They foreshadowed the isothermic lines, which were later shown also to turn north-westwards and which, as will be seen, radically altered popular Canadian perceptions of the HBC territories.

Edward Sabine would have been pleased with considerably less, and he assured Lefroy that whatever else he had accomplished, if he had satisfactorily proved the decrease in intensity in going northwards and its corresponding increase in the return southwards, that would be his 'first feather.' Sabine believed that the full value of Lefroy's observations could be appreciated only once they reached England, to be interpreted at Woolwich and presented to the

Royal Society. 'I fully expect,' he wrote, 'they will be found to reflect the highest credit on all parties concerned, in proportion to the share which each shall have contributed.'[35]

Lefroy's ideas did not always coincide with those of his superior officer. Anxious to make a name for himself, Lefroy hoped to analyse his own observations. His increasingly outspokenness in this regard revealed far-reaching repercussions, both negative and positive, of his British North American magnetic survey. Even before embarking on the survey, he had confessed his 'fancy firmly possessed' by dreams of an overland survey around the globe. In April 1843 George Simpson had helped him plan an extensive magnetic survey whose first phase was to have been the journey of 1843-4. A second phase would have taken Lefroy back to Norway House in the spring of 1845, thence with the Columbia brigade to Fort Vancouver, by HBC ship to the Sandwich Islands, and on to England in 1846. A third phase in the spring in 1846 would have taken him from England to Sitka, across the North Pacific in a Russian American Company ship to Okhotsk, on to St Petersburg, and back by steamer to England late in 1846.[36]

The actual experiences of 1843-4 sharpened Lefroy's vision and narrowed its focus so that, a mere six weeks after his return to Toronto, he submitted an ambitious proposal for a second journey to the Arctic north-west from 1846 to 1848. At Fort Simpson and Fort Good Hope he had observed magnetic fluctuations that varied with both the season and the occurrence of the aurora borealis. These fluctuations suggested causal relationships which Lefroy hoped to verify as scientific fact. Moreover, he pointed out, the HBC was extending its establishments north of New Caledonia into the only remaining portion of the continent whose physiography, natural history, and geography remained completely unknown. Additional scientific advantages could be reaped if a naturalist were sent out to join him in such an expedition, since he and Henry had been too preoccupied to collect flora and fauna, as Sir John Richardson had requested. Lefroy felt responsible for the future progress of terrestrial magnetism in Canada since none of the young officers had shown any real interest in sharing in his plans for a magnetic survey of British North America. Neither did 'my occasional hints as to observations wanted from Lower Canada, or of the field open there, elicit any sort of disposition or ambition to enter it.' In the meantime, Lefroy wrote to Sabine, he hoped to make himself 'thoroughly master of the whole system' by which his observations were computed and reduced. He desperately wished 'to avoid the risk of becoming a mere *observing machine*, and to avail myself of this lesson for my improvement as a magnetician.'[37]

In Sabine's mind, the question of a second survey and that of Lefroy's private ambitions were but two sides of the same political coin. A misunderstanding had arisen between the HBC and the Royal Society over expenses incurred by Lefroy's first excursion. The resulting 'difficulties,' Sabine believed, would continue for some time, and made embarrassing yet another request for help from the company. Moreover, Sabine feared that Lefroy's justification of a second survey might be seen as an admission that his first had failed to determine the position of maximum intensity. Unable 'to speak with confidence of any striking results obtained,' Sabine concluded, 'we are clearly not in the case to talk of new projects.'[38]

What united both men despite a growing rift between them was a common interest in the future of the Toronto observatory, which would be decided in 1845. Sabine had recently presented a paper extolling the Toronto observatory as a model source of data obtained from 'work so steadily persevered in and carefully performed.' He advised Lefroy that as a result 'you could not be better placed than at Toronto working hard to shew that the good you have accomplished is well commensurate to the outlay which has been made.' Their bickering over a future survey continued to July 1845, but in May Lefroy renounced his commitment to the project in favour of the 'mill work' needed to prolong the existence of the observatory. It was wiser, he reflected, 'on an enlarged view of the objects and duties of life, to forego the present advantage for the greater one behind.'[39] His decision was to have far-reaching consequences.

Scientific interest in terrestrial magnetism and meteorology had spread steadily over much of the world during the 1830s and 1840s. By 1844 research in British North America, centred at Toronto, had attained international significance. Henry Lefroy had gone a long way in his gruelling survey to clarify the patterns of northern isodynamic lines. Yet while these sciences were now institutionalized in Canada, the work was highly technical and could not of itself fire the popular imagination or alter Canadian self-perceptions. Like geology, terrestrial magnetism and meteorology would have to take root in Canadian soil.

7

Mutual Attractions, 1845–1850

The matters in which we differ are nothing in comparison of those in which we agree.

Rev. William Bell to Egerton Ryerson (1830)

The practical and political snags which made a second magnetic survey of British North America improbable left Henry Lefroy to face the prospect of a lengthy sojourn in Toronto. Having returned from magnetic studies in the field, he turned his attention to other kinds of attraction. Although he had spent only a few weeks in Toronto before leaving for Lachine in 1843, he had become acquainted with the families of the town's most prominent citizens. While wintering at Lake Athabasca he had written to Younghusband asking to be remembered to the Macaulays, the Robinsons, and the Boultons. James Buchanan Macaulay (1793–1859) was puisne judge of the Court of Queen's Bench; John Beverley Robinson (1791–1863) was chief justice of Upper Canada; and the Boultons, one of the wealthiest Toronto families, included several members in public life and law and provided one of the main centres of Toronto social life at their home called the Grange. All were highly influential members of the Family Compact. It was not just that Toronto society took an interest in the official movements of the British troops stationed there. Many took a personal interest in the twenty-eight-year-old well-educated bachelor as a prime candidate for social invitations.[1] It was, both for Lefroy and for Toronto society, the beginning of a mutually important relationship.

No doubt as a result, Lefroy's own attitude towards Toronto changed. The town had, he felt, 'much altered since I left it.' It was 'a wonderfully growing place' destined to become 'the leading commercial city of Western Canada, a sort of Fresh-water Boston.' Toronto society he found 'dull enough,' but endowed with exceptions which offered a pleasant change after his long and isolated absence. These included the blue-eyed Mary Jane ('Polly') Hagerman, without whose gifted singing voice, he wrote to his sister, 'I should consider evenings spent abroad a great infliction.' Polly Hagerman was the twenty-two-year-old daughter of Christopher Hagerman (1792–1847), a pillar of the Family Compact. Hagerman was a former attorney general and judge of the Court of Queen's Bench, and his wife, Elizabeth, was a sister of J.B. Macaulay. Lefroy had been back at the observatory only a few weeks when he was visited by Mrs John Beverley Robinson, accompanied by her eldest daughter, Emily Merry, and 'the fair P.H.,' all professing an interest in Lefroy's collection of 'curiosities from the north.' Polly Hagerman may have possessed a character too headstrong and ebullient for the more reserved Lefroy, shown several years later when she accepted an offer to sing publicly at a resort in Saratoga Springs and scandalized Toronto society. She married John Beverley Robinson, Jr in 1846.[2]

Lefroy next turned his eyes to Emily Merry Robinson. They had met at Beverley House even before he departed for Athabasca in 1843. In 1845 Lefroy confided 'a sort of liking' for Emily despite her 'highly worldly' – by which he meant money-conscious – character. Feeling increasingly at home in social centres like the Robinsons', Lefroy felt himself 'getting towards that state of philosophy, that the only society I care for is that of a rational family circle.' The Robinsons had seven offspring of marriageable age, and it was natural that Lefroy should be drawn to their home. When the eldest, James Lukin Robinson, married at Woodstock in May 1845, a days-long celebration of picnics, drives, and dances included Lefroy, 'falling more in love all the time.' By November 1845 he was engaged to Emily, having convinced himself that her character and influence would 'an hundred fold enhance' his own usefulness to the magnetic service. 'I have long felt,' he wrote to Sabine, 'that the director of an Observatory in a place where there is much society, ought to be a married man, and long felt it to be very essential to my happiness that I should become one.'[3]

Lefroy married Emily Robinson on 16 April 1846 in a double ceremony with her sister Louisa and Lefroy's friend George W. Allan. Allan, a barrister and future mayor of Toronto, was the son of William Allan, the financial genius of the Family Compact. A third sister had married James McGill Strachan, son of John Strachan (1778–1867), the Anglican bishop of Toronto and pivot of the Family Compact, some months earlier. These marriages brought Lefroy into the network of the Family Compact and consolidated his stature within Toronto society.

Relationship by marriage was not the only link between Lefroy and some of the most powerful elements in Canadian society. The political and social network known as the Family Compact shared with him ideals they had absorbed as pupils of John Strachan. The socially and politically conservative outlook usually associated with the compact included a desire to achieve material progress and social order in Canada through far-reaching economic and educational improvements. While the Geological Survey of Canada refracted economic aspects of this outlook, Lefroy's magnetic survey highlighted the educational.

The roots of the educational views of the Family Compact can be traced to John Strachan. Before emigrating from Scotland in 1799, Strachan studied at Aberdeen and St Andrew's. There he met the Reverend James Brown, who not long afterwards became professor of natural philosophy at Glasgow. Brown offered Strachan an assistantship, mainly to prepare scientific experiments illustrating Brown's lectures. Strachan never attained the position because Brown became ill and retired prematurely. Bitterly disappointed that a 'career of honourable usefulness had been opened in a way after [his] own heart,' Strachan never relinquished his interest in teaching natural philosophy.[4]

Strachan's outlook and experience struck a harmonious chord with Governor John Graves Simcoe in Canada. He too believed in the power of scientific education to promote social order, a 'tone of principle and manners that would be of infinite support to government.' In order to promote this social role of science in education, Strachan, who was then teaching at the Cornwall School, instigated a bill in the Upper Canadian Assembly in 1805 for the purchase of scientific instruments. The act (46 GeoIII, c.3) passed in 1806 appropriated £400 for the purpose, entrusting these instruments to 'some person employed in the education of the youth of this Province, in order that they may be as useful as the state of the Province will permit.' The collection, in Strachan's care, gave several generations of Upper Canadians their 'earliest inkling of the existence and significance of scientific apparatus.' Strachan taught natural philosophy and natural history at the district grammar school at York and conducted a course of popular lectures in natural philosophy there in 1818.[5]

Strachan's students, who included most of the Family Compact, imbibed the conservative ideal of cultural uniformity in religion and education. The ultimate goal in the community they longed for was 'culture,' a quality of life attained through social peace and harmony. Victorian science, in seeming to reveal the harmony of creation and ensure material progress, offered a prime example of this kind of culture. John Beverley Robinson, for example, had known Edward Sabine during the latter's military service in Canada and respected J.H. Lefroy's work in the magnetic and meteorological service as a part of the higher culture he wished to inculcate in Canadian society, especially after the shock of the 1837 rebellions.[6]

Lefroy, devoutly Anglican, highly disciplined, well educated and cultured, more than shared these conservative ideals. His scientific activities seemed to build from the compact's conservative blueprint for a Canadian or even British North American community. Lefroy's extended travels in the north-west of British North America sharpened his understanding of the enormous difficulties related to the vastness of the land, where 'people in one district are sometimes as ignorant of the details of the next, as men in London.' Nor was this to be wondered at, he thought, since the unity of the country under the HBC 'deceives us as to its immense extent.' A keen observer of the Upper Canadian political scene and an ultra-Tory himself, he watched with interest the unfolding of the Clergy Reserve and university questions. Canadian Tories, he reflected, were no longer British Tories. They had become so 'liberalized' that they hardly knew themselves; even Strachan, he noted, had made concessions to 'Dissenters' who clamoured for non-sectarian solutions to both problems. Such complexities could hardly be grasped from London, and Lefroy wished that British legislators would take the trouble to visit Canada for themselves, 'before they throw it away.'[7]

As Lefroy ingratiated himself with the Upper Canadian financial and social élite, he added prestige to his scientific work in the eyes of his status-conscious fellow citizens. Personal contact with political power elevated his magnetic and meteorological pursuits to a degree of legitimacy that they probably would not otherwise have enjoyed. But he also became a symbol of the power of science to act as the cultural adhesive they sought in consolidating a Canadian community. An indispensable prerequisite for both sides of this relationship was, of course, the continuance of the Toronto observatory.

I

In order to demonstrate the value of the Toronto observatory, Sabine made a proud display of volume 1 of its magnetic and meteorological observations for 1840-2, published in 1845. The hope held out to Lefroy for continued support of the observatory was hooked into Sabine's plan to maintain and extend his world-wide chain of observatories. Sabine recruited Sir John Herschel (1792-1851), recently returned from his astronomical studies at the Cape of Good Hope, to address the BAAS in 1845 on the subject. Herschel recognized physical observatories 'as part of the integrant institutions of each nation calling itself civilized.' After priorities such as national defence and justice, he continued, there was no object 'greater and more noble – none more worthy of national effort – than the furtherance of science.' Indeed, there was 'no surer test of the civilization of an age or nation than the degree in which this conviction' was felt.[8]

Ultimately, Sabine and Herschel hoped, colonies where these observatories were established by Britain would emulate the activities of the Royal Artillery officers who staffed them. Local scientists and surveyors might look to the observatories for geographical determinations and for future trigonometric and magnetic surveys. In the final analysis, it was 'the spirit of grouping, combining, and eliciting results' which the magnetic and meteorological observatories were intended to instil: without this spirit of co-operation, 'a man may as well keep a register of his dreams as of the weather.' Sabine used the Toronto volume to justify further observations at intermediate stations between the North American interior and the European continent. Newfoundland offered an important instance of magnetic transition between the two continents, and Sabine accordingly urged its governor, Sir John Harvey, to request an observatory there. He similarly encouraged Sir William Colebrooke at Fredericton, a station 'remarkable for its brilliant aurora borealis.' Future plans also included naval stations at Nova Scotia and Prince Edward Island, but Sabine found his ambitions squeezed between the Scylla of increasing demands on the exchequer by the growing distress in Ireland and the Charybdis of waning enthusiasm for his expansive projects among new members of the Board of Ordnance who succeeded his former allies, Lord Vivian and Sir Alexander Dickson.[9]

For the time being, Sabine and Herschel secured a number of important resolutions at the BAAS meeting in 1845, with the concurrence of both the Royal Society and the Peel government. The Toronto observatory was to be continued for three more years, unless arrangements could be made in the meantime for its permanent establishment. Sabine explained to Lefroy his hope that, once King's College at Toronto had settled its affairs, the observations might be continued under one of its professors if the buildings and instruments were transferred to the college.[10]

Lefroy lost little time in making an official presentation of the Toronto volume to the council of King's College. Since it was clear that the British government would not continue the observatory indefinitely, Sabine and Lefroy discussed ways of inducing the college to undertake the observations and of securing financial support from the Canadian government. In 1845 Lefroy confided pessimistically that the college would be unable for at least three years to undertake the observations. Even more disheartening, the professor of natural philosophy there was 'a very unlikely person.' Henry Croft, the professor of chemistry, seemed to be the only one likely to feel interested in such an enquiry, but it was out of his line. Lefroy despaired in the belief that there were no scientific men, that not a single copy of any scientific journal was taken, and hence that there was a want of general interest in scientific subjects in Canada.[11]

But just as Lefroy's estimate of Toronto society had recently been favourably transformed, so too would his estimation of Canadian intellectual life. Meanwhile, he understood correctly that the university's future lay in doubt and that the institution would undergo considerable structural change. But these uncertainties might in turn brighten the observatory's prospects: both he and Sabine erred in calculating that the college would soon 'get rid' of its professor of natural philosophy and that the prospect of directing a physical observatory might even induce a more capable person to undertake the duties.[12]

The Reverend Robert Murray, professor of mathematics and natural philosophy at King's College, was a Presbyterian minister and former mathematics teacher who had been shunted into the professorship in 1844 in a political coup by Governor Charles Metcalfe. Metcalfe needed a Presbyterian to succeed the previous professor, and at the same time he wished to place Egerton Ryerson in Murray's position as assistant superintendent of education. As reluctantly as Murray accepted his new position, he retained it until his death in 1853. His interests virtually excluded the natural philosophy aspects of his teaching duties.[13]

Sabine and Lefroy agreed, however, to work on influential college council members who were known to be favourable to the observatory. Dr John McCaul was one, and Sabine advised Lefroy to broach the subject casually, making sure to stress the growing prestige of the Toronto observatory and mentioning in passing the possibility of the observatory's reversion to the college. An encouraging sign, as it was to the GSC, was Lord Cathcart's appointment as governor of Canada in 1845. Cathcart, known for his active interest in science, visited the observatory while on a tour of inspection and, according to Lefroy, 'went minutely into the details, apparently with considerable interest.'[14]

Charge of the observatory, even for Lefroy, began to appear more attractive than an extensive magnetic survey into uncharted territory. It now appeared to be the post 'most likely to bring a certain amount of subordinate distinction and credit to the Director.'[15] Lefroy's position in Toronto society, he realized, might stand the observatory in good stead when the time came. But three years seemed hardly enough time to effect the kind of conceptual transformation needed to sell Canadians the idea of their own physical observatory.

II

Yet developments during the 1840s piqued Canadians' interest in the Toronto observatory as a means of comprehending and even controlling the vicissitudes of life in the land they inhabited. Although terrestrial magnetism was by no means ignored, popular interest centred more on meteorology. A major factor which directed public attention to the study of Canadian climate was a series of 'years of visitation,' which brought widespread epidemics of cholera, typhus, and influenza. Their unprecedented destructiveness baffled physicians, whose reputations rested on the discovery of the etiologies of these diseases.

Epidemics had for centuries been linked to malevolent influences of heavenly bodies or unknown elements of the weather, especially when medical knowledge could confirm no direct causes. The spread of European commerce and migration over the world unleashed pandemics of unheard-of proportions. Epidemics of influenza had been recorded in North America during the eighteenth century, from 1830 to 1833, from 1836 to 1837, and again in 1846 and 1847. Typhus reached Toronto in 1847 and was blamed on the destitute Irish who emigrated to Canada during those years. But it was cholera, an Asiatic disease unknown in North America before 1830, that shocked the population and inspired new approaches to explain its virulence, which fitted no known model of contagion. As late as the summer of 1849, Canadian physicians and politicians were divided over whether the disease was contagious, requiring quarantine and tighter immigration laws, or rather epidemic, the product of atmospheric changes and local conditions requiring sanitary reform and poor relief. In this highly political dichotomy medical opinion leaned towards the latter theory, presuming the presence of a 'miasma' which facilitated the spread of the disease overland through changes in the quality of the air.[16]

In Canada this concern reached the public in the first issue of the *British American Journal of Medical and Physical Science*, founded in Montreal in 1845. Its editor, Dr Archibald Hall (1812–68), was a Montreal physician educated at the University of Edinburgh after the cholera epidemic of 1832. Like many of his contemporaries, Hall combined his profession with a personal interest in meteorology and natural science. The first issue of the *British American Journal* recruited meteorological observers to submit monthly meteorological reports for various Canadian towns. Hall guaranteed that a very important object, 'too palpable to require to be specified,' would be secured by such a measure. He solicited charts of thermometric and barometric readings taken three times daily (7 A.M., 3 P.M., 10 P.M.), along with weather descriptions, and hoped that maximum, minimum, and mean temperatures might eventually be established for the whole country. Immediate responses were forthcoming from Montreal and Quebec, where registers were already kept by the military and by local natural history societies. Contributions came also from Dr William Craigie (1790–1863), a Scottish physician whose medical training had instilled in him the habit of keeping thermometric and meteorological registers, first in Ancaster and then in Hamilton, since 1835; and, of course, from J.H. Lefroy. Registers kept at HBC posts were also published when they became available. Whenever possible, Hall appended bills of mortality to the monthly meteorological registers, with the goal of illuminating causal relationships between weather and disease.[17]

Evidence for such a relationship seemed to be on the rise. In 1845 C. F. Schönbein, professor of chemistry at the University of Basel, discovered ozone, an unstable compound of oxygen detected in the atmosphere after discharges of lightning. This discovery galvanized medical interest in possible relationships between epidemic disease and the amount of electricity in the air. The theory seemed reasonable because the nervous system was thought to be powered by an analogous electrical fluid in the body. Hall's *British American Journal* published a number of clever expositions of the ozone theory. This theory postulated that the production of ozone, directly related to the amount of electricity discharged into the atmosphere, might be used to measure the potential for certain diseases to spread. A condition of too little ozone, with its bleach-like qualities, might allow unidentified causes of cholera to flourish in the miasma; too much ozone, known to be an irritant, might somehow provoke the body to produce symptoms of influenza. In the decades before the germ theory was firmly established, the theory was an ingenious rationalization of a mystifying yet crucial problem of diagnostic medicine.[18]

Several scientists in Canada continued to defend the ozone theory long after it ceased to be fashionable. Charles Smallwood (1812–74) made a name for himself by pursuing a relationship between ozone and disease. Smallwood emigrated from England in 1833 to St Martin's, Isle Jésus, near Montreal. There he built a remarkable private meteorological and electrical observatory to measure and record atmospheric changes. Credence was also given to the ozone theory by J.H. Lefroy. Lefroy noted that during the Toronto cholera epidemic in 1849, he was unable to magnetize a metal bar. Attributing this phenomenon to a deficiency in the usual 'electro-magnetic influences' of the atmosphere, he wondered whether a link with disease might be substantiated. A visit to the observatory by the secretary of the Smithsonian Institution, Joseph Henry, who 'disbelieve[d] *in toto* the alleged connection,' did little to encourage Lefroy's investigations. But he continued to contribute to Hall's respected *British American Journal*, creating a widespread impression among its readers of the authority, usefulness, and co-operativeness of the Toronto observatory.[19]

In an age replete with demonstrations of the practical uses of electromagnetism epitomized by the telegraph, popular interest in the subject was bound to increase. Lefroy used the *British American Journal* as a forum for his own attempts to clarify the obscure relationships between electricity, magnetism, and the atmosphere. His northern expedition aroused his interest in the aurora borealis as an electrical phenomenon. North American displays of the northern lights, he felt, 'were always of a more brilliant and definite character' and more frequently observed than anywhere else in the world. In 1848 he arranged for nightly weather registers, including observations of the aurora, to be kept at all military posts in British North America where Royal Artillery officers were stationed. A more general circular inviting communications with the Toronto observatory soon established Lefroy as the foremost authority on the aurora in North America. Accounts of auroral displays were printed in local newspapers, and Lefroy contributed articles to the *British American Journal* that he intended to be understood in connection with his meteorological register. The years 1847-8 were spectacular auroral years, exhibiting the greatest disturbances of the magnetical elements ever recorded. Magnetism, Lefroy explained, offered 'proof of operations or effects not cognizable otherwise, and probably important to a full comprehension of the *physique du globe* in direct proportion to their remoteness from us.' In addition to these Humboldtian benefits, he added, the study of magnetic forces bore upon sanitary questions, particularly the spread of epidemics.[20]

Hall believed that the attention of future historians of science would be arrested by 'the remarkable impulse given to inquiries in terrestrial magnetism within the last few years': 'Equally finding a field for its researches in the snows of the arctic circle, or at the "nether pole", in the most familiar of European countries, and the least known regions of Asia or Africa – at sea or on land, it has yet sprung so suddenly into life, and so suddenly embraced the globe, that the bulk even of intellectual and well informed persons, are scarcely conscious of its existence, and quite unaware of its aims.'[21] Since the area of maximum intensity lay between Lake Winnipeg and Hudson Bay, Hall suggested that Canadians had a special contribution to make to the science of terrestrial magnetism. As much for the country's distinction on the ladder of civilized nations as for the sake of science itself, he hoped that Canada, possessing 'one of the regions of the greatest magnetical interest in the globe,' would be distinguished by observers of its own in physical science and by its own well supported scientific establishments.[22]

This sprouting interest in terrestrial magnetism was nothing compared with popular interest in meteorology. Edward Sabine admitted that meteorological observations were destined to be more popular than magnetic ones, not only because of their more obvious relevance to daily life but also because meteorological theory possessed 'a completeness and fullness not yet attained in magnetism.' An additional reason for growing public interest in organized meteorology in Canada was its importance, particularly after the repeal of the Corn Laws in 1846, in encouraging immigration and efficient agriculture. Both William Logan and his assistant, Alexander Murray, kept meteorological tables during their geological surveys of the region around Lakes Huron and Superior.[23] J.S. McCord of Montreal anticipated that in a short time a continued series of observations would determine the mean temperature of Canada and its maxima and minima.

Such data would not be of merely local or short-term significance. McCord thought they would decide the question of climatic progress and would help to determine whether Canada's climate had changed since the first settlement of the country, and if so, to what degree. Available evidence had convinced McCord that while mean temperatures had not altered much, extremes had moderated so that Canadians were no longer exposed to such intense periods of heat and cold as had earlier been the case. Even Sir Francis Bond Head extolled the cumulative benefits of each farmer's efforts to clear the soil: 'While every backwoodsman in America is occupying himself, as he thinks, solely for his own interest, in clearing his location,' he wrote, 'every tree which, falling under his axe, admits a patch of sunshine to the earth, in an infinitesimal degree softens and ameliorates the climate of the vast continent around him.'[24]

The idea that the climate of the less hospitable north-west might be made more amenable to agriculture found support in Lefroy's observations in 1844 that as far north as the Mackenzie River, the soil thawed to a depth of two feet in open places and only one foot in the woods. At Cumberland House Lefroy had taken an interest in a field of wheat, and he expressed astonishment to find a farm of thirteen acres yielding barley and potatoes at Fort Simpson. 'All this is entirely a new feature in this country,' he wrote to Sabine, 'and adds immensely to the comfort of life here.' The HBC did not directly encourage farming although Lefroy thought it wise to do so. 'Every increase to the productive resources of the country,' he suggested, was 'an increase to the value of the Company's property and interests, while it leaves the native resources more exclusively to the natives.'[25]

Finally, the usefulness of the barometer in tracking movements of large storms over the American continent was recognized from data collected by James P. Espy for the United States navy. The publication of Espy's *The Philosophy of Storms* in 1841 earned his appointment as American meteorologist in 1842 and the expansion of his network of volunteer observers. Once again the Toronto observatory was advantageously located to study continental patterns originating in the north-west. In simultaneous observations with the Americans, Sabine had told the BAAS in 1844, Toronto would 'form a very valuable extensive basis of induction for the movements of the atmosphere over that great continent.'[26]

For several reasons, then, abstracts of the monthly register of observations kept by the Toronto observatory were in considerable demand in both Canada and the United States by 1848, when its term came up for renewal. Sabine was satisfied that the observatory was 'well adapted to stimulate and to give a systematic direction to amateur observation' among Canadians. As Lefroy, however, realized, there was little public awareness that the long-term continuance of the observations at Toronto was in doubt, and enthusiasm was lacking for a permanent commitment from either King's College or the Canadian government. To discontinue the observatory prematurely, Lefroy believed, would be 'a reproach to the philosophy of the age, almost an act of barbarism.' Both he and Sabine counted on the college's option of reclaiming the observatory. They hoped that the new University of Toronto, once established, would acknowledge this inheritance by the end of 1850.[27] The means of accomplishing this, they decided, was to demonstrate the Toronto observatory's indispensability to the best interests not only of magnetic and meteorological science but of Canada as well.

III

The first of these demonstrations took full advantage of Sabine's forte as a master manipulator. Sabine had all along striven to attach the work of the Toronto observatory to that of both British itinerant observers sent occasionally to the north and American observers to the south. As early as 1845, Elias Loomis had written to Sabine that he was 'particularly impressed with the importance of the observatory at Toronto' for its meteorological no less than for its magnetic observations, 'pre-eminent for their accuracy and completeness.' Great winter storms in North America, he noted, had little respect for political boundaries and almost always projected themselves over British North America. 'We want,' he continued, 'a chain of meteorological posts extending indefinitely northward to the furthest outpost of civilization.' There was in Loomis's view 'nothing which would hold out a prospect of so rich a harvest to American meteorology as the establishment of such a chain of posts.' But it could only be effected through the agency of the British government. Loomis stood at the vanguard of a 'general meteorological crusade' that, he told the American Philosophical Society in 1843, would supersede the 'Guerilla [sic] Warfare' which had been maintained for centuries with indifferent success.[28]

The society decided in February 1846 to postpone its application to the British government until its chances of success with the American government had been determined. But Sabine repeatedly advised American scientists that only the co-operation of British observers in Canada and the provinces to the north and west of Canada could give 'full efficacy' to their plans. He assured them that such co-operation was guaranteed if they approached the British government formally through the American ambassador. An American system of volunteer meteorological observers was introduced under the direction of Joseph Henry of the Smithsonian Institution in 1848. Both Henry and François Guyot of Harvard visited Lefroy at Toronto the following year, in order to study self-registering photographic equipment being tested there, and both agreed that the abandonment of the observatory was 'a thing not to be thought of.' Just as Sabine had hoped, they promised to 'put in a plea' for continuance of its operations.[29]

Fortunately for the Toronto observatory at this crucial point in its existence, the Montreal riots of 1849 occasioned the transfer of the seat of government to Toronto in 1850. Influential personalities thus became accessible to Lefroy, who requested an interview with Robert Baldwin. When Baldwin failed to respond immediately, Lefroy became restive and decided to 'take an opportunity of hinting to Dr Henry and others what must be done if the American meteorologists desire the continuance of this establishment.' If they went to the expense of the complete self-registering system he was testing, he reasoned, they would probably take action in support of Toronto. Baldwin's 'satisfactory reply' in May, that provincial funds for the observatory would be forthcoming 'if *absolutely* required,' left Lefroy even more anxious to 'stir up the Smithsonian and Cambridge people.'[30]

Baldwin told Lefroy that, despite the Reform government's embarrassment over cries for retrenchment from within its own ranks, his colleagues were prepared to recommend the sum Lefroy requested (£300) as an annual appropriation. But he warned that the outcome in the House of Assembly was by no means certain. For although they opposed the idea of 'permitting observations of so valuable a nature to fall to the ground,' the Reformers preferred the observatory's eventual transfer to the university. These revelations were confidential, and formal application had to be made through the provincial secretary. The fact that William Logan was experiencing the same difficulty obtaining funds to complete his geological survey convinced Lefroy that 'our best hope lies in the success of an application to be made by the Americans.'[31]

Here Sabine was not above resorting to a little moral suasion and nationalistic flattery. He reminded Joseph Henry that it would tell for much more than the simple continuance of the observatory 'if the request so freely made by the Royal Society for the Establishment of cooperating Observatories in the United States, and which was so well received and responded to on your side the water, were returned on the present occasion by the expression of a hope from the United States that the Toronto Observatory would go on ... Of course,' he added, 'I write on the supposition that the continuance of the Toronto Observatory does really appear to the Body making the communication an object of sufficient importance to justify it by its bearing on the prosecution of *American* meteorology and magnetism.' Finally, he cautioned Henry that 'any proposition coming from your side the water must not appear to have originated in suggestions from this country.'[32]

For his part, Lefroy wrote to Henry, Loomis, Bache, and Bond at Harvard. Henry, he assured Sabine, 'will do all we want.' Indeed, Henry expressed interest not only in co-operative meteorological but also in magnetic and auroral observations. He assured Lefroy that he would incur the support of the Smithsonian, whose chancellor, Millard Fillmore, had recently become president of the United States. He felt encouraged as well by the American Philosophical Society, the American Academy, and plans for the forthcoming meeting of the AAAS at New Haven.[33]

The president of the AAAS for that year was Bache, who referred the issue of the continuance of the Toronto observatory to a special committee of Loomis, Henry, and himself. The outcome was a unanimously adopted series of resolutions recognizing the importance of the Toronto observations, 'so near the focus of maximum magnetic intensity,' as 'of great value to the progress of science, especially as connected with similar operations in the United States.' They invited the British government and directors of the HBC to co-operate 'in united & systematic meteorological inquiries.' The American ambassador to Great Britain, Abbott Laurence, added a personal appeal to Lord Palmerston, assuring him that 'the objects to be accomplished are strictly international.' Copies of the resolutions were sent to the Canadian government as well as to Lefroy, who lost no time in forwarding one to Sabine. With this advantage Sabine began to work on the president of the Royal Society well before the British government could grind out a decision on the formal American request.[34] That Sabine's diplomatic skills achieved their goal was no surprise.

The second task, that of convincing Canadians that the Toronto observatory embodied a Canadian cultural necessity, succeeded beyond even Lefroy's wildest dreams. When the government moved to Toronto, Lord Elgin honoured Lefroy and lent the observatory a certain amount of prestige when he and Lady Elgin visited it and 'appeared to take a very intelligent interest in it.'[35] In addition to the steps already taken to hook the activities of the observatory firmly into the British North American social fabric, Lefroy secured an important token of recognition in 1850.

Lefroy persuaded Egerton Ryerson (1803–82), now chief superintendent of education for Upper Canada, to establish a chain of meteorological stations throughout the province. Ryerson's draft of a grammar school bill in 1850 promoted the study of meteorology in grammar schools to be used as meteorological stations. He pointed out the utilitarian value of climatology and meteorology to Canadian society. It was desirable, he urged, to direct attention at educational institutions to natural phenomena and to encourage habits of observation. Moreover, a better knowledge of the climate and meteorology of Canada would be beneficial to agricultural pursuits as well as scientific enquiries, especially as to whether Canada's climate was improving. Ryerson proposed to charge masters of senior county grammar schools with daily meteorological observations to be submitted periodically to the Education Office. Barometers, thermometers, hygrometers, rain gauges, and wind vanes would be supplied by the Education Department. For unrelated political reasons the grammar school bill was temporarily withdrawn.[36] Undaunted, Ryerson would soon have his grammar school act, complete with meteorological stations. The community of outlook that permitted this unexpected transfer of scientific responsibilities to the Canadian public domain escaped a single word of opposition from a notoriously frugal House of Assembly. Instead it inspired 'a good deal of interest in the subject.'[37] The last years of Lefroy's directorship of the Toronto observatory basked in the light of this common outlook.

8

Science as a Cultural Adhesive, 1850–1853

> Canada ... is here presented in a light on which there can be no clashing of opinion, no discordancy of sentiment: the smiling face of Nature, the harmony and beauty of the works of God, may be turned to by men of all parties as a refreshing relief from the stern conflict of political warfare.
>
> P.H. Gosse, *The Canadian Naturalist* (1840)

During the first five years of his directorship Lefroy constructed a firm foundation for the Toronto observatory as a functioning institution within Canadian society. The last three years consolidated that foundation in order to guarantee its permanence.

Lefroy had found himself professionally and personally in harmony with the Family Compact, but there were other elements in Canadian society whose ideas and outlook also struck a sympathetic chord. An Anglican with strong evangelical leanings, Lefroy had belonged to a small group of young officers at Woolwich who met during the 1830s for weekly Bible readings and prayer meetings. Closely related to his personal piety was Lefroy's intense interest in self-improvement through education. In an age which looked increasingly to popular education to diffuse moral and other values and to mitigate growing social conflict, Lefroy's involvement in a Sunday school for officers' children, his desire to improve himself by enrolling in the Royal Engineers' school to study practical astronomy, and his efforts to found the Royal Artillery Institution to afford officers professional instruction all met with the approbation of military authorities. Lefroy's conception of the value of education in moulding the outlooks of future generations remained essential to his own outlook.[1]

In addition to his pursuit of a career as a magnetician and meteorologist after his return from the north-west in 1844, Lefroy became occupied with examining the relationship of science to education. Reflecting the born-again Baconianism typical of many Victorians, he felt 'more and more convinced that it is to industry and not to extraordinary ability, that the world is most indebted for its acquisitions in the knowledge of Nature.' The contribution of his observations to Sabine's theories seemed to Lefroy 'a very beautiful example of the importance of division of labour in observation – and of interrogating nature upon the subject of each phenomenon there where it may be most favourably exhibited, instead of every body trying to do the same thing. In short, of Scientific *Free Trade*.' Lefroy became frustrated with the mechanical nature of his scientific duties, which hampered efforts at analytical thought. As a result, he complained, the laws they had sought to comprehend remained 'locked up in our folios of figures, and we continue heaping together pearls and pebbles to be sifted hereafter, instead of looking out for the one and leaving the others in their bed.'[2]

Lefroy's intellectual soul mate in Canada was not so much the orthodox John Strachan and his followers, with whom he shared a common conservatism, as the evangelical Reverend Dr Egerton Ryerson. Ryerson's enormous intellectual and administrative contribution to the advancement and organization of science in the Ontario school system warrants some consideration. Like Lefroy and the Family Compact, Ryerson was a social conservative convinced that a healthy and viable society depended upon the harmonious mutual dependence of its various elements. But as a self-educated convert to Methodism, Ryerson spent most of his career as chief superintendent of education for Upper Canada acting upon his belief in the individual value of a popular and practical liberal education. In a lecture to the Mechanics' Institutes of Niagara and Toronto in 1849, Ryerson revealed a Baconian touch that must have appealed to Lefroy. 'The well-being of a community,' declared Ryerson, 'is to be estimated not so much by its possessing a few men of great knowledge, as by its having many men of competent knowledge.' The broad sweep of Ryerson's outlook can be gleaned from his conviction that nothing in the world was really insignificant, that 'all has meaning – all tends to one harmonious whole in the order of creation.' Ryerson sought to instil in Upper Canadian schoolteachers and, through them, in their pupils patient and discriminating habits of observation as students of nature.[3]

Ryerson had superintended the common schools of Canada West since 1844. Never far from controversy, he was usually seen as the outspoken opponent of Strachan's sectarian educational system. Assessed no doubt accurately by Lord Elgin as 'cunning,' Ryerson sustained a keen intelligence with a 'consciousness of partnership with the Divine.' His aura of 'self-sustained virtue' won the respect of even his bitterest political opponents. John Langton confessed, 'He is surrounded as by an atmosphere with such a concentrated essence of respectability, and meets all opposition with such a calm unruffled air of conscious superiority that I do not deny that I am half afraid of him myself.' Langton,

147 Science as a Cultural Adhesive

vice-chancellor of the newly founded University of Toronto and a staunch supporter of the established church, shrewdly assessed Ryerson's strategy in creating a school system in which all power converged on himself. Increasing power permitted Ryerson and his assistant John George Hodgins to disseminate Ryerson's ideas at public expense. Through the official *Journal of Education for Upper Canada* his ideas spread from the Education Office to every school section.[4]

In Ryerson's terms, the *Journal* was adapted 'to the wants of the country, rather than to individual taste.' The difficulty of determining the difference, where Ryerson was concerned, is a measure of the man himself. His advice as principal of Victoria College, that 'It is wise to anticipate what is inevitable,' encapsulates his influential and varied career.[5] Ryerson's genius was his ability not only to anticipate and articulate Upper Canadian developments in the mid-nineteenth century but moreover to harness and lend respectability to their implications.

I

One such development was the growing popular appreciation of science during the nineteenth century. An avid reader while riding his circuit during the 1820s, Ryerson was deeply impressed with the *Natural Philosophy* of John Wesley, prescribed reading for prospective Methodist ministers. Originally conceived as *A Survey of the Wisdom of God in the Creation*, the work was first published in 1763. Through many editions it was constantly updated in a conscientious effort to accommodate new developments in natural philosophy. By the 1830s its title emphasized its purpose as *A Compendium of Natural Philosophy* intended to provide a 'short, full' and above all 'plain' description of the 'visible creation' for readers of modest means and even less formal education. The work reflects Wesley's particular interest in electricity, magnetism, and atmospheric change. Early editions explained the history and uses of barometers and thermometers; the revised edition of 1836 then made a striking attempt to assure its readers that weather forecasting using barometric readings did not reflect a form of superstition, but rather the rational understanding of an effect that 'points out what has already happened[,] not what has still to happen.' Wesley's advocacy of Francis Bacon's 'diligent Search into Natural History' by 'many Experiments and Observations' firmly linked the promotion of science with the Wesleyan ideals of self-improvement and education in the highly effective organizational structures of Methodism.[6]

Ryerson's advanced views on the value of science as part of the educational curriculum at all levels have been well documented, but one passage from his inaugural address at Victoria College in 1842 exemplified the Wesleyan outlook applied to the Canadian context. After quoting Bacon, Anson Green, the president of the Methodist Conference, remarked, 'It is our happy privilege to live at a period when the star of prosperity is dawning upon our land, and the light of science is spreading a brilliant lustre over the civilized world.' Ryerson then pointed to the need in Canada for more practical education, which included physical science:

If one branch of education *must* be omitted, surely the knowledge of the laws of the universe, and of the works of God, is of more practical advantage, socially and morally, than a knowledge of Greek and Latin. How useful, how instructive, how delightful, to be made acquainted with the wonders and glories of the visible creation – the invariable laws, by which the heavenly bodies are directed in their complicated and unceasing evolutions through the amplitudes of space – the structure of the earth on which we move, the materials of which it is composed, the arrangement of its component parts, the revolutions and changes to which its masses have been subjected, the laws which govern their ever-varying compositions and decompositions – the mechanical powers of air and water – the properties of light, heat, electricity, magnetism, &c – and the application of these various branches of physical science to the art of increasing the means of support, the comforts, refinement, and enjoyments of life, of facilitating the intercourse of nations, and of promoting the general happiness of our race![7]

Ryerson, like Lefroy, greatly admired Alexander von Humboldt as 'the Nestor of Philosophers' who had earned a 'permanent place among the princes of the intellectual world.' On his first educational tour through Europe from 1844 to 1846, he made a point of hearing François Arago lecture on aspects of astronomy at the observatory in Paris. Deeply impressed, he sought out scientific lectures on chemistry by Dumas and on mineralogy, electricity, and natural history at the Sorbonne.[8]

Like others of his generation, Ryerson believed that 'by *science* a nation is enabled to profit by the advantages of its natural situation.' His powerful position and strong determination to apply such ideas during the 1840s and 1850s gave particular significance to Ryerson's plans for the development of Upper Canada. Besides teaching physical science starting at the common-school level, the most obvious way to carry out these beliefs was to encourage agricultural education. After 1846 'the change in the commercial policy of the mother country should induce us to put forth extraordinary exertion,' he urged, 'to demonstrate that two ears of corn could be grown where we now raise one.' New challenges seemed to call for the application of scientific principles to enhance the efficiency of agricultural production throughout the province. Addressing audiences on the issue, Ryerson emphasized his personal background as the son of a farmer; but he emphasized that he spoke also as a native of Canada.[9]

In the new 'age of improvement and keen competition,' Ryerson extolled the role of farmers as arbiters of the country's destiny. The American and French revolutions had achieved the cultivation of the land by its people, and this same revolution was being accomplished in Canada by the proposed establishment of agricultural schools and model farms. In educating its farmers Canada would join the ranks of the civilized as opposed to the 'savage nations.' Each farmer, too, would gain a greater sense of his stake in the province's welfare:

In the chemistry of his soils and measures, in the botany and vegetable physiology of his garden, fields and forests; in the animal physiology of his stock and poultry, in the hydraulics of his streams and rivulets, and the geology and mineralogy of their banks, in the mechanics of his tools, and the natural philosophy of the seasons, and the application of this varied knowledge to the culture of his lands, the care of his flocks, and the improvement of his estate, he finds exhaustless subjects of inquiry, conversation and interest, all connected with his own possession, associated with his own home, and involved in his own prosperity.[10]

Ryerson, once employed at the London District Grammar School, had no time for élitists who had become so 'profoundly learned' that 'they can scarcely condescend to look upon a Grammar School at all; nor can they seem to endure any other mode of teaching Agriculture than by a Professor in a University.' Rather he argued that it was more patriotic and farsighted to raise the district grammar schools to the level of importance held by their counterparts in Germany, France, England, Scotland, and the United States. He resolved to knock practical sciences such as agricultural subjects and meteorology down from their ivory towers and into grammar schools, where they could be of most use.[11]

Both the initiative and the approach for the Ryersonian school system derived from his Methodism. Ryerson wanted to preach the gospel of science to the ordinary toilers of the land, the very people he had come to know and understand on his missionary circuit.

II

A first step would be to raise 'a corps of native teachers' trained to diffuse Ryersonian ideals. To this end the Provincial Normal School was founded at Toronto in 1847. Ryerson at first sought British candidates for a lectureship in natural philosophy and chemistry. Instead he was delighted to appoint a recent immigrant, Henry Youle Hind (1823–1908), who had studied at Cambridge and Leipzig. As mathematical master at the normal school, Hind taught natural philosophy (heat, electricity, magnetism, mechanics, hydrostatics, and pneumatics), astronomy, animal and vegetable physiology, and agricultural chemistry. An avid gardener, Hind conducted agricultural experiments on the grounds of the school, with special reference to Canadian soil, climate, and productions. These experiments sparked considerable interest in Upper Canada; the 'able and energetic,' not to mention ambitious, young master soon acquired 'strong admirers' on the Council of Public Instruction as well.[12]

Nor did Hind escape the attention of Edward Sabine and Henry Lefroy. In March 1847 he published a scientific account of parhelia (mock suns) he had witnessed from his home in Thornhill. Hind's knowledge of the subject was deemed by Lefroy to equal his own. Lefroy wrote to Sabine that while he knew nothing of Hind, he had learned from Toronto booksellers that 'there are scattered about on Farms in this country many men of University education,' one of whom even read works like the classic *Mécanique céleste* (5 vols., 1799–1825) by Pierre Simon, Marquis de Laplace (1749–1827). Lefroy determined to meet more of these people, whom a short while before he had believed nonexistent in Canada.[13]

As a corollary to his talents as a teacher, Hind soon served both Ryerson and Lefroy in his capacity as a publicist. His inaugural speech on the importance of teaching natural philosophy and chemistry in an 'age of enterprise and improvement' was reprinted in *The British Colonist*. Thereafter a number of articles penned by Hind, particularly on agricultural education, were published in the *Journal of Education*. Harmonizing with Ryerson's views on the subject, Hind declared: 'The advance in wealth and importance of a country so situated [as Canada] rests entirely upon the national character of its inhabitants. With an energetic and improving population, who are not afraid of competition, and are willing to relinquish ancient forms and prejudices in favour of improved methods and advanced ideas, such a position is the one most likely to ensure a real and continued progress.' Like Ryerson, Hind emphasized the increasing competition against Canadian agriculture brought on by improved transportation, the development of new agricultural areas, and the 'social and commercial revolutions' that Britain and Europe had lately witnessed. His lectures on agricultural chemistry, published in 1850 and reprinted a year later, urged Canadian farmers to sharpen their competitive edge through increased agricultural efficiency.[14]

Hind's most important contribution as a publicist of the cause of meteorology and climatology espoused by Lefroy and Ryerson was an expanded section of his lectures on the climate of the western Canadian peninsula. Published in 1851 as *A Comparative View of the Climate of Western Canada*, the pamphlet drew chiefly upon meteorological data assembled by Lefroy at the Toronto observatory. Hind's stated object was to draw attention to the climate of Canada West in its bearing upon agriculture. His aim was to counteract potential immigrants' false impressions of 'intense and almost unendurable winter cold, together with a hot and fleeting summer, which scarcely affords the agriculturalist time to secure his harvest ... FROM THE PECULIAR SITUATION OF THE PROVINCE AMONG THE GREAT LAKES,' he stressed, the climate of Western Canada instead presented agricultural potential unsurpassed in North America. Hind reiterated that he wrote not out of a 'romantic spirit of patriotism,' or because 'there is no place like home,' but because he believed his argument to be reasonable and verifiable.[15]

Like Sabine, Lefroy, and Ryerson, Hind admired Alexander von Humboldt's 'cosmical' approach to science. A full seven years before the publication of Lorin Blodget's *Climatology of the United States*, Hind understood from Humboldt's use of isotherms to explain climatic variations over continental areas that latitude was by no means the main determinant of climate. Foreshadowing techniques he would utilize in his own analysis of the agricultural potential of the British American north-west later in the decade, Hind divided the province of Canada into geographical regions. He recognized the impossibility of correctly assessing climate solely from annual means of temperature without considering the overall amount and distribution of heat, light, and water. At the same time Hind differed from Humboldt in clinging to the older view that clearing the land would moderate its climate. He predicted that the disappearance of forests 'before the rapid encroachment of the settler' would raise minimum temperatures of spring, summer, and autumn nights. Late spring and early autumn frosts, too, would 'probably become rarer, as the country becomes more cleared.'[16] This combination of advanced scientific awareness with unproven traditional assumptions of climatic progress formed an early matrix for Hind's excessively optimistic assessment of the north-west's climate.

For the time being, Hind's conclusion that the Upper Canadian peninsula was destined to become one of the great granaries of the world was sure to earn appreciative reviews. The *Journal of Education* called Hind's *Comparative View* 'an investigation as original as it is patriotic,' successful in its 'extensive induction of facts' to ascertain and interpret 'the language of nature.' A vice-president of the York County Agricultural Society thought the pamphlet marked 'a decided epoch in the history of a country when it begins to produce works of its own' and wrote 'Nothing can serve so well to point out to other nations its progress in real civilization.'[17] Hind's success as a publicist derived largely from the fact that his own optimism fell in with that of his time. He proved particularly adept at providing the Canadian public with what they most wanted to hear.

III

While H.Y. Hind secured a place in the public eye as someone who could bring science to Canada's social, economic, and even cultural benefit, Egerton Ryerson exposed the ideological significance of the growing interest in collecting meteorological information in Canada. Once again 'anticipating the inevitable,' Ryerson linked his struggle for a public educational system in Upper Canada to the promise of efficiency and progress through science. He appealed to values that few public men of his age would deny. Hoping to diminish public controversy, Ryerson concentrated on such areas of agreement whenever possible. He used the widespread desire for economic and agricultural improvement to heal rifts that religious differences had torn in Canadian society. When Lord Elgin donated two agricultural prizes to the provincial normal school, Ryerson used just this kind of argument to persuade John Beverley Robinson to distribute them. Ryerson and the Family Compact differed over the role of an established church in education, but they did not do so over centralization, efficiency, and loyalism as desirable fruits of a Christian educational system. Nor did they differ over the important role of science in the achievement of these ends. It was, after all, a 'thinking age' in which 'to study, plan, alter, improve, invent and develope, seem[ed] to be the prevailing passions of all classes.'[18]

Nor did these similarities hold only for political conservatives. Despite his major break with Reform leaders in 1844, Ryerson's ideas about education and science found sympathy in the Reform government's inspector general, Francis Hincks (1807–85). Hincks's father, Thomas Dix Hincks, an educator and founder of the Royal Cork Institution in Ireland, lectured on chemistry and natural philosophy. Francis's brother William, a botanist, taught natural philosophy and natural history at both York and Cork before being appointed to University College, Toronto, in 1853.[19] Interest in both science and education ran in the family, and Francis's sympathy was no anomaly. Hincks met Ryerson privately in 1850 to discuss details of Ryerson's grammar school bill. He introduced the bill in June 1850, after circulating questionnaires throughout the school system to secure the support of those 'best acquainted with the subject.' He defended the bill until late in the session, when he was forced to withdraw it from committee of the whole.[20]

Like Lefroy, Ryerson saw a direct connection between continuing the Toronto observatory and providing a system of meteorological stations in every county of Canada West. The former offered focus, direction, and prestige; the latter justified the increasing stake of taxpayers in the public school system. When viewed as an 'extensive machinery' for simultaneous observations over a wide area, the grammar schools seemed to promote scientific enquiry for the benefit of society; storms could be forecast, agriculture improved, and diseases eradicated. This proposal for meteorological stations overlapping the grammar school system was a clever technique to satisfy critics of the grammar school bill, who objected that too few would benefit directly and that it was therefore unjust to tax the whole community for its support.[21]

The idea of ordering headmasters of grammar schools to take meteorological observations had its roots in both the British and the American experiences. David Brewster's observations in Scotland for the BAAS in 1839 had actually been made by a local schoolteacher. As early as 1825, the Board of Regents of New York State had begun meteorological observations at all local schools under its jurisdiction. Both Lefroy and Ryerson were aware of this system and hoped to adapt it to Canada. As a British meteorologist explained, the appointment of responsible and educated observers invited the assumption not only that they were competent to record meteorological phenomena but also that their character impressed the returns with a stamp of trustworthiness and authority.[22]

For Ryerson these first attempts to organize a province-wide scientific inventory held a deeper significance. They helped pave the way towards the sense of Canadian and even of British North American community that he felt was lacking. Ryerson's sense of a British North American community derived from his religious experiences and paralleled that instilled by revival meetings and dramatic personal conversion, which he himself had experienced. In a land where cultural diversity on the level of language and religion was the norm, he hoped, science might be used to bridge these gaps. A pet project of Lefroy's was suggested by the *Journal of Education* in 1851: if British North Americans could extend to Newfoundland the telegraph system already linking Sarnia to Halifax, they could create 'a storm alarm on a stupendous and magnificent scale,' and 'susceptible of indefinite expansion towards the west.' Agriculture and the 'commercial safety of our maratime [sic] sisters' would at last become common interests.[23]

Similar outlooks in Britain have been attributed to a 'Methodist synthesis' of liberalism, order, and national mission. This intellectual synthesis mediated between liberty and order and between the traditional sectarian oligarchy and the modern national-state. Ryerson, like Strachan, never challenged the preoccupation of Upper Canadians with improvement and material progress. He participated in the process, disseminating 'useful scientific knowledge' and organizing scientific inventory to which even those with little education or background could contribute. As 'practical' inventory science, meteorology and even terrestrial magnetism exemplified collective pursuits in the public interest to which not even Roman Catholics or French Canadians objected.[24] By affirming for all Canadians the Christian foundations of scientific inventory, and by teaching 'the wisdom of God in the creation,' Ryerson was emulating John Wesley.

Ryerson saw in organized scientific inventory a bridge from the individual to the community and from the community to the nation. These ideas formed the foundations of an incipient nationalism, which, very much like a religious sect, drew its inspiration from the future. 'Let public attention,' wrote Ryerson in 1849, 'be directed to common interests, rather than to party interests, – to unity rather than to division, – to the practical rather than the speculative, – to the future rather than to the past.' The *Journal of Education* in 1850 lamented 'the absence of a *true Canadian feeling* – a feeling of what might be termed *Canadian nationality*, in contradistinction to a feeling of mere colonial or annexationist vassalage.' This regret mirrored Ryerson's own view: 'It cannot be too strongly impressed upon every mind that it is on Canadian energy, Canadian ambition, Canadian self-reliance, skill and enterprise, – in a word, on Canadian patriotism – that depends Canadian prosperity, elevation, and happiness.' Indeed, it concluded, God never meant that the brotherhood of man should be 'broken by territorial boundaries, or limited by expedients of trade.' Yet 'none, but those who have gone mad upon remote generalisms and unities, will deny that kindred, vicinage and organized reciprocity impose peculiar obligations.'[25]

Just as the scientific spirit infiltrated intellectual disciplines with its ideals of universalized method and centralized effort, so too did it expand the concept of community articulated by Egerton Ryerson. Ryerson believed that if Canada was to survive and prosper, a stronger sense of community had to be built. The only common ground to which he could appeal was the promise held out by practical scientific endeavour. Since this endeavour was ancillary to his own Methodism, Ryerson found inventory science a highly acceptable substitute for religion in bringing social harmony first to one and eventually to both sections of the province. In this he found many supporters.[26]

IV

Physical science bridged political and religious differences in the formation of the Canadian Institute, which attained a royal charter in November 1851. Conceived originally in 1849 as an exclusively professional association of architects, surveyors, and engineers, the institute was nevertheless heir to earlier attempts to found a scientific society at Toronto. In 1851 it was redefined as 'a Society for the encouragement and general advancement of the Physical Sciences, the Arts and Manufactures.' The institute's net was cast more widely in order to attract as many members as possible. Yet some of its founders also believed that the larger aims with which they identified their professions must be echoed by all who had the best interests of the province at heart. These aims included opening up the wilderness and preparing the country for agriculture, adjusting property boundaries, improving transportation and communications, developing resources and industry, and 'otherwise smoothing the path of Civilization.' This latter task included collecting specimens to 'promote the purposes of Science and the general interests of Society.'[27]

Memberships in the institute more than quadrupled, at least on paper, from 64 in 1850 to 263 in 1852. They embraced the entire political spectrum, from George Brown and Robert Baldwin on the left to John Langton, W.H. Draper, J.B. Robinson, and G.W. Allan on the right. Baldwin professed himself 'a lover of science' despite being no 'scientific man.' The institute's *conversaziones* attracted, furthermore, men of stature like John Strachan and Egerton Ryerson. As the *Journal of Education* aptly noted in another context, 'Science is honoured by rank, when rank itself seeks honour from science.'[28]

In order to earn prestige and set a high scientific standard for the Canadian Institute, care was taken to enlist William Logan as its first president. Logan agreed, but since he was based in Montreal and frequently away in the field, the position remained largely a nominal one. An obvious solution was to solicit Henry Lefroy as vice-president. From March 1852 to his election as president later that same year, Lefroy chaired monthly meetings in Logan's stead. He 'set the Institute going' by modelling the proceedings on similar meetings in Britain. More than one source credits Lefroy, rather than Sandford Fleming, with actually founding the institute.[29]

In the Canadian Institute Lefroy found himself in familiar scientific company: H.Y. Hind, now professor of chemistry at Trinity College, was already a member; and J.B. Cherriman, who had replaced Robert Murray as professor of mathematics and natural philosophy at University College, and his colleague H.H. Croft, professor of chemistry, joined soon after Lefroy. Hind edited the institute's *Canadian Journal of Industry, Science and Art*, having at last found a forum for his meteorological ideas. In April 1852 he addressed the first *conversazione* on the climate of Upper Canada.[30] Lefroy, as chairman, not only offered his endorsement but also followed up some of Hind's ideas in a paper subsequently published in the *Canadian Journal*.

Explaining why it was of the 'greatest interest' in Canada to obtain accurate thermometric registers from as early a date as possible, Lefroy for the first time linked meteorological theory explicitly with the idea of territorial expansion. By referring to thermometric registers, he pointed out, we learn 'whether we can bring about changes of climate by human agency: whether such changes are always beneficial, and therefore in harmony with the design of the Universe, or sometimes noxious, and therefore in favour of the opinion that there are preordained bounds to the extension of civilized man over the Globe.' Data were needed to complete isothermal maps depicting equal monthly temperatures on the North American continent. Part of this gap, Lefroy declared, would be filled by observations at grammar-school stations. Still, he cautioned, it was important to remember that the land had 'not yet reached its permanent climatic condition as older countries' had. Lefroy's subscription to the theory of climatic progress was followed by a still bolder analogy in an unsigned article a year later:

Two or three centuries ago the Rhine used to be frozen, and the animals, the natives of the northern regions, were abundant on its banks – how different is the case now? It will be so in British North America, with this difference, that the improving climate will keep pace with the vastly accelerated movements, and more rapidly increasing numbers of the New World settlers.[31]

While interest in determining the truth about climatic progress spiralled to its peak, the perennial issue of the Toronto observatory's continuance rose yet again.

V

After the Smithsonian Institution submitted its memorial to the British government requesting the continuance of the Toronto observatory in 1850–51, a decision was made that subsequent communications would instead be addressed to the governor general of Canada. Lord Elgin invited Henry Lefroy to outline the grounds for continuance as the basis for Elgin's application to the British Parliament on behalf of the observatory. Lefroy's immediate reply listed not only the obvious contributions of an unbroken series of observations to both magnetic and meteorological science. It also enumerated special and local reasons for continuance: first, the observatory was the only public establishment in British America devoted to any branch of physical research. It alone possessed instruments for studying even meteorology 'in the manner which the present state of that science' demanded. Canadian colleges and universities were struggling with financial problems, against public opinion too uninformed to assess the importance of physical and mathematical pursuits. 'At no distant time,' he trusted, the provincial university would take responsibility for the observatory. Meanwhile, continued support was necessary to maintain the unbroken series of observations.

Second, Canada's social development could benefit greatly from such 'higher intellectual pursuits.' 'Nothing less than this,' Lefroy insisted, 'will satisfy what science may reasonably claim from a country of the growing wealth and importance of Canada. Nothing less will place the native Canadians, seeking education at the colonial universities, upon a footing of equal advantage with the youth of the neighbouring states.' Third, surveyors would benefit from astronomical calculations at the observatory, as would the growing trade between the upper Great Lakes and the provinces on the Atlantic. Fourth, Canada's geographical position 'adapts it especially for the application of the Electric Telegraph to Geodesic purposes,' namely for the study of storms, barometric fluctuations, and the aurora borealis, in tandem with similar American studies. In fact, American scientists looked to the Toronto observatory as 'the Colonial centre for extensive enquiries' into such problems. Elgin forwarded these arguments with his concurrence.[32]

In 1852 the Toronto observatory was voted financial support from the British treasury for one final term ending 31 March 1853. While Sabine could 'scarcely anticipate' the possibility of a transfer to colonial authorities by then, Lefroy began a final campaign to impress upon Canadians the importance of their assuming responsibility after that date. From his executive position in the Canadian Institute, Lefroy took the authority to speak out on what was essentially a civil matter and hence of no official concern to military personnel. In a memorandum to the Canadian government, he reiterated the contribution of the observatory to magnetic and meteorological science. As proof, he added that Sabine was about to publish a second and then a third volume of Toronto observations.

But his main thrust was an appeal to Canadian interests. The buildings of the observatory would, after all, revert to University College upon the withdrawal of the military contingent. 'There should continue to be an Establishment devoted to Physical Enquiries and observations in Canada,' he argued, because 'the intelligence of the community appears to demand it,' as did growing scientific interest in the United States. Moreover, he drew a conceptual link between scientific and educational institutions for which he and Ryerson strove. Lefroy categorized the abandonment of such an establishment already in existence as 'hardly reconcileable with the high standing which the Educational and other Institutions of the country have attained.' Alluding to his auroral and meteorological observations, he argued further that although the original objects of the observatory had been attained, 'others of great interest have been subsequently added to them,' requiring a longer span of attention. Lefroy suggested that the instruments and library be maintained at Toronto, to be utilized in a 'Physical Observatory' not only to continue magnetic and meteorological observations but also to serve larger educational and astronomical purposes. The establishment could form a training school for skilful and accurate observers and 'give practical bearings to instructions in Mathematics and Natural Philosophy' at the university, with a professor in charge. It should, he stated, be connected to the electric telegraph. Finally an annual report should be submitted to the governor general by a board of visitors who included the chancellor of the university, the chief commissioner of public works, a judge, the chief engineer of the Board of Works, professors of mathematics and natural philosophy at each college of the University of Toronto, and the president of the Canadian Institute.[33]

Edward Sabine, meanwhile, ordered the withdrawal of both the detachment and the instruments to Woolwich. This order became Lefroy's lever to pry action from the governor general, the provincial secretary, the University of Toronto senate, and the Canadian Institute. The interest in terrestrial magnetism and meteorology that the first two had been pleased to express, he warned, led him to believe that they would not permit his military orders to take steps 'which the well wishers of the progress of scientific pursuits in this Colony, may afterwards regret.'[34] In his presidential address to the Canadian Institute in January 1853, Lefroy again played up the question of honour and the weight of public opinion.[35]

By early February 1853 Lord Elgin had received memorials on the subject not only from the Canadian Institute but also from the Natural History Society of Montreal, the Literary and Historical Society of Quebec, and the Smithsonian Institution. The Canadian Institute expressed great concern with regard to the discontinuance of 'the only observations made systematically and upon a large scale on any class of natural phenomena in British North America.' Besides the practical agricultural and medical advantages of making Toronto a 'centre of reference and comparison' for the study of the 'gradual change of climate which Canada is supposed to be undergoing,' it would, they argued, be seen as 'a reproach to a Country so populous as Canada, of so large a public revenue, and possessing a University so largely endowed' if the observatory were to be discontinued.

The Natural History Society of Montreal added that it was 'in expectation of ere long setting on foot, in connection with, and under the superintendence of, the Director of the Magnetic Observatory a valuable and interesting series of simultaneous meteorological observations all over the province of Canada.' The Toronto observatory was important 'not only as regards the Natural & Physical History of Canada, but of the whole World: and at the same time highly creditable to this Country, whether regarded in a Provincial or National point of view.' The Literary and Historical Society of Quebec called the Toronto observatory, with its peculiar geographical position, the best situated in the world for researches in terrestrial magnetism, the aurora, and similar phenomena. It was, moreover, 'universally acknowledged to be the most perfect of its kind.'[36]

The most striking non-scientific appeal for provincial maintenance of the Toronto observatory was an editorial by H.Y. Hind in the *Canadian Journal*. Science was at the foundation of all national prosperity, argued the writer, and such an establishment as a national observatory held the elements of the truest claims to national respect. 'Shall the British colonies acquiesce in the sentiment,' he asked rhetorically, that a colony was by necessity a place 'where the refinements of life, the pleasures of the intellect, and the pursuits which lead to other distinctions than that of wealth, can never be naturalized? ... To be respected abroad,' he countered, 'we must respect ourselves, and seize with no timid or reluctant hand each occasion, as it arises, for displaying an enlarged and enlightened public feeling.' In order to earn social equality with the mother country, Canadians had to prove their right 'by measures which the consent of the civilized world receives as true indices of the advancement of a community.' In this larger sense, the Toronto observatory could be 'obscure or distinguished, – a vigorous mainspring to a thousand scientific impulses, – or a mere machine for tracing a tame routine,' depending upon its director and especially upon 'the measure of public liberality dealt to it.' But Canada did deserve to have an observatory; it could maintain and appreciate one. Its success would take time, but no delay in the production of important scientific results could detract from 'the credit which will be justly due to the Canadian public for the formation and maintenance of such an establishment.'[37] The Canadian Institute at the same time attempted to stir up public interest by offering one of two prize medals for the best essay on 'the physical form, climate, soil, and natural productions of Canada.'[38]

Elgin's resulting sense that there was 'a strong feeling in the Province' in favour of continuing the observatory with colonial funds was understandable, and it was echoed in an Executive Council decision to buy time by requesting a delay in the transfer of the instruments to Woolwich. All of this was accomplished even before the formal petition of the Canadian Institute was considered in the House of Assembly. As A.N. Morin, the provincial secretary, commented to the secretary of state for the colonies, 'The provincial authorities seem to have been taken by surprise, they appear to have a vague feeling that it is incumbent upon them to follow in the ways of Science that have been traced out to them.'[39]

But it was not only scientific societies who pressured the government into these 'vague feelings.' The Toronto *British Colonist*, quoting the *Quebec Morning Chronicle*, argued that it was the duty of the legislature to undertake responsibility for the physical observatory on behalf of the province. Citing practical navigational and medical reasons, the papers argued that regardless of cost, Canada must have an astronomical observatory as did the United States, England, Prussia, Russia, and France. For there were 'signs in the heavens' for 'seasons, for years, and for days, the study of which is not without impunity to be neglected either in Canada or anywhere else.'[40]

The indeterminate state of the observatory in the spring of 1853 strengthened the case for an alternate system of meteorological observations in the province. In May the House of Assembly passed the Act to Amend the Law Relating to Grammar Schools in Upper Canada (16 Vict., c. 186), which established Ryerson's system of meteorological stations at the senior county grammar schools. Once again, no one questioned these clauses in debate.[41]

When the necessary series of transatlantic communications had finally been exchanged to countermand the orders to withdraw after 31 March, Lefroy had already departed for headquarters and had to return temporarily to aid in the retention of the instruments. On 13 June, after his final departure, the House of Assembly voted £2,000 'for the reorganization and temporary maintenance of the Scientific Observatory' for one year. Direction of the observatory, which the Canadian Institute felt had, under Lefroy, 'caused the name of Canada, and of Toronto in particular, to be honoured in all parts of the world where science is cultivated,' and which had 'earned a reputation second to none throughout the world,' was given provisionally to J.B. Cherriman. Edward Sabine considered these arrangements 'very liberal' and congratulated Lefroy 'that so strong a sense should be felt and so handsome an acknowledgement should be made of [the observatory's] usefulness.' The provincial government, the university, and almost every public voice in Canada had apparently been convinced that 'if Captain Lefroy had lighted the lamp of Science in the Province, it was their duty sedulously to supply and trim it.'[42] These tributes were well justified because Lefroy had done much to assure the Toronto observatory of a firm foundation for others to build upon and even to extend.

9

Encompassing the North

> I try in vain to be persuaded that the pole is the seat of frost and desolation; it ever presents itself to my imagination as the region of beauty and delight . . . There snow and frost are banished; and, sailing over a calm sea, we may be wafted to a land surpassing in wonders and in beauty every region hitherto discovered on the habitable globe. Its productions and features may be without example, as the phenomena of the heavenly bodies undoubtedly are in those undiscovered solitudes. What may not be expected in a country of eternal light?
>
> Mary Shelley, *Frankenstein; or, the Modern Prometheus* (1816)

Mary Shelley's classic novel anticipated the Victorians' ambivalent responses to the Arctic. With the disappearance of Sir John Franklin's expedition in 1845, British society shrank from the horrors on the dark side of the sublime. In contrast, Canadians during the second half of the nineteenth century permitted an almost unbounded optimism to guide their northern approaches.

Of the two inventory sciences institutionalized by the Canadian government after Lefroy's departure in 1853, meteorology attracted more public attention and was more influential in shaping ideas about Canada. Meteorology had more obvious practical applications than terrestrial magnetism and it was conceptually more advanced, as even Sabine admitted. J.H. Lefroy too had admitted in a public lecture to the Toronto Mechanics' Institute that governments had expended vast sums of money assisting the magnetic researches of Sabine, Hansteen, Humboldt, and Gauss, 'and one vast expedition has now been lost to the civilized world, for many years, in a region of ice.' It would be asked, he realized, 'what has been done by all these expenditures of time, labour, and treasure in the cause of science?' Collected observations filled more than twenty-seven quarto volumes of figures, which had yet to be compared and reduced. After all that, no confirmed theory of terrestrial magnetism had yet been established.[1]

By the 1850s meteorology had benefited from both technology and theory to become a more exact science. Expanding telegraph networks inaugurated the use of synoptic charts in the study of continental weather systems. Thermodynamics clarified the relationship between heat and motion and found important applications in explaining adiabatic processes which produce precipitation and storms. These developments increased public interest in weather forecasting. It has been argued that a resulting shift from descriptive meteorology with longer-range climatological purposes to shorter-range forecasting and storm-warning services prevented greater progress in meteorological theory during the latter part of the century.[2] In Canada the new emphasis upon meteorological service splintered basic aims, resulting in confusion of purpose and two different orientations in the meteorological inventory of the country.

I

Promoters of meteorological service envisioned the Toronto observatory as the hub of a navigational storm-warning network along the entire St Lawrence system, drawing in Canada East as well as the maritime colonies. Administrative changes at the observatory introduced by the Canadian government in 1855 strengthened this orientation.

While J.B. Cherriman and Egerton Ryerson were busy searching for meteorological instruments simple, accurate, and inexpensive enough to distribute to the senior county grammar schools, a chair of meteorology was established by the government at University College in 1855. The professor of meteorology was also to direct the Toronto observatory, with one third of his salary paid by the province and two thirds by the university. Ryerson heartily approved this linking of his grammar school stations to the observatory. Cherriman at first took the new position, with the chair of natural philosophy to be filled by his brother-in-law, George Templeman Kingston. Kingston, a former midshipman with the Royal Navy who graduated from Cambridge in mathematics, had been principal of the Nautical College at Quebec since 1852. However for reasons of health Cherriman did not take up the meteorological chair, and the appointments were reversed when Kingston arrived in August 1855. Instead of a natural philosopher who leaned towards long-term descriptive studies, the chief meteorologist of Canada was now a man inclined towards the navigational uses of the science.[3]

This blurring of basic aims in the pursuit of meteorological knowledge in Canada was aggravated by petitions to the Canadian Institute from Maj. Robert

Lachlan. Lachlan had served with the British Royal Leicestershire (17th) Regiment in India; he settled in Colchester Township in Upper Canada in 1836. He had been sheriff of the Western District and president of both the Western District Agricultural Society and the Western District Literary, Philosophical, and Agricultural Association at Amherstburg. In 1853 he was president of the Natural History Society of Montreal.[4]

In March 1854 the major addressed the Canadian Institute on 'The Establishment of Simultaneous Meteorological Observations, etc., throughout the British North American Provinces.' He maintained that his proposal had been sent ten years earlier to Dr William Craigie, the amateur meteorologist at Hamilton. In a written discourse Lachlan had suggested simultaneous meteorological diaries be kept in different parts of the country to ascertain 'the various shades of *climate* in different quarters.' As president of the NHSM he had joined with other amateur meteorologists to approach J.H. Lefroy at the Canadian Institute. Lachlan, Charles Smallwood, and J.S. McCord hoped Lefroy would direct extended meteorological observations, 'with the lines of electric telegraphs acting as magically powerful assistants.' Lefroy agreed that such stations ought to stretch east-west every hundred miles and north-south along the Ottawa, but he regretted that their establishment would take more time than he had left to spend in Canada. He advised Lachlan to contact the Smithsonian Institution for the leverage to recruit 'an efficient corps of zealous and accurate *volunteer* observers.'

Lachlan hoped to include officers of the Hudson's Bay Company, educational institutions, the military and medical professions, scientific societies, and public institutions (harbour-masters, lighthouse keepers, and customs collectors) in his network. His goal was a thorough knowledge of climate 'in all its relations, and of its variations in the same and in different localities.' Temperature and seasonal variations, the limits of vegetation, the progress of epidemics, and the direction and motion of storms were all to be recorded. Lachlan appealed to the governments of New Brunswick, Nova Scotia, and Prince Edward Island, 'invoking their cordial cooperation in the patriotic undertaking.'[5]

Lachlan's scheme was noticed by *The Globe* in Toronto but received no official reply from the Canadian Institute until early the following year. In his presidential address John Beverley Robinson agreed that no one questioned the scientific and social value of Lachlan's suggestions. But he added that it was premature for the institute, with its limited resources, to undertake the project. Either the number of stations or the range of observations would have to be curtailed. Robinson did perceive an urgency in that in Canada the opportunity to study the effects of cultivation was rapidly vanishing, especially in the northwestern part of the province.[6]

The institute's 'neglect' of his scheme provoked Lachlan's 'Supplementary Remarks in Behalf of the Establishment of a Provincial System of Meteorological Observations.' Forced to acknowledge that the 'enlightened foresight' of Egerton Ryerson already supplied meteorological stations, even though there was no plan to link the grammar school stations to the telegraph system, Lachlan countered that Ryerson's jurisdiction, for all his good intentions, was too limited. If £2,000 were forthcoming from the provincial legislature to expand the observations to Lower Canada, Canada would be setting an example 'to the advancement of so laudable a National Work' and recommending 'a similar line of conduct to the favourable consideration of the Sister Provinces.' Mustering the support of a scientific argument penned by Joseph Henry, Lachlan linked the proposed meteorological stations to a North American storm-watch system: 'No greater favour could be conferred on the science of Meteorology than the establishment of a series of observations in the British possessions in North America. If this were done, all the phases of a winter's storm could be noted from the moment of its rise through all its changes, until its disappearance; and for want of data of that kind, the observations now made within the boundaries of the United States are of much less value than they otherwise would be.'[7]

Lachlan was barking up the wrong tree, and was told as much by the editing committee of the *Canadian Journal*, including Hind and Cherriman but, interestingly, not G.T. Kingston. In a report to the Canadian Institute, the committee concluded that at least until Ryerson's grammar school stations were functioning, Lachlan's application was both unnecessary and impracticable. They accused Lachlan of misleading his listeners when he denied that meteorological observations received considerable attention in Canada. G.T. Kingston, they pointed out, was training teachers and others in the use of instruments and the scientific applications of the results of observations. The committee added that although staff and funding might eventually allow the establishment of a comprehensive meteorological scheme in British North America, they believed it was not for the institute but for the observatory to take the initiative. All the Canadian Institute could do was publish results and recommend a similar grammar school system in Canada East. This final recommendation was followed up in the pages of the *Canadian Journal*, where the editors expressed the expectation that scientific and educational institutions all over British North America would be stimulated to action by the Upper Canadian example and that the results of a 'great chain of philosophical researches in Physical Geography and Magnetism' might some day be published in a meteorological and magnetic journal of British North America.[8]

The probability of realizing some part of what G.T. Kingston termed 'a great scientific movement now in progress in Canada' increased the following year when Kingston proposed to pursue the main aim of Lachlan's scheme. Kingston planned to employ the electric telegraph to predict storms in a chain of meteorological stations east of Sarnia to Halifax and Fredericton. The system would help to prevent shipwrecks on the St Lawrence route, and costs would be minimized by requiring captains and shipowners to consult meteorological stations as part of their insurance policies. A meteorological committee struck by the Canadian Institute in 1856 and chaired by Kingston recommended that the institute emulate the Smithsonian by enlarging the grammar school system of stations and providing instruments for observers.[9] But the report was still deemed over-ambitious by the institute and at cross-purposes with the institute's interest in accumulating knowledge more than providing a public weather service.

Indeed, even Ryerson's grammar school stations, already passed into law, encountered unforeseen difficulties. It was not clear to what end, whether long-term or short-term, these stations were actually being established.

II

Egerton Ryerson and John George Hodgins, his assistant, planned thirty chief grammar school meteorological stations, with possibilities of adding forty-three more as circumstances warranted. Each senior county grammar school was to be supplied with an aneroid barometer, wet and dry bulb thermometers, self-registering maximum and minimum thermometers, a hygrometer, a rain gauge, a wind vane, a manual (John Drew's *Practical Meteorology*), and standardized forms for daily observations and monthly abstracts. The government would pay half the cost of this equipment if the municipalities would pay the other half. Thermometers shipped from London had to be numbered and compared with standards set at the Toronto observatory. Preparations were finally completed in 1858, and public response was surprisingly enthusiastic. The plan, thought one newspaper, 'would make of Canada one vast observatory; we should arrive at facts, and might thence deduce natural laws, a knowledge of which would be of incalculable advantage to Canada and the world.'[10]

Some papers demonstrated a clear understanding of Ryerson's use of science as a unifying social force. The Barrie *Northern Advance* noted: 'Whatever difference of opinion may exist amongst the gentlemen composing our Municipal Council on questions of a debateable character, we are proud to say that in matters of this kind a liberal and enlightened spirit has ever characterized their proceedings.' The Hamilton *Spectator* reminded readers that 'not only the master, but the pupils also, shall learn to make the observations requisite; and the youth of the country be thus indoctrinated with a taste for natural science.' Canada would attune itself to 'every enlightened country in Europe' engaged in similar enquiries. In addition to 'truths of the highest value and importance in scientific research' that had already been revealed, the stations were thought to provide 'great practical importance to a new and but partially settled country.' They encouraged uniform exactness and accuracy 'throughout the length and breadth of the land,' making every observer aware of his meaningful part within the whole system.[11]

Reality, of course, was quite unlike the rhetoric which the meteorological stations inspired. By December 1860, instruments had been purchased by municipal councils of only sixteen out of thirty-one counties with senior grammar schools. Of these, half submitted their abstracts at least one third of the time, and only five did so most of the time. Only six schools usually submitted well-prepared abstracts. The Education Department sought ways to enforce the law at the remaining schools, offering monetary incentives to those who complied.[12]

Part of the problem was an understandable reluctance among headmasters to perform forty extra hours of work per month, including observations three times daily at inconvenient hours at the schoolhouse (7 A.M., 1 P.M., 9 P.M.), with no remuneration. Ryerson offered only laudatory assurances to those who did the job well. 'If you continue to progress in the same manner,' he wrote to one headmaster, 'you will do much for the advancement of science in this country.' Although the public revenue allotted an extra £100 annually to each senior county grammar school, none of the money was used to reward headmasters for performing this part of their duties. Ryerson hid behind the law when one headmaster threatened to interrupt his observations until he received compensation: 'As to the taking or not taking such observations,' he wrote, 'I have no authority to suspend the law, to which I am as much subject as you are.' But 'it might be gratifying to many and might create an interest in the subject amongst intelligent persons in your county, if you were to prepare and publish monthly in one of the local papers, the results of your observations.' Periodic extremes in the weather did excite attention throughout the province, but newspaper editors in general were reluctant to make room for weekly meteorological abstracts which interested few readers and were often criticized for their inaccuracies.[13]

In 1865 a revised statute granted an abbreviated list of senior county grammar schools $15 a month in exchange for satisfactory abstracts of observations reduced to twice daily. But the overall result was still not encouraging. Headmasters found their 'confinement' in a single spot for the whole year irksome and inconvenient. When one dared to suggest that an increase to $20 would not be 'running much risk of overpayment,' Ryerson began withholding grant money from schools whose returns were judged incomplete at the observatory. Some headmasters used space reserved for miscellaneous observations to give free rein to their more whimsical poetic urges, but others deepened their interest to the point of building their own observatories and vehemently defending the accuracy of their own instruments when challenged by the Toronto standards.[14]

Papers such as the Toronto *Leader* felt gratified that steps were being taken to remedy Canada's lagging behind other countries in meteorological observations. But contrary to Ryerson's ambition that the system be 'more complete than in any other part of America,' he admitted in 1862 that, other than providing northern outposts for the Smithsonian system, it had not yet yielded practical results in Canada. By 1865 Ryerson boasted that returns from the grammar schools had aided the Standing Committee on Colonization and Emigration. The committee used tabulated returns in 1864 as 'proof' that Canada's climate was 'adapted to every kind of agriculture' and compared favourably with the genial climate of European growing districts.[15]

Indeed, the only use of the grammar school stations even remotely felt by the Canadian public by the 1870s was in providing climatological statistics for prospective immigrants. Canadians were still confused as to the actual purpose of the stations. The Toronto *Leader* erroneously viewed them as Canada's version of the meteorological warning services provided in other countries by telegraph.[16] It was nevertheless apparent from G.T. Kingston's continued attempts to establish a larger telegraphic storm signal service that the director of the Toronto observatory did not see the grammar school stations as part of such a scheme.

III

As director of an expanding network of observatories in British North America, G.T. Kingston judged it an appropriate duty for a 'National Observatory' to collect, classify, and reduce the grammar school observations. Kingston emerges as a diligent, almost plodding worker strikingly unlike H.Y. Hind in his refusal to colour his work with speculation. Kingston made few attempts to set his observations into broader context or to participate in the growing debate over the interpretation of meteorological data from the north-west. He defined the primary object of the observatory as to furnish to the scientific world 'the data necessary for evolving the Laws that regulate the Magnetic and Meteorological Phenomena of the Earth.' Unlike Lefroy, he seldom ventured beyond these bounds to try his hand at analyzing the information he collected. Kingston took pride in the publication of his magnetic observations up to 1862 before the imperial authorities had published those up to 1852, and he was satisfied that his work in the Toronto observatory supplied 'a valuable Canadian contingent to that common intellectual property' in which the whole human family shared and towards which it was 'the bounden duty of each nation to contribute according to the opportunities afforded by its geographical position and physical peculiarities.'[17]

While others around him weighted various elements in assessing the climate of a region, Kingston would state only that more detail was needed to ascertain the geographical limits of monthly mean temperatures. The climate of a given locality with respect to any one meteorological element, such as temperature, Kingston hypothesized, could be characterized by the mean annual value of that element. But he added vaguely that annual and diurnal periodic variations also had to be considered. Even while the Palliser and Hind expeditions were boldly reshaping traditional views of the climate of the British American north-west, the dedicated Kingston remained hunched over his data sheets, concluding with a touch of irony that the north-west had dominated Toronto's prevailing winds in recent years and that weather disturbances originating there warranted further study. Kingston pursued these minute studies until 1871, when he solicited dominion government support for his telegraphic storm signal system.[18]

In 1860 influential critics of the University of Toronto became disillusioned by increasing expenditures and diminishing returns from both the chair of meteorology and the Toronto observatory. They included Egerton Ryerson, who had earlier seconded the establishment of the chair. John Langton, vice-chancellor of the university, believed it had been a financial error to connect the observatory to the university in the first place. A select committee of the assembly agreed, recommending that the chair be funded entirely by the government because meteorology was too limited a subject to form an exclusive chair and attracted too few students. Instead of stepping more deeply into the University of Toronto's financial quagmire, the government took on the support of two other observatories. The Kingston observatory, established by private subscription in 1855, was mainly astronomical and only incidentally meteorological. But the observatory at Isle Jésus was mainly meteorological and magnetic and its builder, Charles Smallwood, had been professor of meteorology at McGill University since 1858.[19]

Petitions for provincial meteorological observatories at Montreal or Quebec had collected in the House since 1856. Smallwood's observatory was touted as the potential nucleus for such a provincial observatory, a centre for the advancement of meteorological science in Canada East. The work of this "Canadian merlin [sic]" was appreciated beyond the limits of Canada, and J.W. Dawson urged the 'intelligent public' to aid his meteorological investigations, if only 'by studying and expounding' his conclusions. Smallwood's petition for financial aid found sympathy not only in the NHSM and McGill College, but also in the bishop of Montreal and several other Roman Catholic bishops. The McGill College observatory, which transferred and incorporated Smallwood's observatory in 1864, included a storm-warning system connected to the telegraph and to all the church bells in Montreal. Smallwood's ambition grew with his prestige, and in 1866 he requested facilities for astronomy, celestial chemistry, and spectrum analysis.[20]

The great aim of modern science, Smallwood believed, was 'unity of purpose.' Sciences like meteorology thrived over large areas such as the Dominion of Canada. But Confederation did not bring about the centralization, standardization, and synchronization that Smallwood believed necessary to achieve such unity of purpose. He urged the government to do its share and supported Kingston's proposed storm-warning system. French Canadians like Abbé Provancher, it was noted in the *Canadian Naturalist and Geologist* in 1869, had begun to contribute regular meteorological observations, and it was hoped that these would be increased all along the St Lawrence.[21]

A small astronomical observatory for time-reckoning, under the direction of Lt Edward Ashe of the Royal Navy, was publicly supported at Quebec since 1849. By linking the observatory, temporarily located in the Quebec citadel, to the telegraph, it would be possible to establish exact longitudes of places in Canada and to determine exact local times. Ashe proposed in 1857 to take advantage of Quebec's position as both a seaport and an inland town to establish a 'National Canadian Astronomical and Meteorological Observatory'; he was supported by a petition of the Canadian Institute. It would not, declared the institute, 'consist with [Canada's] dignity to lag behind in the march of scientific research.'[22]

Nor was interest in establishing observatories limited to the province of Canada. In 1855 Ashe commenced chronometric telegraph operations with W.B. Jack at Fredericton. Equipped with the largest telescope in North America, Jack's observatory had been enthusiastically encouraged by the governor, Edmund Walker Head, who was by 1855 governor general of Canada. Jack strongly favoured a central observatory at Montreal. 'I suppose the Observatory at Quebec is maintained by the British Government,' he wrote to William Logan, 'but such a country as Canada should have a sort of national Observatory at which, besides other important work performed, the shipping resorting to its Ports could have their chronometers partly rated.' He believed that Nova Scotia 'ought to take some pains to have her geographical position accurately defined, seeing that, – thanks to the exertions of Mr. Dawson, – so much is known of her Geology.' Finally Jack pointed out that the recently laid ocean cable should be utilized to ascertain longitudinal differences between Newfoundland and Greenwich: 'Could you do nothing to forward it on this side of the Atlantic?' he asked William Logan. 'If we do not attempt it, it is certain the Americans will, and it would be somewhat disgraceful to be forestalled by them on our own ground.'[23]

Eventually the Institute of Rupert's Land was founded at Fort Garry in 1862, to collect and make known information on the condition and resources of the HBC territories. The membership included some competent to take charge of a magnetic observatory if the government or some scientific society would supply a building and instruments, declared its secretary, J.C. Schultz. Members scattered from the Pacific to Lake Superior and towards the Arctic Sea, if given instruments, would also transmit meteorological information. The effects of such an institute on the character of the Red River settlement would, it was argued, certainly be positive: 'The eyes of the world will be turned toward us. We will be judged by a new and loftier standard – it will be the means of inducing scientific men to visit us and of bringing to light much that is not generally known.' Success would 'show the world that we were rising not falling in the scale of civilization.'[24] The 'spirit of research' was of course not new to the HBC lands; but news that it had now taken root in these 'remote confines of civilization' was welcomed by the Canadian Institute as evidence of an 'intellectual dawn' that heralded the period 'when states and empires of the great northwest are to claim their place in the world's commonwealth of nations.'[25] While G.T. Kingston focused meteorology at the Toronto observatory on an eastward-looking storm signal and time service, others turned in the direction of Canada's prevailing wind system, the north-west.

IV

By the mid-1850s meteorological data invited attempts at generalization; nowhere was this more true than with regard to the north-western region of British North America. Only a decade earlier the north-west had been labelled *terra incognita* on Alexander von Humboldt's isothermic maps. On these maps, lines of equal temperature deviated from the latitudinal grid and challenged the long-held assumption that temperature varied inversely and rigidly with latitude. Humboldt's isotherms measured annual as well as seasonal temperature means, facilitating the comparative climatology he pursued as part of his larger study of the *Kosmos*. By 1849 the usefulness of isolines as a climatological research tool was enhanced when Wilhelm Dove at Berlin narrowed their scope from annual to monthly means.[26]

It was not only that these developments challenged some old scientific assumptions; they reshaped still others to lend scientific authority to expansionist ideologies. One example was the idea that climate improves as one proceeds westwards on a large continental mass. This widely held notion was a colloquial version of an eighteenth-century hypothesis formulated by the German naturalist and geographer Georg Forster. Forster (1754–94) accompanied James Cook on his second circumnavigation of the globe from 1772 to 1775. In comparing the European with the North American continent, Forster noted two things. First, considerable temperature and climatic differences separated the west and east coasts of each continent. Second, the west coasts of both resembled each other in the relative mildness of their climates. Forster became Alexander von Humboldt's mentor and prodded him to investigate the significance of these observations.[27]

English immigrants who settled in eastern North America also accepted as self-evident the adage that climate improved 'by westing.' They relied on it to assess the agricultural potential of North America. It surfaced in the well-known and influential 1847 address by Robert Baldwin Sullivan entitled 'Emigration and Colonization,' which inspired a reconsideration of the possibilities of the great north-west:

Just take the map of Canada – but that will not do – take the map of North America, and look to the westward of that glorious inland sea, Lake Superior. I will say nothing of the mineral treasures of its northern shores, or of those of our own Lake Huron, but I ask you to go with me on to the head of Lake Superior, to the boundary line: you will expect a cold journey, but I tell you the climate still improves as you go westward. At the head of Lake Superior, we ascend to the height of land, and then descend into the real garden of the British possessions, of which so few know anything.[28]

The idea attained its full ideological application in the testimony of convinced expansionists a decade later before the Canadian select committee on the HBC territories. William Logan mentioned it in his GSC report of 1845. It was accepted even by non-expansionist scientists such as J.H. Lefroy and Lt. Thomas Blakiston, the magnetic and meteorological surveyor who accompanied the Palliser expedition in 1857.[29]

Closely related to the idea that climate improves 'by westing' was the theory of climatic amelioration, a phenomenon that as we have seen, was believed to result from clearing and cultivation. Published expressions of this theory date back at least to 1544, and it was taken up again in Pierre François Xavier de Charlevoix's influential *Journal*, published in 1761. When Canada was found by early settlers to deviate from European climatic norms at comparable latitudes, the obvious differences in population and cultivation seemed reasonable explanations. Other eighteenth-century writers, including Abbé J.B. du Bos (*Réflections sur la Poésie et la Peinture* [1719]), Simon Pelloutier (*Histoire des Celtes* [1740]), and David Hume (*The History of England from the Invasion of Julius Caesar* [1754–62]), studied ancient descriptions of the climate in Germany. They theorized – from accounts that the Rhine and the Danube often froze over with ice thick enough to support invading barbarian armies and that reindeer had frequented the forests of Germany and Poland – that Europe had formerly been much colder. It seemed to follow that the cold had been diminished by clearing woods and draining swamps, improvements which permitted the rays of the sun to penetrate and warm the earth. As the soil was cultivated, the earth became more temperate.[30] It did not take long to draw what seemed an obvious analogy with the climate of North America.

The notion that the climate of Canada would be ameliorated with increased deforestation and cultivation received the stamp of authority from Edward Gibbon's *History of the Decline and Fall of the Roman Empire*, first published in 1776. Relying on Charlevoix, Hume, du Bos, and Pelloutier, Gibbon wrote: 'Canada, at this day, is an exact picture of ancient Germany. Although situated in the same parallel with the finest provinces of France and England, that country experiences the most rigorous cold.' Gibbon's *History* became a literary institution even in his own day, and his interpretation of Canadian climate was perpetuated in the minds of generations to come. Nearly every educated person in the English-speaking world read Gibbon's *History*, which fixed a certain impression of the climate of Canada.[31]

The theory of climatic progress helps to explain why early settlers attacked the forest with more than the necessary vigour. Despite William Kelly's protestations, most new settlers resembled Catharine Parr Traill, who believed that the harsh climate of the backwoods would gradually be moderated as the forests were cleared. J.H. Lefroy, as we have seen, induced Egerton Ryerson in 1850 to organize observations at the grammar school stations with this problem in mind. As late as 1855 the chief clerk in the Crown Lands Department, William Spragge, testified before a select committee of the House of Assembly that the climate of Upper Canada had indeed, 'with the progress of improvement, and of opening up the country become wonderfully ameliorated, and seems to be approaching to the character of the same Latitude in Europe.' He inferred that a milder climate would be felt in Lower Canada under corresponding circumstances.[32]

That same year, first prize was awarded by the Paris Exhibition Committee of Canada to John Sheridan Hogan for his essay on Canada, which included the observation: 'Since 1818 the climate has greatly changed, owing principally, it is supposed, to the large clearings of the primeval forests.' Abbé J.B.A. Ferland addressed the question in his *Cours d'histoire du Canada*, published in 1861, and the Canadian Institute interviewed G.T. Kingston on the matter in 1862. In 1864 the New York *Evening Post* claimed, 'The severity of the cold season has also much abated since the forests have been cut into; the winters of New Brunswick have, it is affirmed, been shortened two months by this one cause.'[33]

Yet enough meteorological data had been accumulated and reduced by the 1850s that some observers doubted the validity of the theory of climatic progress. Ferland, Kingston, and Lefroy joined those who saw little scientific evidence that the climate of Canada had indeed improved with cultivation. When asked outright by the British parliamentary select committee on the HBC in 1857 whether his experiences in British North America could support the impression created by Gibbon, Lefroy replied in the negative.[34]

V

Although the question of climatic progress was by no means yet laid to rest, by the 1860s the authority of Humboldt's isotherms relaid the entire theoretical basis of climatology in British North America. Isotherms freed temperature variability from rigid ties to latitude and permitted more northerly latitudes to be seen in a new light. If, indeed, the climate of the province of Canada could not be expected to improve significantly over time, the north-western territories of the HBC might be found to promise what Canada could not. As the Crown Lands Department saw it, 'the easier livelihood that is to be earned by cultivating the prairies of the west, where the plough can be immediately used, and great crops be obtained with comparatively little labour,' would be most attractive to both prospective immigrants and sons of Canadian farmers who might otherwise move to the United States.[35]

Much has been written on the intellectual contribution of Lorin Blodget's *Climatology of the United States and of the Temperate Latitudes of the North American Continent*, published in 1857. But the impression created by the amount of attention his work received requires an understanding that Blodget's name and authority served as shorthand for a body of meteorological knowledge accumulating at least since the founding of the Toronto observatory. Lefroy's observations in the north-west had been noted by Sir John Richardson, who inspired Blodget and publicized the agricultural possibilities of the north-west. These observations, as we have seen, also lent scientific authority to H.Y. Hind's assessment of Canadian climate in 1851. Hind's opinion that Canadian agricultural potential was unsurpassed in North America was in turn imbibed by Alexander Morris and urged upon the Canadian public in his prize-winning essay *Canada and Her Resources* in 1855.[36]

In reducing and translating accumulated observations Hind, appointed by the Canadian government to report upon the climate and agricultural potential of the HBC lands in 1857 and again in 1858, skipped the necessary step of collecting the data as objectively as possible. He was neither a trained explorer nor a meteorologist but an avid student of scientific literature whose enthusiasm towards the north-west was well known before his departure. Meteorological instruments for the expedition were provided by Egerton Ryerson from his supply intended for the grammar school stations. Ryerson's *Journal of Education* also reprinted parts of Joseph Henry's paper on the 'sterile belt' of the north-western United States, which Hind predicted would force the progress of settlement north to the valley of the Red River and the Saskatchewan. Eager to move the summer isotherms of the north-west still higher than Blodget's he relied on only a single rapid rise in temperature noted by Lefroy at Fort Simpson in 1844. Given Hind's previous publications, his hyperbolic assessment of a 'fertile belt' from Lake of the Woods to the Rocky Mountains was not surprising: 'No other part of the American Continent,' he enthused, could even approach 'this singularly favourable disposition of soil and climate.' Notwithstanding its rigour during the winter season the climate 'confers, on account of its humidity, inestimable value on British America south of the 54th parallel.' Hind's *Narrative* of his expeditions shared something of the style of Francis Parkman's popular *The Oregon Trail*, published in 1849, as well as some of its key concepts, including the westward movement of peoples and the 'fertile belt.'[37]

It is interesting to note the number and diversity of people who took his judgments literally. In Montreal, not nearly as rabidly expansionist as Toronto, Hind's *Narrative of the Canadian Red River Exploring Expedition* was favourably received as a credit to the province that belonged in all public libraries and should 'be carefully studied by those who interest themselves in the prosperity and extension of the Province to the Westward.' By the same token, when Blodget's *Climatology* was reviewed by G.T. Kingston, an outsider to the expansionist movement, he ignored its propagandistic final chapter. Nevertheless, Kingston came away with a new climatological conception of the North American continent; the whole region from the Atlantic to the Mississippi Valley now appeared as one meteorological region, different from that west of the Rockies. For Canadians the significance of Blodget's work was not so much that his isotherms transcended the north-south political boundary but that these new meteorological concepts drew the eyes of 'civilization' north-westwards beyond the pale.[38]

The strength of the isothermal argument put forward by Hind and, through him, by Blodget was that it made unnecessary the questionable theory of climatic progress. It also disposed the uncommitted to accept more readily its expansionist implications. Hind premised his conclusions on the older idea that climate improves 'with the westing,' an assumption which Thomas Blakiston, one of the few scientists to call Hind to task for inaccuracies and hasty conclusions, shared: 'There have been and possibly still exist,' wrote Blakiston, 'more particularly in Canada, most erroneous opinions concerning the country and climate of the Red River and the west.' He had heard it described as superior even to the south-western peninsula of Canada, an absurd idea 'in the face of 5 degrees difference of latitude between even the central part of that peninsula and the most southern limits of the interior[;] *notwithstanding the westing*, this was most improbable' (italics added).[39]

Dismissing frequent analogies between the climate of the north-west and that of north-central Europe, Blakiston instead asked, 'Why does Europe differ in climate so greatly from North America?' His answer was the Gulf Stream, for 'in comparing the climate of the two continents, we should rather contrast Europe with the Pacific side of North America, the east or greater portion with Asia ... with the exception, perhaps of Siberia containing no equivalent to Hudson's Bay.' Blakiston by no means opposed settlement in the north-west; he believed it to be 'against the laws of humanity to offer any obstacle to the progress of civilization.' But his more balanced critical assessment of the HBC territories was obscured when he broke from the Palliser expedition in August 1858. His report lies buried among the expedition's papers, and his meteorological register was never recovered for use in Palliser's final report.[40]

Support for the isothermic interpretation of the potential of the north-west, with its geopolitical corollaries, was of course not limited to H.Y. Hind. William McDonell Dawson of the Crown Lands Department testified before the Canadian select committee on the HBC charter. His arguments closely resembled the climatic arguments and European analogies put forward in the report of the commissioner of Crown lands for 1856. Dawson and others like him were irritated by J.H. Lefroy's testimony before the British select committee on the same subject. Lefroy stated that agricultural settlement 'can make but very slender progress in any portion of that region' and moreover that 'the natural affinities' of the Red River rested with the valley of the Missouri. It was, he stated, 'going against nature if we try to force it into the valley of the St Lawrence.' For those in Toronto who recalled Lefroy's favourable impressions after his northern magnetic survey in 1844, this was heresy, and all the more mystifying because years later Lefroy recalled that behind the belief that frequent summer frosts prevented the north-west from ever becoming an agricultural country lay a misapprehension that it would be fatal to the fur trade. It was natural for expansionists who rejected George Simpson's denial of the fitness of the territory for settlement to turn to Hind's report for a strong antidote.[41]

While developments in the science of meteorology bestowed legitimacy upon a growing Canadian expansionist movement, they also challenged Canadians' views of themselves as a mere colonial people. The optimistic interpretation of the north-west encouraged by Hind and Palliser, combined with growing meteorological evidence by the late 1850s that Canada's climate would never resemble that of western Europe, proved fertile soil for the idea of Canada's 'northernness' as connoting an appealing manliness. As Gibbon wrote, in ancient times the rigorous northern cold of the north and 'the keen air of Germany formed the large and masculine limbs of the natives, who were, in general, of a more lofty stature than the people of the South; gave them a kind of strength better adapted to violent exertion than the patient labourer, and inspired them with constitutional bravery, which is the result of nerves and spirits.' The severity of a winter campaign 'that chilled the courage of the Roman troops, was scarcely felt by these hardy children of the North.'[42]

In 1836 Catharine Parr Traill noted 'the spirit and vigour infused into one's blood by the purity' of the winter air in Canada. Early interpretations of the 'dry, cold and elastic' north-west wind as favourable to health and longevity were borrowed from Hind by both John Sheridan Hogan and Alexander Morris for their prize essays in 1855. By 1858 Morris had constructed a full-blown image of a 'Great Britannic Empire of the North,' with its 'goodly band of Northmen from Acadia, and Canada, and the North-West, and the Columbia, and the Britain of the Pacific.' They constituted 'a noble army of hardy spirits encased in stalwart forms.' He invited Canadian audiences to 'consider the energetic character inherited by our people, which the fusion of races and the conquering from the forest of new territories' had fostered, and which climate had 'rendered hardier.' The result, he predicted, would be one nation, a 'harmonious whole – rendered the more vigorous by our northern position.'[43]

Edward Sabine promoted a world-wide chain of magnetic and meteorological observatories in the hope that he could explain the physical causes of terrestrial magnetism in an *opus magnum* modelled on Newton's reduction of gravitation to a series of mathematical laws. Since this great work never appeared, the 'greatest scientific undertaking which the world had ever seen' ended in failure. Although daunting mountains of observed data did lead to some minor conclusions, essentially they revealed the magnetic qualities of the earth to be 'so peculiar and complicated' that the project reached a point of diminishing returns.⁴⁴ One of the sticking-points was the need to develop a mathematical field theory more powerful than Newton's scalar theory or even Gauss's vector theory, a long-range difficulty which even quantum theory could not later surmount. But Sabine had also placed more emphasis on terrestrial magnetism in relation to meteorological processes, and less on its relation to electrical currents within the earth; his decision to exclude frequent anomalies, or 'disturbances,' in the data was recognized within his lifetime as erroneous. Subsequent geomagneticians spent decades defining efficient ways to collect meaningful data, and only in the 1930s did new techniques revive widespread interest in geomagnetism. Despite Sabine's efforts, then, and partly because of them, the basis of terrestrial magnetism remained very much a mystery.

As a dedicated soldier in Sabine's campaign to raise terrestrial magnetism to the level of completeness achieved in astronomy, J.H. Lefroy expressed a certain cynicism over the defeat. His personal manuscript journals having been stolen en route to England in 1846, no account of his magnetic survey of the north-west reached the public during his lifetime. Lefroy revealed in his autobiography a regret that his tireless observations at Lake Athabasca had failed to attract much interest and had only 'swelled the volume of wasted labour,' for no one 'ever tried to sift them or deduce comprehensive results.' A grand theory would have compensated for the hardship and tedium of collecting countless minutiae of magnetic and meteorological information. While taking heart that the published numerical results of his observations showed 'that I tried at least to do my own share of this work,' Lefroy concluded bitterly that it was never noticed because 'the interest of the whole inquiry was largely factitious.'⁴⁵

But to infer from this failure of terrestrial magnetism to find its Newton in Gauss, Faraday, or Sabine that the Humboldtian sciences of geomagnetism and meteorology had no lasting influence in Canada is to miss an important point. The main characteristic of nineteenth-century physical science, as of other facets of Victorian thought, was its tendency towards an increasingly unified world-view.⁴⁶ This world view found scientific expression in Faraday's theory of electromagnetic induction and in the unity of physical forces underpinning the laws of thermodynamics. Both terrestrial magnetism and meteorology as Victorian inventory sciences imbued Canadians with an outlook characterized by this unified world-view. They linked together apparently isolated scientific facts, abstract ideas, and even ideologies which had not coalesced before. This they accomplished in several ways.

First, the policies pursued by the imperial officers responsible for the Toronto observatory, in their anxiety to preserve the observatory from oblivion when imperial financial support was no longer forthcoming, offer an interesting example of Lord Durham's philosophy at work. J.H. Lefroy, and through him

Edward Sabine, strove to instil in Canadians a collective sense of pride, ambition, and cultural advancement which would impel them to assume the observatory's scientific functions. They appealed to the national uses of internationalism by calling upon prestigious American scientists to persuade Canadian Reformers and Tories alike that Canada was ready to occupy the important place allotted to it by geography on the scale of 'civilized' nations.[47]

These policies are not well accounted for by the traditional Whig interpretation that Canadians struggled to wrest political responsibility from Great Britain during the period of the Union. Nor do they fit into the more revisionist interpretation of British policy towards Canada at mid-century, which claims it was intended to kill an incipient nationalism with kindness.[48] Whether Lefroy intended it or not (and his testimony before the British select committee on the HBC suggests that he did not), his policies altered Canadians' self-image through their perceptions of the land they inhabited. One means to this end was to establish the Toronto observatory as a centre of pride and authority whose influence reached far beyond the local and provincial, even to the intercolonial and ultimately 'national' levels. The degree of nationalistic rhetoric which accompanied these changes in perception was second only to that associated with the achievements of the Geological Survey of Canada.

Second, geomagnetism and meteorology were observational sciences unencumbered by the materialism associated with geology in the search for mineral wealth. They were easily cast in a subtle ideological role by social conservatives like Egerton Ryerson, the Family Compact, most legislators of the day, and even some French-Canadian bishops. Ryerson clearly saw these sciences as a means to achieving Methodist goals. They were used as a social and cultural adhesive to cement the foundations of a society Ryerson feared would shatter on the shoals of its own diversity.

Third, ideas of expansionism and nationalism began to flourish in Canada during the 1850s partly because of the authority lent them by scientific developments in meteorology, climatology, and terrestrial magnetism. In the light of these developments the British American north-west came to be viewed as a habitable land and moreover as an agricultural one to be coveted as part of Canada's manifest destiny. In the same vein, Canadian meteorologists interested in forecasting helped tie the maritime provinces into a meteorological network in which Canada formed the core.

Fourth, attention upon Canada as a northern country was focused by the discovery of the north magnetic pole near Hudson Bay. Observatories in Toronto and Montreal participated as northern outposts of co-operative British and American systems of magnetic and meteorological stations. A peculiar alchemy of old climatological myths and new meteorological theories after mid-century helped first to forge and then to sharpen conceptions of a Canadian character which transcended differences in origin or region. This new character was to be tempered by a healthy and invigorating climate; such conceptions later combined with Spencerian notions that the main ('fittest') centres of civilization had survived in the cooler climates. These ideas formed the intellectual tap-root of the 'northern myth' seized upon by Canadian imperialists and other nationalists late in the nineteenth century. 'The directive magnetic force that controls the mariner's needle,' wrote Charles R. Tuttle in 1885 after accompanying a government expedition to the north,

is not a more attractive problem than is the not less unerring north-westerly trend of human progress. Westward and northward have the marching orders been, until the people of the present generation must look southward and eastward for the homes of their ancestors. The greatest deeds have always been accomplished in high latitudes, because the highest latitudes produce the greatest men. And yet, strange as it may seem, the north is always underrated ... Here on this continent the trend of all material progress is north-westerly. The flow of immigration is north-westerly, and the Great Creator, as if to make way for the advance, has pushed back as it were, the cold of the Arctic nearer to the Pole, and spread out the vast fertile belt of the North Temperate Zone from the Great Lakes to the Mackenzie River; so that may not this England of the New World yet become to the Western Hemisphere all that the England of the Old World is to the Eastern?

For many years Canada has held an obscure place among the countries of the globe. Our borders have been pictured as the abode of perpetual snows, and our people as indifferent, easy-going, indolent. But a change is taking place. The narrow, little, rugged country on the margins of the St Lawrence has extended its borders from Atlantic to Pacific, and to the Arctic Circle of the north; the harvest-patches of Western Ontario, once the pride of United Canada, have blossomed into boundless fertile prairies, stretching away toward the setting sun, and pushing their golden fields far above the fifty-fifth parallel. With these changes have arisen national questions of trans-Pacific and transcontinental trade, and Canada is putting on the garments of preparation to enter the race of nations.[49]

Humboldtian science emphasized the value of comparison, continuity, and systematic co-operation. As a result, earlier work in these sciences was perceived as 'for the most part desultory, independent,' and indeed 'worthless.' The spirit of the Humboldtian method infused even politics with its practical corollaries of standardization, universalization, and centralization, all in the name of efficiency and progress. British North American legislative assemblies, normally bitterly divided by the dichotomy between expenditure and retrenchment, were convinced by the 1850s of the provincial and even national importance of supporting observatories and meteorological stations. Humboldt's chain of observatories of the 1830s had led, in fact, to a proposal during the 1850s for a 'Meteorological Confederation' over the whole of Europe. In a similar way Canadians, especially in Canada West, began to be conditioned during the 1850s to political union as a natural corollary of growing scientific, economic, and technological links among the colonies. In the absence of 'a positive agitation' on the subject of a union of the British North American colonies, wrote the Toronto *British Colonist* in 1853, 'the public feeling in relation to it is not dead, but only slumbers': 'We believe the feeling of Canada, and especially of Upper Canada, is strongly in favour of it, and only wants a fitting opportunity to manifest itself, in a very decided form.'[50]

Terrestrial magnetism and meteorology did not originate the idea of a political union to form a community of Canadian people. But they helped create the intellectual climate which made such ideas appear sensible and perhaps even inevitable. 'Such an establishment,' said W.H. Draper of the Toronto observatory in 1858, 'is worthy of the rising character of this fast-growing community, and affords to foreign countries one of the best proofs of our real advancement ... It is a thing of a world-wide character, designed to co-operate with all other nations engaged in similar researches ... which seeks to benefit as well future generations as our own.'[51]

Two years later Daniel Wilson observed that Canada possessed at least two institutions worthy of the inherited duties of 'a people sprung from the old stock that gave a Bacon and a Newton to the world.' More important than material triumphs as the highest and most enduring moments of a nation's progress were 'those two great departments of scientific labour on which this Province has hitherto chiefly concentrated its intellectual energies: the Geological Survey and the Magnetic Observatory,' in relation to which Montreal and Toronto were 'named with pride wherever science is cultivated and knowledge revered.' An 'indefatigable Provincial phalanx of workers,' continued Wilson, looked through geology with a 'wise retrospection' into Canada's distant past, and from the vantage-point of terrestrial magnetism and meteorology into 'the unseen truths of a great future.'[52]

Magnetic and meteorological inventory, of course, could not create the Canadian community which some people hoped it would help to do. It was a means by which those like Egerton Ryerson attempted 'to make ourselves and sister provinces a good deal more respectable in size than we have hitherto been made to appear' on any map. By the 1860s a scheme like Confederation was unthinkable without the aid of science: 'For the prospect before us,' declared Oliver Mowat in 1865, 'increases immensely the importance of every agency that is fitted to advance the reputation or mould the character of the people of this new nation.[53]

PART III

BOTANY

10

Adventitious Roots

> Where vast Ontario rolls his brineless tides,
> And feeds the trackless forests on his sides,
> Fair Cassia trembling hears the howling woods,
> And trusts her tawny children to the floods.
>
> Cinctured with gold while ten fond brothers stand,
> And guard the beauty on her native land,
> Soft breathes the gale, the current gently moves,
> And bears to Norway's coasts her infant-loves.
>
> Erasmus Darwin, *The Botanic Garden* (1791)

Closely related to the study of the soil and climate of British North America was the study of its vegetation. Erasmus Darwin's widely read scientific poem intertwined three main functions served by the botanical inventory of Victorian Canada two generations later. First, Darwin explained the prevailing system of plant taxonomy. He located *Cassia*, a flowering plant, in the Linnaean system of classification according to the number of stamens. Second, he recognized the growing relationship between botany and geography. *Cassia*'s emigrant seeds, given up to the currents of the St Lawrence and carried across the Atlantic to Norway, highlighted the geographical distribution of plants, an object of intense interest to Victorian botanists. Third, Darwin foreshadowed Canada's increasing prominence in the development of Victorian botany. Like geomagnetism and meteorology, botany came from sources outside British North America to take root in the great northerly expanse of the territories.

Botany flourished with the participation of British North Americans in the investigation of the flora around them. In 1836 Catharine Parr Traill, a Canadian amateur botanist, wondered whether the country's endless forest could promise its settlers a secure future. The same thought must have crossed the minds of many immigrants to Canada during the early nineteenth century. For this reason, botany could be considered the most fundamental of the four main inventory sciences, for it could shed important light upon the agricultural capabilities of the new land, adding new evidence to theories proffered by geology and meteorology regarding soil and climate. A science that identified and cultivated plants in inclement northerly latitudes was of obvious utilitarian importance to an incipient agricultural economy. Botany became an important auxiliary to Canadian development.

Like the geophysical sciences centred at Toronto, botany was organized around close links with British imperial interests. Imperial botany found a counterpart to Edward Sabine, co-ordinator of the magnetic observatories, in Sir William Jackson Hooker, director of the Royal Botanic Gardens at Kew after 1840. Tropical regions of the earth may have been valued for the lush exotic vegetation which was both attractive (rhododendrons) and useful (cinchona, cacao, coffee), but Canada held attention as a source of the white pine most suitable for masts on the ships which carried these products to the mother country.[1]

Developments in botany during the Victorian period influenced the ideas of Canadians as well as of imperial botanists. They added to the stock of theoretical and practical knowledge of northern vegetation, important for the future development of British North America. In addition to scientific knowledge, increased awareness of British North America's position as a northern territory propagated a whole northern mythology imbued with botanical metaphors and analogies. Canadian nationalist ideology, clothed in an aura of scientific authority, served up an enticing vision that embraced all of British North America and encouraged Canadians to reconsider their place among the nations of the world.

No obvious institutional focus for a botanical inventory comparable to the geological, geomagnetic, and meteorological surveys presented itself readily. The idea of a botanical survey of Canada did not attain public support before the Geological Survey of Canada was expanded to include natural history in 1877. Botanical inventory did not earn the international prestige reaped by the GSC or the Toronto observatory. Nor were its major achievements embodied in the career of a single pioneering botanist. Only James Barnston and George Lawson held promise in the way that William Logan and Henry Lefroy personified their respective fields. Instead, theoretical and practical complications discouraged a centralized botanical inventory of Canada. Botany everywhere suffered from the difficulty of establishing a satisfactory system of classification with which to organize and comprehend specimens. The ensuing confusion invited adherents of different systems to undervalue or even disparage one another's work. Local rivalries, personal animosities, and untimely departures, including those of Barnston and Lawson, combined further to scatter efforts in Canadian botanical research.

Yet despite fundamental organizational difficulties, botany shared in ideological currents that informed the other inventory sciences. At the same time, its peculiar permutations of McGee's 'construction, extension, permanence, grandeur and historical renown' underscored the uniqueness of botany's intellectual contribution. Before this contribution can be considered, it is necessary to sketch important trends in botanical science to the early nineteenth century. A distinctive feature from the outset was the relative lateness of botany's emergence as a modern science. Geologists operated within the Lyellian framework, and geomagnetists and meteorologists within the Newtonian; botanists and other naturalists were touched by developments in both, but by the mid-nineteenth century they were still groping for a solid theoretical foundation.

I

Plants have been valued from ancient times as sources of food, medicine, and raw materials, giving botany strong utilitarian roots. By the seventeenth century the discovery of the New World and the influence of the Reformation broadened the conceptual horizons of collectors, encouraging them to seek out plants other than those already known in classical herbal descriptions. Yet botany long remained mainly a branch of medicine, with practical utility its essential organizing principle.[2]

By the nineteenth century this strict utilitarian attitude towards the natural world changed both in England and on the Continent. By 1826 botany could accordingly be subdivided into three branches. Physiological botany studied the internal structures and processes of plants, making great strides in plant histology with J.M. Schleiden (1804–81) and T.A.H. Schwann's (1810–82) discovery of the cell by the late 1830s. Systematical botany compared external structures of plants, distinguished species, and arranged them into successively larger categories of genera, orders, and classes. Economic botany studied uses of plants, either by observation or by deduction from those already known.[3] British botanists, with vast uncharted colonial territories and a strong Baconian belief in the power of science to improve the quality of life, pursued mainly systematical and economic botany.

Systematists' thorniest problem was to devise a universal system for classifying specimens. A major issue was the basis for such a system. Botanical taxonomy received its classical treatment from Carolus Linnaeus (1707–78), the

Swedish botanist whose binomial Latin nomenclature classified specimens according to genera and species. For Linnaeus, plant classification was botany's only real task. His main interest, flowering plants (phanerogams), formed the core of his classificatory system. The Linnaean system grouped plants by the number and arrangement of their pistils and stamens; as a result, large numbers of non-flowering plants (cryptogams, so named because their method of reproduction was still a mystery), including ferns, algae, lichens, mosses, and fungi, were lumped together in one large undifferentiated group. But the Linnaean system had definite advantages. Whereas traditional herbal catalogues merely listed plants unsystematically in alphabetical order, Linnaeus ordered them as in a city directory. The simplicity which permitted anyone to categorize and identify virtually any specimen gained the Linnaean system international acceptance for a century after his *Systema naturae* was published in 1735. The system was particularly popular in Britain, with its strong amateur naturalist tradition.[4]

Linnaeus admitted that the basis of his system was artificial. It revealed no fundamental relationships among species, since any other feature might as well have replaced the reproductive parts as the classificatory key. Linnaeus agreed that a more natural classification expressing degrees of interrelationship was needed. But such a system, he was convinced, was not yet possible given the level of knowledge of plant physiology and function. In this sense the Linnaean system marked less the beginning of a new era in botany than the 'tidying up' of an old one. Both Linnaeus and his botanical precursors drew upon Aristotelian principles in comparing natural with logical forms. These borrowed assumptions were reflected in the close structural analogy between the Linnaean system and formal class logic.[5]

The elusiveness of the key to a natural system of classification did not prevent contemporaries of Linnaeus from insisting upon the inadequacy of anything less. Linnaeus's popularity in England, which inspired disciples to collect, describe, and name specimens from all over the world, contrasted with a strong reaction against artificial systems in France. This reaction was led by the Comte de Buffon (1707–88), whose multivolume *Histoire naturelle* did much to shape public understanding of the natural world for the century after 1749. The reaction against Linnaeus followed John Locke's philosophical attacks against the Aristotelian tradition in classification. Locke opposed the arbitrary selection of certain characteristics as indicative of natural affinities.[6]

In botany this reaction found expression in the revolutionary work of Bernard de Jussieu (1699–1777) and his nephew Antoine Laurent de Jussieu (1748–1836). Their *Genera plantarum*, the first workable example of a natural system of botanical classification, appeared in 1789. The work made a great conceptual leap in classifying plants according to the totality of their organization. The Jussieus began with the lower, more simple plant forms, the cryptogams, and worked upwards to the phanerogams. This upward approach permitted them to discern a unity of plan in the structures of plants which Linnaeus had not recognized.[7]

The Jussieus' achievement was furthered by the publication in 1813 of *Théorie élémentaire de la botanique* by the Swiss Augustin-Pyramus de Candolle. De Candolle (1778–1841) established the principle that comparative plant morphology, the study of symmetry in plant forms and structures, provided a more complete key to taxonomy than Linnaeus's reliance upon numbers of reproductive parts. This morphological principle formed the basis of plant taxonomy for the next half-century and indeed still influences contemporary systematics.[8]

The idea of symmetry in plants derived from current developments in crystallography, particularly the theories of Abbé René-Just Haüy. Haüy linked constancy in plant shapes to regular crystal shapes, for which he sought a natural law. After Haüy's *Traité de la minéralogie* was published in 1801, de Candolle sought a similar law for the plan underlying plant families. In de Candolle's mind, such a law would advance an artificial 'analytic' approach to classification, which emphasized differences, towards a more natural 'synthetic' approach, which sought resemblances among plant forms. By arranging plants according to the symmetry of their essential organs, de Candolle replaced Linnaeus's 'directory' with a weblike conception of plants reorganized into natural families. Still, no precise rules for determining the symmetry of plant forms were forthcoming. It was up to the botanist to study the plants' life history and to note the presence of organs and their position, number, size, shape, function, consistency, colour, odour, and taste.[9]

By the 1820s British botanists too realized that on the one hand no natural classification could yet identify an unknown plant as well as the Linnaean system could. But on the other hand, natural affinities could no longer be overlooked by botanists who strove to 'contemplate the Vegetable Kingdom with any degree of philosophical attention.' Indeed, as one observed in 1832: 'Systems, and branches of systems, sprang up over the whole of this ample field, each aspiring to eminence and distinction above its neighbours.' All 'armed themselves plentifully with thorns of offense, as well as defence, by which they hoped finally to prevail over their numerous competitors.'[10] De Candolle's natural system accommodated innumerable exotic relatives of European specimens which the Linnaean system had failed to anticipate. But by the 1830s theoretical botanists realized they still lacked sufficient knowledge to perfect this natural system, the highest goal of their science.[11] The difficulty, it seemed, could be overcome only with the accumulation of more information.

The rapid acquisition of new botanical information from all parts of the world promoted new approaches to the study of plants. In particular, the voyages of Capt. James Cook renewed interest in the geographical distribution of plants and in climatic influences upon vegetation. Arctic voyages financed by the British government in search of the North Pole in the early nineteenth century, including those with Edward Sabine as the scientist on board, returned with new and interesting specimens of Arctic flora.[12]

Reports of exploratory voyages formed a new body of scientific literature on plant geography. After a voyage to South America, Alexander von Humboldt and Aimé de Bonpland published their *Essai sur la géographie des plantes* in 1805. It was a pioneering attempt to typify plant forms in relation to climatic regions. Then in 1820 de Candolle's 'Essai élémentaire de géographie botanique' gave credence to Buffon's Law, which held that different species of the same plants populate even ecologically similar places. In an effort to explain this anomaly, Buffon compared the flora and fauna of the Old and New Worlds. He noted that northern animals in Eurasia and North America resembled one another more closely than did North and South American specimens. Buffon speculated that European animals had migrated over a land-bridge between Asia and North America. After some generations these migrants had 'degenerated,' or fallen away, from their earlier types into a new (and often smaller) mould. Buffon attributed these physical changes to climatic influences and concluded that conditions in the New World somehow hampered the physical development of Old World forms.[13]

Many North Americans were disturbed by the ominous implications of Buffon's speculations for North America's immigrant society, but Euro-centred responses concentrated more upon the dissimilarity of Old and New World species which Buffon had pointed out. Charles Lyell's *Principles of Geology* summed up the significance of Buffon's Law: he agreed with de Candolle that these geographical differences suggested the unicentric origins of species. His own uniformitarian theories supported the notion of species as historical entities: 'A species, like an individual,' he wrote 'cannot have two birth-places.' Each species had therefore spread from its place of origin to other geographical locations. Evidence compiled by Humboldt and de Candolle on the geographical distribution of plants elevated botany to a higher level than zoology in this regard, and Lyell was prepared to extend Buffon's Law to the entire natural world. The final proof, he believed, lay in the vegetation of northern regions. Scandinavian botanists had already scrutinized northern Eurasian plant geography, but little had yet been done in British North America, the other end of Buffon's hypothetical land-bridge.[14] In this way Buffon's Law gave an added fillip to botany in the northern reaches of the British North American territories.

By the early nineteenth century those who studied plants, whether their taxonomy or their geographical distribution, shared certain basic assumptions. First, species constituted the basic units of nature, despite the difficulty of distinguishing them from some varieties. Second, taxonomy and plant distribution rested upon the same classical epistemological foundation. The critical question for Linnaeus, as for de Candolle and Buffon, was to define a family of species, or genus, by grouping plants with obvious structural similarities and factoring others out. This continued emphasis upon similarities thus retained a link to Aristotelian essentialism.[15]

But the geographic approach of de Candolle and Buffon, so influential in Lyell's uniformitarian theory, introduced a counter-emphasis upon nature's continuity. The synthesis of differences in nature implicit in Lyell's *Principles of Geology* became the key to one main strand of British botanical thought during the nineteenth century, one focused upon the geographical distribution of plant forms. This geographic approach to botany 'took possession' of Victorian Canadians, shaping their perceptions and thereby their behaviour.[16]

II

While attempts to classify plants raised theoretical questions, economic botany promoted by the British government harked back to the science's utilitarian roots. Under the influence of Sir Joseph Banks (1743-1820), and later of Sir William Jackson Hooker (1785-1865) and his son Joseph Dalton Hooker (1817-1911), botany rose dramatically in economic importance to the British Empire. As a large part of that empire, British North America was destined to play a considerable role in this expansion of imperial botany.

Banks had been stimulated by Linnaean taxonomy to undertake botanical explorations farther afield; he became England's most prominent Linnaean and the system's most enthusiastic advocate. In 1766 he visited Newfoundland and Labrador aboard a fishery protection vessel and returned with the most comprehensive documented collection of botanical and zoological specimens ever brought from North America. The voyage enhanced Banks's enthusiasm for botanical exploration, and two years later he accompanied James Cook on his first voyage around the world. In 1772 Banks became scientific adviser to the Royal Botanic Gardens at Kew, and from 1778 until his death some forty years later he was president of the Royal Society of London.[17]

From this dual vantage-point Banks wielded enormous influence in the promotion and direction of botanical exploration. He encouraged the study and collection of economically useful plants from all corners of the empire. Banks became a virtual 'department of state,' and Kew Gardens 'the great exchange house' of the empire. Staff at Kew Gardens studied, acclimatized, and transferred exotic plants to plantations in other colonies. This imperial aspect of British botany had been evident long before Banks brought it to prominence and had focused largely upon North America. But the loss of the American colonies deflected attention away from plantations in the southern states to more equatorial climates.[18]

Besides applying plant taxonomy to record and control plants accumulated and redistributed within the empire, Banks encouraged agricultural improvement at home. The Industrial Revolution created demand for exotic vegetable products such as rubber and hemp and reinforced the Baconian idea, particularly in northern England and Scotland, that scientific principles could induce a parallel agricultural revolution. In 1799 Banks helped found the Royal Institution in London to pursue means by which such agricultural improvements might be achieved. The sciences of chemistry and botany were considered among the most important in the long run, but agricultural depression and the threat of Napoleon's blockade induced a short-term preference for economic solutions in the form of the imperial Corn Laws. Rapid developments in chemistry and botany after 1815 permitted the repeal of those same Corn Laws by the 1840s.[19]

The growing interest in scientific cultivation among the British landed gentry was not limited to the field, as botany flourished also in the garden. Cook's voyages opened British eyes to exotic fruits and ornamental plants which could be acclimatized and even improved in Great Britain. Plant nurseries burgeoned as collectors sold specimens to an English syndicate for commercial distribution. The Royal Horticultural Society of London, founded in 1804 by Joseph Banks and others, increased British stocks of fruits, vegetables, and ornamental plants. The society specialized in imports that could be grown in Britain, leaving larger-scale economic enterprises to Kew Gardens and theoretical problems to the Linnean Society, which Banks also helped found in 1788.[20]

Through the Royal Horticultural Society, plant and vegetable products of British North America became better appreciated in Britain. The society's influential secretary during the 1820s was Joseph Sabine, elder brother of Edward. In 1823 the society sent David Douglas, a Scottish gardener recommended by William Jackson Hooker, to collect fruit trees and other plants in North America. After visiting New York and Pennsylvania, Douglas made his way across Lake Erie to Amherstburg, where his brother had social connections dating from the War of 1812. From the Canadian side of the Detroit River

Douglas collected apples, pears, plums, peaches, and grapes surpassing in size and flavour any he had yet seen.[21]

Plant collectors like Douglas required settlements with transportation facilities to serve as bases in a new territory. Species were located in spring and summer, and seeds and roots harvested in autumn and winter. Douglas's laudatory report on orchards in the Western District of Upper Canada promoted an image of the south-western peninsula of the colony as a fertile Eden suitable for British immigration. His expedition was deemed such a success that the society underwrote a second journey by Douglas in 1825, via Hudson Bay to the mouth of the Columbia River.[22]

Douglas's travels as a plant collector exemplified the economic aspect of British botany in the 1820s, but many botanists and collectors had preceded him to British North America. The earliest came as European explorers, their activities reflecting needs of European botanical gardens rather than local needs. Later botanists lamented that early Canadian botany had depended for its development entirely upon 'aliens.' It was not until the 1850s that inhabitants of Canada began to take a major part in the development of the science.[23]

Yet it must be realized that the earliest history of botany in Canada antedated Linnaeus and helped lay the foundations of modern systematics. French botanical collectors sent by the Jardin des plantes at Paris were particularly important. Jacques Cornut published his *Canadensium plantarum* as early as 1635. Joseph Pitton de Tournefort (1656–1708) commissioned a number of collectors for the Jardin, the best known of whom was Dr Michael Sarrazin (1659–1734), the Médecin de Roi who died in an epidemic at Quebec. For despite the known mineral wealth of New France, plants were thought to constitute the country's true wealth and potential for settlement and trade. In addition to food, medicine, and raw materials extracted from plants in New France, the French sought valuable herbs. They believed that the colony's latitudinal and topographical similarities with Asian Tartary might make New France rich in ginseng. This herb, which grew wild near Montreal, sold for enormous profits to China until bad business methods collapsed the trade in the eighteenth century.[24]

Pioneer European botanists after Linnaeus investigated northern parts of North America to enrich taxonomy, plant geography, and horticulture. Peter Kalm, a Swedish student of Linnaeus, visited Canada in 1749 as the first field botanist to study the Canadian flora. In 1803 the Frenchman François André Michaux published his *Flora Boreali-Americana*, the first attempt to trace the northern tree-line. In 1814 Frederick Pursh from Saxony published his influential *Flora Americae septentrionalis*. Pursh came to Canada seeking ornamental plant specimens and was deeply impressed by the range of new species and varieties he discovered.[25]

The British too investigated the flora of North America. The Hudson's Bay Company had a long-standing tradition of supporting scientific investigations of the Royal Society of London. In hopes of discovering another lucrative vegetable like ginseng, the HBC instigated an inventory of plant resources, requiring company surgeons after 1730 to include botanical descriptions and specimens of native plants in their regular reports. Orders to collect every species of wild plant were reiterated after 1760 but with little response. Company factors were urged to create more permanent settlements around forts by planting large gardens for local consumption.[26]

Both the HBC and the North West Company engaged fur traders who were also amateur naturalists. Andrew Graham (173?–1815) was among a number of pioneers who collected and described botanical specimens. These they transmitted to Britain less for material gain than for the 'peace and rational pleasure of cultivating the mind.' Indeed, this phrase encapsulated the overriding eighteenth- and early nineteenth-century motivation for studying natural history. Officers in the fur trade habitually kept journals, cultivating powers of observation which were frequently turned to scientific benefit.[27]

The increased involvement of the British government in Arctic exploration after 1815 forged a further connection between the government's botanical advisers, William Jackson Hooker and Robert Brown, and botany in British North America. In 1817 John Goldie, a gardener at the botanical garden directed by Hooker at Glasgow, embarked on a botanical journey to North America from Montreal around Lake Ontario to Pittsburgh. Most of Goldie's patiently assembled collections perished at sea, but Hooker reported some of Goldie's discoveries to the Edinburgh Philosophical Society in 1822.[28]

In another widely reprinted paper given in 1825, Hooker emphasized the importance of pursuing botanical investigations in North America, which for its extent and interest, he believed, had 'scarcely any parallel in the world.' What rendered the botany of North America peculiarly interesting to the British naturalist, Hooker explained, was that a large proportion of its vegetable productions could be assimilated to Great Britain's climate. These productions included valuable oaks and firs, in which Hooker was particularly interested. He pointed out links between botanical products of the American Arctic and those at the summits of Scottish mountains, suggesting a subject in need of further study. Yet such scientific investigations could hardly be expected of Canada, he concluded reluctantly, since he believed a certain degree of 'political and mental improvement' to be the precondition to a proper appreciation of science. Still, many valuable specimens had been contributed from British North America to Hooker's renowned herbarium. He resolved that if no individual more competent to the task came forward, he would personally undertake to publish a flora of British North America.[29]

III

It was not only botanists outside of British North America who recognized the territory as a vast and unique botanical field. Three types of arguments preoccupied Canadians interested as well in a botanical inventory of the colony after 1820: intellectual, agricultural, and botanical.

The intellectual argument was best articulated by David Chisholme, editor of the *Canadian Review and Magazine* in Montreal, who also called for a geological survey of Canada. Chisholme believed it was high time to turn from sensational features of the Canadian landscape, 'the impressive grandeur of its scenery, the solitude of its woods, the roarings of its cataracts, and the unsuspicious security and freedom of its inhabitants,' to more legitimate objects of research. The sole object of investigation and description, Chisholme held, was a thorough inventory of the country, including its flora, in preparation for permanent settlement. Other writers reiterated the necessity of such an inventory among the first efforts of a country 'arising from a state of semi-barbarism into a comparative position in the scale of nations.' Canada was now 'verging from the dark and gothic gloom,' and should 'commence to examine into its interior, and find how far she is capable of being independent of other nations for her resources.'[30]

This sense of moving from darkness into light was to be taken literally to some extent. Life in the backwoods was widely seen as detrimental to the health and even to the physique of European immigrants, who were likened to plants stifled by forest overgrowth: 'Buried in the depth of a boundless forest, the breeze of health never reaches these poor wanderers; the bright prospect of distant hills, fading away into the semblance of clouds, never cheered their sight; they are tall and pale, like vegetables that grow in a vault, pining for light.' Receding forest and rising towns represented the physical and mental freedom to address oneself to questions larger than those of basic survival. Most settlers accepted without question the rule of thumb that a quick inventory of the types of plants and forest trees in a district could determine the correct choice for a fertile and prosperous homestead.[31]

The growth of towns permitted natural history societies like those at Quebec, Montreal, and Halifax to become focuses for local botanical inventories. An important inspiration for botanical collectors in the Literary and Historical Society of Quebec after 1820 was Lady Dalhousie, herself an amateur botanist. In Montreal the Natural History Society resolved in 1828 to establish a museum to display botanical specimens. In particular, a local herbarium initiated by Dr A.F. Holmes formed the core of the McGill University collections. Similarly in Nova Scotia, the few with leisure to pursue botany usually did so in the neighbourhood of larger towns like Halifax. The most extensive catalogue of Nova Scotia plants published before 1830 was assembled by Dr William Cochrane, vice-president and professor of languages at King's College, Windsor. Cochrane, a Linnaean systematist, inspired generations of students throughout the lower provinces with his passion for natural history.[32]

A second type of argumentation for a botanical inventory came from the movement for agricultural improvement. This movement was epitomized by John Young (1773–1837), a Scottish immigrant to Nova Scotia. In 1818 Young published a series of open letters to the Halifax *Acadian Recorder* under the pseudonym 'Agricola.' He aimed to draw public attention to the backward state of Nova Scotia agriculture, vulnerable to the vicissitudes of world trade after the peace of 1815. Agricola hoped to convince Nova Scotians that poverty was not inevitable and that grain could be successfully harvested in Nova Scotia. The fault, he countered, lay not in the soil and climate but in local farmers' lack of scientific knowledge.[33]

Young followed the example of Sir John Sinclair's movement for agricultural improvement in Britain. Sinclair, well-known author of the *Statistical Account of Scotland*, believed that the first key to improvement was the collection of useful information. His understanding of statistics was rooted in the etymological origins of that word: the collection of information useful to the state. The second key was the application of scientific principles to agriculture. Important parts of Sinclair's preliminary inventory were questionnaires sent to individual farmers and the subsequent founding of the Board of Agriculture. The duties of the board were to survey and report upon the farming of the country, to research applications of botany and chemistry to farming, and to educate the public in these applications.[34]

Like Sinclair, Young believed that land reclaimed from the forest was 'the Alpha and Omega of national grandeur and wealth.' A preliminary to determining the best use of this land was to appoint 'enlightened and scientific men' to prepare accounts of the various districts of the province and to collect evidence on the nature of the soil and the plants that suited it. An appointed Board of Agriculture, he assured readers, would then be able to 'overthrow every barrier to improvement.' Young attracted a circle of influential 'improving farmers,' including the science enthusiast Lord Dalhousie, who promoted progressive policies in the colony as its lieutenant-governor. Young then served as secretary to the new Board of Agriculture of Nova Scotia in 1819.[35]

Interest in agricultural improvement through inventory and scientific applications was not restricted to Nova Scotia during the 1820s. It flourished throughout British North America, stimulated by the changing position of the colonies within the empire after the Napoleonic Wars. A dominant economic factor in the British North American colonies before 1821 had been the sudden dependence of the British navy upon British North American timber during the Napoleonic blockade. Duties made permanent in 1816 gave British North American timber a preferential position in British markets. But after the war's end British merchants who depended on the former Baltic trade determined to regain their competitive edge. They secured a reduction of imperial preferences in 1821, widely seen to foreshadow the demise of the entire imperial tariff structure. Evidence of Britain's declining reliance on Canadian timber included the decision no longer to reserve all Canadian white pines for the Crown.[36] If Canadian forests were no longer vital to the empire, an obvious alternative was agriculture, since the colonial grain trade still enjoyed a British preference. In order to survive within the imperial economic system, it behooved Canadians to determine the extent and nature of their country's agricultural potential. Botany was soon called upon to aid in this enormous inventorial task.

A third type of argumentation was more strictly scientific. In 1833 William Jackson Hooker published the first volume of his promised *Flora Boreali-Americana; or, the Botany of the Northern Parts of British North America*. This event, more than any other, propelled botanical study in British North America to higher levels of sophistication. Assessed by one author as Hooker's greatest work, the book constituted the most complete summary of Canadian flora for decades to follow. Hooker's compendium was based mainly upon collections by John Richardson and Thomas Drummond on Sir John Franklin's first two polar expeditions (1819–22; 1825–7) and by David Douglas in travels across northwestern North America in 1825–6. It noted great strides made in Canadian botany and apprised an international audience that British North America had produced many native botanists. Hooker acknowledged and documented contributions by British North American collectors from Lake Huron to the Atlantic, including Lady Dalhousie and her social circle, William and Harriet Sheppard and Anne Marie Percival, all of Quebec; Dr A.F. Holmes and Messrs Cleghorn, Chandler, and Buckingham, all of Montreal; John Goldie and James McNab in Upper Canada; Dr Todd of the HBC, at Lake Huron; and a Mr Kendal of New Brunswick. Hooker's Glasgow herbarium became a major clearing-house for colonial collectors, a source of strength to those who would otherwise 'lose heart in the want of sympathy.'[37]

Nor was Hooker the only British botanist who recognized British North American collectors. Catharine Parr Traill's botanical activities gained an added sense of purpose when a professor at the University of Edinburgh – possibly

Robert Graham, who held the chair of botany from 1819 to 1845 – asked her to contribute native specimens to his collection. Immigrants like Traill who had not encountered botany through medical studies were acquainted with the amateur naturalist tradition as part of their cultural heritage. She discovered that the country opened a wide and fruitful field to the botanist. 'I now deeply regret,' she confessed, that 'I did not benefit by the frequent offers [my sister] made me of prosecuting a study which I once thought dry, but now regard as highly interesting; and the fertile source of mental enjoyment.' The latter point rang especially true to those who, living in the bush, were denied the pleasures of a large circle of friends and the distractions that a town or village offered. Traill discovered to her delight that one of her few neighbours, Frances Stewart, also relished botanical collecting. The only botanical guide available to them was Pursh's intimidating *Flora* with its Latin nomenclature. This difficulty inspired Traill to model her more popular Canadian works on Gilbert White's beloved *The Natural History of Selborne*.[38]

Both the Literary and Historical Society of Quebec and the Natural History Society of Montreal soon noted that many persons, 'natives or long resident in the Provinces,' had turned their attention to botanical pursuits. They strove to encourage 'one of the most generally acceptable of the natural sciences' by offering popular lectures on botany. These lectures increased popular awareness of native trees, shrubs, and flowers. The societies offered prizes for essays on the plants of Canada which they judged qualified to bear foreign scrutiny. A boost came unexpectedly in 1837 in the form of a dispatch from Lord Glenelg, the colonial secretary, to Sir Francis Bond Head. Glenelg requested that Canadian authorities accord the utmost consideration to a request from the British Museum for specimens of the colony's natural history.[39]

In addition to this imperial attention, there was a growing effort in the United States to collect and classify North American plants. These collections were based only upon type specimens held in American rather than European herbaria. John Torrey (1796–1873) and Asa Gray (1810–88) gained prominence as leading American botanical taxonomists in 1838 with the publication of the first instalment of their influential *Flora of North America*, which marked a shift in American botanical taxonomy from the Linnaean to the de Candollean system of classification. In British North America it attracted the attention of teachers of botany such as James Robb at the University of New Brunswick, who used Gray's *Elements of Botany* (1836) as a text. Torrey and Gray, like Hooker in Britain, acknowledged contributions by several Canadian collectors. Specimens from Montreal and Quebec City marked the northern limits of the flora under their scrutiny. But the interior of Upper Canada also attracted American botanists, who were convinced that its ancient forest had not changed 'since the deluge.' 'No straggling foreign plants or naturalized exotic,' enthused one American explorer of the Ottawa Valley, 'is ever found vegetating there.'[40]

Appreciation of British North America as a northerly botanical region offering unique glimpses into botanical history flourished during the 1830s for several reasons. First, Sir William Hooker intensified interest in cryptogams, which were known to inhabit the rugged Canadian Shield. Second, the availability of achromatic microscopes permitted even amateur botanists to specialize in these mysterious plant forms. Third, although scientific societies were fully occupied in collecting local specimens, they began seeking out plants from the Canadian interior. The NHSM looked to a government expedition to the St Maurice and Ottawa Rivers, and to the HBC, for specimens. It also invited essays specifically on the natural history of the Ottawa Valley.[41] For beyond the local botany of Montreal and the Eastern Townships, the society evinced interests more practical and expansive. It invited practical essays on the commercial and manufacturing value of Canadian forest trees, on hemp and flax for rope and linen, and on plants used for pigments. More far-reaching themes included changes in exotic plants cultivated in the northern parts of America and the comparative adaptation of prairie and forest lands to settlement in a new country.[42]

At York founders of both the Literary and Philosophical Society of Upper Canada and the Mechanics' Institute pursued the grandiose goals of 'investigating the natural and civil history of the Colony and the whole interior as far as the Pacific and Polar Seas.' More realistically, in 1835 Dr William Rees petitioned Lieutenant-Governor Sir John Colborne for assignment to the Baddeley-Carthew expedition north of Lake Huron. Rees hoped to explore the natural history of the country, especially its medical botany. Although his request was denied, Rees typified Upper Canadians who concentrated upon botany in order to discern Canada's aptitude for settlement and agriculture. John Rae, for example, included botany in his personal inventory of Canada during his eighteen-year sojourn in the colony. But like Rae's projected 'Outline of the Natural History and Statistics of Canada,' Rees's proposed medical botany of Canada was never published.[43]

By the time the second volume of Hooker's *Flora Boreali-Americana* appeared in 1840, British North Americans called for a closer investigation of plant life around them. What is more, the study of plants broadened the geographical horizons of even amateur botanists. A recent book of Nova Scotia wildflowers induced the *Quebec Mercury* to recommend it to all the British provinces, as the same flowers and plants seemed to be indigenous to the whole tract of country encompassed by them.[44]

IV

After 1830 botany provided much-needed information about British North America. First, newly founded botanical gardens facilitated the successful cultivation of fruits and vegetables. One such garden was established in Halifax after 1835. In Toronto Tiger Dunlop, Charles Fothergill, and William Rees petitioned the legislature in 1836 for a provincial natural history museum, including botanical and zoological gardens. Land near the military barracks was set aside for the project, which was never completed. In Bytown the *Gazette* advocated the establishment of an experimental farm in the Ottawa Valley. 'One well attested fact,' averred the editor, 'is in agriculture worth a thousand hypotheses.'[45]

Second, local horticultural and agricultural societies reoriented themselves to meet recent developments in botany and chemistry. Such organizations emphasized improvement of method and production rather than more traditional social interests. The Western District Agricultural and Horticultural Society was founded in 1837 to increase the district's agricultural production and efficiency. Montreal's Horticultural Society, founded during the eighteenth century, foundered and was re-established along similar lines during the 1840s. The Toronto Horticultural Society, founded in 1834, introduced and cultivated new varieties of garden plants and fruits. It found a generous patron in Sir John Colborne, who donated land surrounding Government House for a botanical garden. Much like its counterpart in Montreal, the Toronto society relied upon financial assistance from affluent citizens interested in cultivating their park lots as fruit and flower gardens. These same people could afford to indulge in the fad of building private greenhouses to introduce more exotic plant species.[46]

The Toronto Horticultural Society was welcomed by still others who believed it could serve a deeper social purpose. Thomas Dalton, editor of the Toronto *Patriot and Farmers' Monitor*, suggested that the society would render temperance meetings superfluous. Drawing from the experience of industrializing cities such as his native Birmingham, Dalton declared optimistically that horticulture could keep the working classes out of alehouses. 'No temptation,' he held, 'could lure to the Tavern, while the prize cauliflower, the giant asparagus, the huge gooseberry, the gay tulip, or fragrant carnation demanded attention.'[47]

Dalton no doubt overstated his case. But a relationship between botany and the idea of progress in Canada transgressed well beyond the bounds of the Toronto Horticultural Society. In 1834 Dalton published a series of letters on gardening by Alexander Gordon of Pickering. Gordon extolled the progress he witnessed around him. In the settled parts of Canada, he observed, the pioneer era was drawing to a close. To that era, Gordon reasoned, the idea of improvement naturally succeeded. Where, he asked rhetorically, 'can the results shine more conspicuously than in the effects which art will realize in gardening operations?' Gordon noted benefits realized by artificial means to improve fruits to the point where they no longer resembled the original. Plums from the 'savage sloe'; apples in hundreds of varieties from the 'worthless crab'; pears from a 'hard, stony, astringent fruit'; all represented clear 'cases of improvement, resulting from time and skill, patiently and constantly in action.'

Gordon hoped ultimately 'to create a desire in the minds of *all gardeners* for exploring and combining the scientific with the practical' and to direct attention to auxiliary sciences that could 'accelerate their knowledge on the just principles of gardening.' To this end he urged farmers and gardeners to recognize botany as 'one of the most useful, interesting, and instructive sciences extant' and, moreover, as 'of the most essential service to the practical gardener.' If a gardener had no knowledge of botany, Gordon warned, he had 'a great barrier between him and all real improvement, which he cannot surmount without exploring the depths of that science.'[48]

The question of botany as a means of improvement in British North America reverberated in issues other than increasing production. More than just the improvement of plant varieties, some Canadian cultivators believed they witnessed the transmutation of species in their own wheat fields. One backwoodsman wrote to the *British American Journal* in 1834 that more than fifteen years previously, he had made 'the great discovery' that chess, a kind of grass, could be cultivated to produce wheat, and vice versa. The farmer supported his claim by appealing to the Mosaic account of wheat as a grass cultivated by man to produce grain. The issue persisted for more than a decade as other farmers insisted that through proper techniques in cultivation oats, wheat, and barley could be transformed into one another. Thomas Dalton addressed the question in an editorial, challenging botanists to investigate the farmers' hypothesis. At best, he prodded, they could discover that 'other kinds of valuable grain may still be hid in the forms of grasses.'[49]

What Dalton left unsaid was that, at worst, the farmers' claims were considered by religious orthodoxy to be heretical. The increasing manipulation of nature through cross-fertilization and artificial selection to improve plant varieties, and even to create new ones, left some amateur gardeners, especially clergymen, uneasy about the implication of this process for the sacred dogma of the immutability of species. Many took comfort in the fact that man-made varieties inevitably reverted to their uncultivated states if left to themselves. The very limits indicated by such changes, declared Charles Lyell in 1833, demonstrated 'fixed and invariable relations between the physiological peculiarities of the plant, and the influence of certain external agents.' They afforded 'no ground for questioning the instability of species, but rather the contrary.'[50]

There was as yet no theorist in Upper Canada to whom such claims could be referred. Before 1853 they might have been handled by Henry Croft, professor of chemistry at University College, who lectured to the public on Canadian soils and plants, or by J.B. Hurlburt, who taught science at Victoria College and wrote scientific articles for newspapers. But even they could only urge farmers to reconsider the premises of their argument. The editor of the *British American Cultivator* advised the errant farmers to abandon 'the false assumption' that 'the change of wheat was possible'; for their forefathers too saw ghosts 'where only natural appearances present themselves.' The need for scientific authority in this and more practical agricultural questions found expression in petitions to the Canadian legislature for a chair of agriculture at King's College, Toronto, and for model farms in each district.[51]

V

In the meantime, authority rested by default on a third means of gathering and disseminating botanical information, the agricultural press. Improving farmers who immigrated from Britain during the 1830s continued the tradition by using newspapers such as the *British American Cultivator* to foment an 'intellectual revolution' among Canadian farmers. Farmers were to be educated not only to respect agriculture as a science but also to appreciate natural science, including botany, as its handmaid. Fully in the tradition of Sir John Sinclair and 'Agricola,' the *Cultivator* named an important preliminary to developing the agricultural resources of British America. This prerequisite was to employ competent persons to collect information from the most successful farmers. Even the varieties of vegetables and fruit cultivated would be noted in such a statistical survey: 'What we want in this country,' urged the *Cultivator*, 'is to learn how, in the shortest time and at the least expense, to produce the greatest quantity of food and other necessaries of life ... without permanent injury to the soil.'[53]

A twofold agricultural crisis during the 1840s stimulated Canadian interest in agricultural botany. The first half of this crisis involved vegetable pathology, the onset and rapid advance of plant disease throughout the province. Potato blight reached Canada West in 1843 and ruined crops repeatedly until the 1860s. Called the greatest calamity of the age for labouring classes, the potato murrain, which devastated populations on both sides of the Atlantic, was all the more frustrating because its cause eluded even scientists. Meanwhile H.Y. Hind and others in Canada searched both the scientific literature and the results of their own experiments for potato substitutes.[54]

The devout interpreted the potato blight as just punishment of those who had opened nature's Pandora's box with the keys of overcultivation and artificial selection. Experiments showed, warned one farmer, that it was not the potato alone, but all seeds, grains, and grasses that had 'run out.' These fears had a solid basis; a lecture to the Royal Institution of London attributed the blight to selective overcultivation: 'While we almost entirely suppressed certain properties in plants cultivated, and encouraged others to grow to excess, we impaired the vigour of the plants' constitution, and thereby rendered them liable to disease.' After years of intensive yet fruitless investigation, British botanists still favoured inventory techniques to collect systematic observations. They sought further clues by circulating printed questionnaires to determine the parameters of the problem more accurately.[55]

The second half of the agricultural crisis followed the repeal of the imperial Corn Laws in 1846 and was less economic than psychological in its effects. While not as immediately devastating to farmers as to millers and forwarders, the repeal led Canadian farmers to question the future of their agricultural economy and with it their role in the British Empire. Some saw the repeal as a 'breach of faith,' since the Corn Laws had given 'ten-fold strength to the connection of Canada with the Mother Country.' Now other means would have to be adopted to keep Canadians abreast of 'the age of improvement and general civilization.' It was high time, argued the *British American Cultivator*, for Canadians to forge their own economic destiny. Imperial trade preference had given Canadians practical reasons to refrain from 'taking any decided action, or adopting enlarged and liberal views in encouraging manufacturing enterprises, and in developing the abundant resources of wealth in which this country abounds.' As a result, the inhabitants of British North America possessed 'a small share of national feeling, – their education and habits have engrafted upon their minds a strong prejudice to any innovation upon their early prejudices, – and any movement approaching to an enterprise, would at the outset be met with terms of disapprobation.'[56] One remedy was to diversify from wheat production to other important cash crops, such as fruit, to diminish Canada's reliance upon Britain and the United States. Domestic industry could also be strengthened by cultivating hemp and flax as lucrative sources of cordage, linen, and linseed oil. In certain parts of Canada soil and climate so favoured hemp that it grew 'wild and self-sown, so as to become a troublesome weed.' An agricultural and botanical inventory could help to pinpoint such locations.[57]

Threats posed by epidemic vegetable diseases and infestations and by the repeal of the Corn Laws generated considerable public discussion. Agricultural associations in both Upper and Lower Canada, uniting local societies under larger sectional umbrella organizations, were incorporated in 1847. Their purpose was to provide expert information about the progress of agriculture in the province. The idea that such provincial agricultural associations 'should be truly national in character and all [their] bearings' was reflected in their peripatetic meetings and marked a distinct change from earlier local attitudes. Lord Durham had noted in 1839 that Upper Canada lacked a great centre linking all the separate parts in sentiment and action. Nor did he find 'that habitual intercourse between the inhabitants of different parts of the country' which made a people 'one and united, in spite of extent of territory and dispersion of population.' On the contrary, he cited striking examples where attempts to acquire agricultural information in one district about another were answered not only with 'very gross attempts' to deceive, but also with wildly false information given in perfectly good faith. The provincial agricultural associations were intended to remedy these problems, first by a preliminary scientific stock-taking of agricultural capabilities. The rewards, urged Lord Elgin in 1850, would be more immediate and abundant in Canada because the province enjoyed access not only to the capital of older countries but also to their 'treasures of knowledge.' 'When the nations of Europe were young,' he reasoned, 'science was in its infancy.' But Canada sprang from the cradle 'into full possession of the privileges of manhood.' On utilitarian grounds alone, agreed the *Cultivator*, the mind of the agriculturalist 'should be irradiated with the beams of science,' including botany, through 'an entire new order of practical and scientific education' in the rural districts.[58]

By mid-century the importance of botany and chemistry to agriculture, and of agriculture to Canada's future, was reiterated in the *Canadian Agriculturist*, founded in Toronto in 1849. Owned and edited by George Buckland and William McDougall, the paper continued in the tradition of the *British American Cultivator*. Buckland (1804–85) was a prominent English agriculturalist who toured the United States and Canada during the early 1840s. In 1847 he emigrated to Toronto in the hope of becoming professor of agriculture at King's College. McDougall (1822–1905) was born at York, Upper Canada, the descendant of Scottish Loyalists. Raised in Vaughan Township and educated at Victoria College, Cobourg, McDougall studied law and was called to the bar in 1847.

While continuing to emphasize the importance of science to agricultural improvement, Buckland and McDougall drew out the importance and pleasure of studying natural history. As the 'alphabet and grammar of the language of nature,' natural history entailed 'not just hard names and intricate classification.' Classification merely provided a means to appreciate 'the beautiful adaptation of parts; and of the whole to the place it has to occupy,' both ecologically and taxonomically in the general scale of being. Just as natural science aided farmers, the reverse was also true. For it was 'scarcely out of the power of any one to contribute something to the general stock of knowledge.' Agriculture, like natural history, advanced first through government-supported fact-gathering surveys, like that conducted in New Brunswick and Nova Scotia by the Scottish agricultural chemist J.F.W. Johnston. Failing such a survey, no opportunity to become acquainted with native plants should be foregone, urged Buckland and McDougall. For with botany came knowledge, power, and enjoyment.[59]

The *Canadian Agriculturist*, 'exclusively devoted to Agriculture and Science,' purported to transcend party politics. The word 'Canadian' in its title expressed 'our desire that the work should assume a distinct and *national* character.' The editors promised 'to do everything in our power for the agricultural advancement of our own country. Canada and Canadian interests, so far as they are comprised within the legitimate and professed objects of our paper, shall have our first consideration.' It was, they suggested, non-partisan to improve the process of colonization, 'the new impulse reserved for our century' and 'the pioneer of civilization.' The applications of science to solve modern problems 'all converged to one high purpose, the mastery of the Globe.'[60]

By the 1840s interest in the botany of Canada converged from several different directions and ranged from ostensibly detached scientific observation to the determined species imperialism practised by immigrant farmers.[61] Although the participation of the Canadian public was still fragmented, these various outlooks were linked by at least one common aim. That aim was to gather information about the vegetable products and potential of British North America. Whereas before 1830 scarcely anyone perceived the necessity of a botanical inventory of Canada, by mid-century this had changed. The seeds of early botanical investigation had fallen on fertile soil and sprouted interesting intellectual plantlets, not all of them on the progressive edge of botanical thought.

11

The Metamorphosed Leaf

> It is that same organ of the plant, the leaf, which unfolds along the stem to assume its various forms: expanding into petals, retracting into reproductive parts, and blossoming once more to shape the fruit.
>
> Goethe, *Metamorphosis of Plants* (1790)

By the 1840s botany had essentially emerged from two different approaches to the study of plants and plant life: the one taxonomic, attempting to classify members of the vegetable kingdom systematically; and the other geographic, attempting to explain patterns of plant distribution over the earth's surface. The two approaches reflected 'morphological' and 'genetic' views of nature. While the morphological view visualized nature as static, seeking what was typical in groups of plants, the genetic view sought to capture nature's dynamism by studying change in those same plants. The one, like geometry, studied similarity through form; the other, like calculus, studied similarity through relationship. The more dominant morphological view was gradually superseded by the genetic, culminating in Charles Darwin's *On the Origin of Species by Means of Natural Selection* in 1859. Darwin anchored the most compelling aspects of his theory, the geographic evidence, firmly in accepted botanical tradition.[1] As a result, his theory found early support from botanists, especially J.D. Hooker in Britain and Asa Gray in the United States. Among botanists in British North America, the theory evoked distress only for those who, often for religious reasons, adhered exclusively to the morphological tradition. Those whose work was guided by the study of plant geographical distribution merely resolved more dispassionately to seek out more evidence one way or the other.

J.H. Lefroy was not alone in his observation during the early 1850s that southern Canada, particularly around Toronto, offered few new challenges to natural historians of either persuasion. Botany, like geology, attracted more collectors along the lower St Lawrence in the older garrison towns of Montreal and Quebec and along the Ottawa Valley, where British officers were stationed. Plants in the flat lands around Toronto were not significantly different from specimens already collected on either side of the border in that latitude. Catharine Parr Traill recognized that if Canadian violets lacked the fragrance of European varieties, the greater range of their colours in Canada more than compensated for this deficiency. Yet some Canadian plants still appeared as poorer versions fo those known in Britain. Lefroy believed that the collection and documentation of the entire flora of the region could be accomplished in a year or two. For this reason, he wrote to W.J. Hooker in 1852, he found it strange that Toronto, with its twenty-seven thousand inhabitants, did not yet boast of a society 'at whose reunion one may successfully start an enquiry of this kind ... People have no leisure in a young country,' he concluded ruefully, 'except to quarrel.'[2]

Although Canadian botany before 1850 could aptly be characterized as 'just the study of floras and the classification of plants in accordance with then prevalent methods,' theoretical advances and new approaches after mid-century turned the attention of botanists increasingly towards British North America. First, post-Linnaean systematics had exhausted itself in the classifying mania generated by de Jussieu and de Candolle. Botany, boasted Edward Forbes, professor of botany at King's College, London, in 1843, had become 'the most advanced and the most pursued' of all the natural sciences. Forbes (1815–54) led a vanguard of naturalists who sought to establish laws for the living realm of nature. These 'philosophical naturalists' practised Humboldtian science by seeking patterns in the geographical distribution of species. A hitherto unrecorded locality of a species became as great an addition to science as the discovery of a new species.[3] Biogeography incorporated information assembled by both botanical and zoological collectors, but for obvious reasons it was much easier to work with plants than with animals in this geographical approach to natural history.

Second, this emphasis upon phytogeography entailed increasing scientific interest in northern vegetation. Joseph Dalton Hooker noted that the northern hemisphere supported two distinct arctic floras, the one American and the other European. The most abrupt line of demarcation was due north of Canada on the shores of Baffin Bay and Davis Strait. Hooker, like Charles Lyell before him, expected great discoveries from his North American botanical colleagues.[4]

Third, the growing botanical interest in the distribution of plants in space was accompanied by a similar interest in their distribution over time. Edward Forbes and others studied fossils to theorize on species distribution as palaeobotany, the study of the geological distribution of fossil plants, gained recognition as a new branch of science. In 1846 Sir Henry De la Beche appointed Joseph Dalton Hooker to relate the flora of Great Britain to the country's geological structure, particularly to fossil ferns found in coal-beds. In North America J.W. Dawson and the staff of the Geological Survey of Canada made original contributions to this new merger of botany and geology. Dawson drew world-wide attention to British North America as a source of palaeobotanical information and deepened the understanding of botanical history during the Carboniferous and Devonian periods.[5]

Fourth, North American botanists advanced their own investigations by leaps and bounds. British North American botanists anticipated the completion of Torrey and Gray's *Flora of North America*, organized on the de Candollean system, as 'every thing that could be desired' to help further their own studies. As fascicles of the *Flora* became available, they were sought after as far northwest as the HBC territories as an organizational framework for growing numbers of private herbaria. Gray's *Manual of the Botany of the Northern United States* (1848) became 'almost as familiar as the Bible, Shakespeare, or the cookbook' to educated North Americans of succeeding generations.[6] Yet since Torrey and Gray included no specimens from north of Quebec City, it was only a matter of time before British North Americans would discover gaps and call for a flora of species in their own collections not found in Torrey and Gray.

The new geographical and geological dimensions of botany revitalized the science after mid-century. Plant geography posed new questions and highlighted the north as a source of new botanical information. It also offered new premises from which Canadians envisioned their destiny as a transcontinental nation. But the complexity of developments in botany, involved as it was in the quest for the origins of species, implied that the fate of such ideas would not be any less tortuous.

I

During the 1850s botany became a teaching subject at Canadian universities. With obvious practical applications to agriculture, horticulture, medicine, and industry, it had long attracted attention. But the predominance of agriculture in the Canadian economy while the nature of a provincial university was being hotly debated lent a unique twist to the relationship between agriculture and natural history. Agriculture and botany were divided into two academic chairs at the University of Toronto. This separation of 'applied' from 'pure' botany differed from the structure of the discipline at McGill and Queen's, where botany embraced an agricultural program on the model of Edinburgh and Yale.[7]

Agricultural education at Toronto was further fragmented by Egerton Ryerson's conviction that agriculture and botany should rather be taught at the provincial normal and grammar schools. To this end he set aside three acres on the normal-school grounds for agricultural experiments and two acres for a botanical garden. Ryerson intended the normal school, with its contemplated educational museum, to serve as an agricultural college, as well as to instil a taste among Canadians for natural history. At the same time it was to be 'as Canadian as possible,' in order to acquaint the people of Upper Canada with 'what is curious and beautiful, ingenious and valuable in their own country, and to aid them in turning it to account.' Even the school's ornamental plants and shrubs served twin purposes of utility and national pride. The grounds included a botanical, fruit, and vegetable garden, a rotation grain and grass farm, as well as a small arboretum that distinguished native from foreign specimens. These would, it was hoped, become identified 'with the Country at large, and such as the people may especially call their own.'[8]

Unlike Ryerson, Robert Baldwin and others favoured the establishment of a professorship of agriculture at the University of Toronto. Two chairs, one in agriculture and the other in geology and mineralogy, were guaranteed by university statute in 1844. Additions to the Faculty of Arts in 1851 included a professorship of natural history (botany and zoology). In contrast, the curricular design of Queen's College at Kingston in 1842 followed Robert McGill's prescription that botany and agriculture should constitute one chair, and natural history and its cognates another.[9]

The appointment of George Buckland as professor of agriculture at the University of Toronto in 1852 formed part of a larger plan for scientific agriculture in Canada West. In 1850 the Agricultural Association of Upper Canada advocated the creation of the Board of Agriculture, a ten-member council including the inspector general and the future professor of agriculture. The resulting legislation (13–14 Vict., c. 73), written by William McDougall, envisioned the board as a popular responsible body with powers to give grants to local societies, to plan experimental farms, and to promote agriculture education.

The board's most important task was to take inventory. In an age of increasing competition, argued Buckland and McDougall, it would not do for Canadians 'to fold their arms in listlessness, and to stand still' while the rest of the world moved rapidly ahead. 'Not a moment ought to be lost. We must be up and doing; bringing willingly to our aid whatever science or experience can suggest for increasing the fertility of our fields, and for developing those great natural resources of wealth and enjoyment, which a bountiful Providence has placed within our reach.'[10] An experimental farm would conduct part of the inventory of Canada by testing the adaptability of its diverse soils and climates to various crops. An agricultural museum would provide a permanent record of this inventory by displaying roots and plants, both wild and cultivated, to illustrate the geology and botany of the country. The professorship of agriculture, in conjunction with the Board of Agriculture, was supposed to coalesce the divided Canadian population by highlighting 'the untold blessings' in store for Canada whenever 'an enlightened, united and vigorous national feeling' supplanted the 'distrust and jealousy' engendered by antipathies of race and party. After the creation of the Board of Agriculture, which Buckland served as secretary, the agricultural association strongly recommended him for the professorship of agriculture. Buckland's only rival, H.Y. Hind, was asked by the selection committee to withdraw in deference to Buckland's qualifications, which they deemed 'much superior both in number and value.'[11]

Buckland failed to attract many students to his agricultural program at the University of Toronto. Of six who registered in 1853, only one or two graduated, and by 1855 the program was 'almost tenantless.' Part of the explanation was that many farmers still evinced a hostility to 'book-farming.' The charge that agricultural chemistry had 'placed nothing within the reach of the everyday farmer' had some validity, since the science could offer no solutions for major problems like rusts, blights, smuts, and mildews. In addition, Buckland was not a good professor; he was too preoccupied as a journalist, farmer, and administrator to spend enough time at the University of Toronto. Buckland's agricultural tours of inspection offered no miracles, but neither could he convert farmers without something positive to compensate his call for patience, practice, and observation before science produced results. By 1860 a legislative select committee on the University of Toronto ran out of patience and recommended the chair of agriculture be discontinued after Buckland's tenure.[12] But this defeat did not end the study of botany at Toronto.

In 1852 the responsibility fell to the new professor of natural history, the Reverend William Hincks. Hincks's appointment was controversial not only because he was the brother of the solicitor general but also because it bypassed Thomas Huxley's application for the same position. There were two factors besides patronage that induced the selection committee to name a known quantity to the chair. First, a 'great uproar' had recently resulted in the House of Assembly over Thompsonian or herbal medicine. Known, unfortunately for botany, as 'botanic' medicine, the method rejected traditional 'mineral'-oriented medicine in favour of plant-derived tonics. To many Canadians this implied quackery, and it could not have strengthened public faith in botany as science.[13]

Second, the university still smarted after the curious case of 'Doctor' Naphegyi. A self-proclaimed assistant professor of chemistry and botany at the Royal University of Pest in Hungary, Gabor Naphegyi applied for a position at the University of Toronto in 1851. The entire Faculty of Medicine met with the university president to consider both the application and a botanical garden for the medical faculty. To their astonishment, Naphegyi possessed no documentation that even a Hungarian translator could decipher. Nor was he able to discuss botany intelligibly in any of the languages they suggested. The apparent fraud excited public interest in Toronto and set the faculty on edge regarding future appointments. While a subcommittee continued to plan for a botanical garden, new faculty problems delayed its actual establishment.[14]

Onto this scene William Hincks (1794–1871) arrived in October 1853. Hincks had studied for ministries in both the Presbyterian and the Unitarian churches and had taught mathematics and philosophy at Manchester College, York, England, and natural history at Queen's College, Cork, Ireland. The decision to appoint him to Toronto was heartily approved by the Board of Agriculture, who hoped also that a botanical garden might finally be established at the university.[15]

When Hincks took up his Toronto duties he was fifty-nine years old and intellectually quite rigid. One of his first public lectures, 'On the Relations of Natural History to Agriculture,' reiterated the usual arguments with which readers of the *Canadian Agriculturist* were already familiar. He taught plant and animal taxonomy as well as vegetable physiology. Although his lectures utilized the microscope and emphasized the natural system of classification, Hincks's students remembered his ideas as 'antiquated' and his teaching methods as leaving 'much to be desired.' Hincks demanded the recall of material by rote in outmoded form and forced students to tailor their papers to his 'peculiar taste.' Nor did he contribute to the establishment of a botanic garden, still anticipated in 1855 by George Buckland as 'an ornament to Toronto, and an honor to Canada.'[16]

Hincks's botanical approach was antiquated in two main respects, both derived from a zealous and uncritical adherence to the views of John Lindley, his former teacher at the University of London. Hincks imbibed too deeply the unsound theoretical foundations of Lindley's practical contributions to botanical systematics. Lindley (1799–1864) was a distinguished botanist and horticulturist who popularized de Candolle's natural system of classification on both sides of the Atlantic after 1830. But the culmination of his published work, *The Vegetable Kingdom* (1846), ended a series of vacillations on Lindley's part by rejecting de Candolle's principle that structure was more important than physiology to botanical classification. What might otherwise have been Lindley's masterpiece took a retrograde step in the attempt to clarify botanical relationships.

Hincks also followed Lindley in subscribing to the Neoplatonic idealism revived among philosophical naturalists in Britain during the 1820s. Rooted both in the Platonism prevalent at Cambridge during the seventeenth century and in German *Naturphilosophie*, the approach of these philosophical naturalists was transcendental; they sought ideal patterns in nature and often based their investigations upon the a priori assumption that individual specimens were deviations from some abstract ideal form.[17]

Hincks revealed his idealist inclinations in two ways, the first through his adherence to the quinary or circular system of classification. Quinarianism was the brain-child of William Sharp Macleay (1792-1865), a retired English civil servant and zoologist who published his *Horae entomologicae* in two volumes from 1819 to 1821. After Macleay's theoretical basis was lost in a fire, its explication was taken up by a colleague in the Linnean Society of London, William Swainson (1789-1855). Swainson promoted quinarian principles in a contribution to John Richardson's *Fauna Boreali-Americana* (1829-1831) and in his own works, *Treatise on the Geography and Classification of Animals* (1835) and *On the History and Natural Arrangement of Insects* (1840). The system assumed that a single plan organized the entire natural world, and that this arrangement could be reduced to three main axioms. First, natural groups of plants or animals (from kingdom on down through class, order, tribe, family, genus, and subgenus) could be enclosed in circles because their members were related by 'affinities,' patterns of close structural similarities that led from one member to the next and ultimately back to the first. Second, some definite number regulated the members of every natural group, both large and small; Macleay believed this number to be five. Third, natural groups also displayed 'analogies' to other natural groups, in that structural parallels linked members of one group to corresponding members in other groups. Such analogies were discerned even between groups of plants and animals, and quinarians worked to increase their understanding of these complex patterns in nature. In 1838 Hincks addressed the BAAS on vegetable monstrosities which he analysed in five 'classes.' His repeated references to 'circles' of parts suggest the depth to which the quinary system dominated his thinking. Although this system elicited much criticism in its heyday and was outdated after the 1830s, Hincks continued to impose his circular arrangements and diagrams upon students of natural history at Toronto to the end of his lengthy career.[18]

Hincks's interest in teratology, the study of abnormalities and monstrosities as a tool for understanding normal forms, reflected the second aspect of his idealist approach. Throughout his career, Hincks shared John Lindley's admiration of the botany of *Naturphilosophie* inspired by J.W. von Goethe. Goethe laid the foundations of comparative morphology, and his *Die Metamorphose der Pflanzen* was its first compendium in the field of botany. The key to Goethe's *Metamorphose* was that each part of the plant should be understood as a metamorphosed leaf. De Candolle understood plants to consist of separate functional units; Goethe instead abstracted the concept of 'leaf' as an idealization from which all parts supported by the root and stem developed. Even after Asa Gray in 1858 denied the usefulness of visualizing petals, stamens, and carpels as derived from a state of foliage, Hincks persisted in scrutinizing vegetable anomalies for evidence of such a metamorphosis. Relying upon analogical reasoning to prove that germ-producing organs in the flower were essentially modified leaves, Hincks ignored the increasingly accepted idea of adaptive change in plants conditioned by the environment.[19] Moreover, he proceeded from principles which some of the best minds of the age had rejected as a promising path for future botanical research. Of the two main approaches to nature taken by philosophical naturalists, the idealist lost momentum after mid-century while the geographical gained enormously. Botany in Canada may have increased its institutional strength at the University of Toronto, but its thrust into the mainstream of botanical thought was not likely to originate there.

II

The idea of a botanical inventory of Canada flourished in the relative economic and social stability of the 1850s, among rural as well as urban populations. In 1852 a farmer from Middlesex-Elgin noted this correlation between rational methods and material progress: 'The present age,' he wrote, 'is preeminently one of practical benefit, where facts, figures, and material prosperity, stand out in bold relief with the sentimentalism of the last century, and seem destined, in a happy combination with the cultivation of the human mind, to elevate the great mass of society, in the enjoyment of material comfort, and the refinements of civilization.' Gardens and gardening represented the victory of these rational methods over the 'immense forest wall round the fortress' of civilization. In 1855 John Sheridan Hogan observed signs of such progress on a typical Canadian homestead which had been worked for only seven years:

Luxurian Indian corn had sole possession of the place where the potatoes had so hard a struggle against the briar and the under-wood. The forest – dense, impenetrable though it seemed – had been pushed far back by the energetic arm of man. A garden bright with flowers, and enclosed in a neat picket fence, fronted the house; a young orchard spread out in rear.[20]

A major contribution to such rapid progress, Hogan believed, lay in scientific knowledge disseminated by agricultural societies. The spread of such information incidentally served a political purpose as well, the assimilation of French Canadians. Hogan anticipated that scientific methods would force even the age-old habits of the *habitant* into conformity: 'Competition indeed shames error out of its follies,' he reasoned, 'for no person, however dogged he may be, will face the ridicule that is attached to clinging to absurd customs in the midst of universal improvement.' C.P. Treadwell of the Board of Agriculture agreed that an important part of such improvement was intellectual; in his view Canada had reached 'that state of advancement which would justify the outlay necessary to the publication of its natural history.'[21]

British botanists assumed that British North American botany east of the Rockies was well enough known since the publication of Hooker's *Flora Boreali-Americana*. In 1843 Hooker dispatched Joseph Burke, an under-gardener at Kew, across the HBC territories to collect certain remaining specimens. Burke's travels via York Factory were facilitated by the company's co-operation in J.H. Lefroy's magnetic survey that same year and by the botanical interests of company officers. But unlike Edward Sabine, who arranged Lefroy's passage, Hooker had limited knowledge of the British North American territories, and the result for Burke's mission was chaos and confusion if not outright failure.[22]

Yet real progress rewarded amateurs in British North America who continued to gather specimens assiduously for British, American, and their own growing collections. 'Natural history is a religion,' quoted the *Canadian Journal*, 'and Botany is one of its sects.' It reprinted instructions prepared by the Smithsonian Institution for collecting and preparing botanical specimens, and it provided a forum for the publication of Canadian collectors' findings.[23]

Writing in this tradition, Edward M. Hodder (1810–78), professor of obstetrics at Trinity College, submitted an essay on poisonous plants indigenous to the neighbourhood of Toronto. Hodder, who had emigrated from England in 1834, botanized the area between the Humber and Don rivers. As a physician, he appreciated the benefits of the investigation of native plants as a 'sovereign panacea for the cure of many of the ills which the flesh is heir to in this thriving city.' But as a naturalist he was interested in larger botanical questions. The geological relations of plants, he explained, threw light on the laws of vegetable distribution in a new country like Canada. The geology of a given region indicated whether a plant was indigenous, or 'coeval with the soil'; if the plant was introduced, geology helped to explain how this had occurred. The means of introduction included human processes like 'arts or commerce, agriculture or manufacture, superstition or medicine,' and natural processes of plant diffusion. Hodder differed from William Hincks both in his focus on geographical distribution and in following the popular de Candollean system to classify his specimens.[24]

But he agreed with Hincks on the necessity of teaching more botany in Canadian schools. Others too recognized the significance of botanical studies for an accurate understanding of the country's potential. William Winder, librarian to the Legislative Assembly, was fascinated by the direct relationship between climate and vegetation in various regions of North America and of Canada in particular. An anonymous writer, likely the editor of the *Canadian Journal*, H.Y. Hind, was astonished by the high quality of fruits, flowers, and vegetables exhibited at the provincial agricultural show in 1852. He judged that these displays 'exhibited in a marked manner the extraordinary adaptation of the climate of this country to all the purposes of Horticulture.' More pointedly, in 1853 Hind claimed first prize from the Board of Agriculture for an agricultural report on York, Ontario, and Peel counties. Under the suggestive motto 'Coming events cast their shadows before,' Hind postulated that the relation of climate to agriculture could 'very satisfactorily' be determined by an inventory of the periods of leafing, flowering, and fruiting. Foreshadowing his own report on the agricultural potential of the north-west, Hind declared that such a botanical inventory might help to evaluate new regions in British North America.[25]

Hind probably authored two unsigned articles which pursued this subject further. The first urged systematic observations of the leafing and flowering of plants throughout Canada. Similar observations in the United States, it pointed out, had produced 'curious results.' Since only a comparative approach could eliminate local 'errors and variations,' the author volunteered the Canadian Institute as a depot for information on the times of frosts, animal migrations, and the leafing, flowering, and defoliating of plants. The second article offered to advise persons commencing such observations and emphasized their contributions to studies of larger climatic and meteorological patterns. Such a proposal assumed the importance of expanding the geographical horizons of its co-ordinators much beyond the merely local scale. The key to its success was co-operation in comprehending a much larger regional and even continental picture, since differences of soil and position frequently caused two, three, or even more days' difference between epochs of the same event in plants of the same species.[26]

As might be expected, responses to these proposals came from those already interested in meteorology. They included Dr William Craigie of Hamilton, who submitted a list of plants indigenous to Hamilton. Dr Charles Smallwood at St. Martin's, Isle Jésus, and the grammar school meteorological network established by Egerton Ryerson later in the decade also collected and submitted similar information. Just as important as the names and life cycles of Canadian plants was their geographical distribution. The *Canadian Journal* asked its widely distributed subscribers to collect such information as the most northern and most eastern townships in which the cactus was found; the limits of the black walnut and Spanish chestnut; and in what districts of Canada West white woods were found.[27]

There were practical as well as theoretical reasons for such questions on plant distribution. One writer observed that it was important to know the plants of the northern as well as the temperate regions in preparation for the day when man reached towards the pole. And in a more practical vein, another writer noted that very extensive sources of revenue were neglected through ignorance of valuable species of wood that did not appear in lists for exports. The Canadian Institute, he suggested, should collect types of wood for its museum, with duplicates sent to the Sydenham Crystal Palace, where the British public 'may be made familiar with the numerous and extensive, but comparatively little known, treasures of our forests.' So little was known of the geographical distribution of plants in Canada that published studies still relied on Samuel Strickland's pioneer descriptions of Canadian forests.[28]

In 1854 Elkanah Billings, curator of the Bytown Mechanics' Institute, future editor of the *Canadian Naturalist and Geologist*, and future palaeontologist to the Geological Survey, exposed the problem of the scarcity of information about Canadian plants. In a prize-winning report to the Board of Agriculture Billings reasoned that while soil, climate, and geography gave his county enormous agricultural potential, its actual state of development was only 'very ordinary.' The missing link between potential and reality, Billings purported, was a more widespread knowledge of natural science. There could be no doubt, he held, that a country inhabited by people educated in the botany, geology, and natural history of their own localities stood a much better chance of being well cultivated than would otherwise be the case. What was required to correct this deficiency was a set of school-books prepared especially for the country; one book would deal with botany, another geology, and a third zoology. Billings acknowledged that specialized information about the natural history of Canada was difficult to come by. Elementary works could be had everywhere, but their general nature prevented students who walked out into the fields from recognizing 'by its specific characters, a single stone, or plant, or insect, except perhaps, a few of the most common.'[29]

Billings's arguments increased their relevance with the success of Canadian agricultural and forest exhibits at Paris in 1855, which showed vast improvements since 1851. In imitation of William Logan, organizers arranged specimens to provide 'an object of *study* for the economic botanist, no less than an object of commercial interest to the merchants and artificers of Europe.' The resulting pride, thought George Buckland, inspired the effort needed to earn Canadians intellectual as well as industrial equality with the foremost nations of the world. Buckland told the Toronto Mechanics' Institute that such an application of science to agriculture served also to broaden the mind, as 'physical things, and the sciences which relate to them, begin to be invested with a garment of meaning and of purpose altogether new ... The drained morass, the fresh-turned fallow, the waving cornfield, the meadow, with its herbage interspersed with flowers,' no longer stood separately 'as things of mere labour, utility, or beauty, our relation to them the accident of a day.'[30]

III

Canadian farmers' need for botanical and other scientific information intensified during the later 1850s when natural disaster threatened Canadian agriculture anew. The 'wheat destroyers' – the Hessian fly, midge, weevil, wheat fly, and other pests – had infested Lower Canada and the eastern Untied States as early as 1828. They moved steadily westward, alarming Upper Canadian wheat farmers by 1850. In the summer of 1856 the insects ravaged the north shores of the Great Lakes, and public confusion as to the exact nature of this new 'visitation' deepened the widespread sense of helplessness.

Two schools of thought proposed remedies. One interpreted the problem as botanical, and the other saw it as entomological. Both agreed that more information was needed before a permanent solution could be found. The crisis induced a growing consensus that government aid and interference were rapidly becoming imperative. The minister of agriculture in the Conservative Taché-Macdonald government, Philip M. Vankoughnet, accordingly offered a prize of $500 for the best essay on the problem.

This attempt by a brilliant lawyer to 'catch the weevil through a committee of inquiry' elicited ridicule from afflicted farmers. But Vankoughnet did grasp the dual nature of farmers' difficulties: on the one hand they relied too exclusively upon wheat to the point of overcropping; they had to be persuaded to diversify. On the other hand aggressive measures were needed to halt the advancing insect hordes. Only a minority of farmers recognized that too much wheat of one type was being cultivated in Canada. Short of abandoning the lucrative crop altogether, the solution was a variety adapted to Canadian latitudes that ripened before the insects hatched. The majority opted for a more straightforward entomological solution. What was needed, they argued, was systematic observations and experiments for several years in each township. The result of such an inventory would be submitted by local agricultural societies to the board and to the Bureau of Agriculture. Entomologists, it was assumed, could then identify the problem and solve it. While most agreed that a preliminary inventorial assessment of wheat crops was indispensable, they also felt that Vankoughnet's essay contest had not gone nearly far enough. If Sir William Logan and a numerous staff had been employed for a number of years in investigating the mineral wealth of the province, and excellent service had been rendered by 'the good knight' to his country, then could not the same rule hold for agriculture, 'decidedly the most in need of scientific aid?'[31]

Vankoughnet's competition closed with the selection of three essays from twenty-two submitted. William Hincks and J.W. Dawson awarded first prize to H.Y. Hind, second to Rev. George Hill of Markham, and third to Abbé Léon Provancher of Montmorency, writing under the pen-name Emilien Dupont. Both Hind and Provancher detailed vegetable and animal parasites known to infest wheat crops. Since plants were more widely studied and better understood than insects, both recommended changes in planting methods and improved drainage to permit plants to attain the vigour which the harsh Canadian climate demanded. Aside from the application of absinthe suggested by Provancher, it appeared from these essays that nothing could effectively eliminate plant pests and disease. Although both Hincks and Dawson commended the patriotism that had motivated Hind's essay, Dawson noted that the paper was based upon scanty primary research in Canada and relied instead upon information culled from abroad.[32]

One Canadian naturalist agreed with Dawson and with the farmers' criticism of the entire exercise. William Couper (d. 1886) found it difficult to conceal his contempt for the prize-winning 'compilations,' which he believed were 'of no more value than waste paper'. Couper's specialized interest in botany and entomology had been piqued by the role of wind and water in the geographical distribution of plants and insects. Studies of geographical distribution convinced him of the need for more widely ranging botanical and entomological inventories of Canada. He believed that the task could best be accomplished by encouraging collectors to take to the field. Even those too diffident to be seen in proper Victorian society with the necessary paraphernalia, he urged, could find clever ways of disguising nets and vials.[33]

By the late 1850s, the need for a scientific inventory of the plant life of Canada had increased in urgency for several reasons. The separation of 'pure' and 'applied' botany at the University of Toronto and the philosophical idealism

217 The Metamorphosed Leaf

underlying the work of William Hincks made it unlikely that any co-ordination of such an inventory would arise at Toronto. Hincks differed from Couper in envisioning himself as a higher-ranking theorist too busy to spend much time on the lower task of collecting specimens. Moreover, he imposed his anachronistic taxonomical outlook upon amateurs in the south-western part of the province, including William Saunders and James Fletcher. Hincks's predominance as the professional botanist of the Toronto region guaranteed that the task of organizing a botanical inventory in Canada would remain for other centres of botanical study.

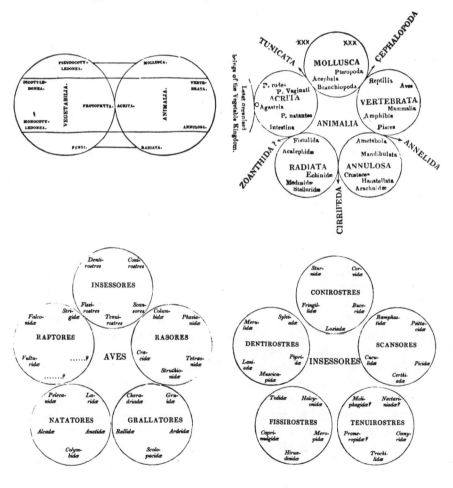

The Quinary System of Classification

12

Fragile Stems, 1857–1863

The varying direction of animal life, the return of seasons, the orbits of the planets, even the eccentric course of comets become defined, and familiarized with our ideas of time, by the inquiring and science of man; but the tree still rears its head towards the heavens in defiance of his research, while tradition and conjecture alone mark the span of its existence.

George Head, *Forest Scenes and Incidents in the Wilds of North America* (1829)

Unlike the University of Toronto, other Canadian universities acquired professors of botany trained in the active research favoured by William Couper. During the later 1850s both McGill College at Montreal and Queen's College at Kingston preferred the botanical tradition of the University of Edinburgh, appointing former students of John Hutton Balfour. Balfour (1808–84) had been professor of medicine and botany and keeper of the Royal Botanic Garden since 1845, when he had edged out Joseph Dalton Hooker for the position. Trained in systematics, Balfour nevertheless developed Edinburgh's tradition of laboratory research using microscopes in vegetable morphology and physiology. He also extended the tradition of field botany. A charismatic teacher, Balfour routinely conducted botanical excursions through the Scottish countryside to procure specimens for the laboratory and herbarium and to stimulate interest in the study of plant forms in their ecological relationships. He persuaded students that such excursions constituted 'the *life* of the botanist,' enriching him with both practical knowledge and the companionship that bound members of the field party by 'ties of no ordinary kind.'[1]

Balfour's other forte was his organizational and expository ability. During the 1830s he had helped found the Botanical Society of Edinburgh and the Edinburgh Botanical Club, both of which inspired similar Canadian organizations later. His popular textbooks also conveyed a deeply religious frame of mind.

Balfour's religious orthodoxy surfaced on the issue of the transmutation of species. Oats, he declared, could not be converted into rye artificially by repeated cutting before flowering. Balfour abhorred the 'very erroneous views' which held up such claims. He imparted to students his lifelong conviction that species varied only within narrow limits, that seeds perpetuated a typical form, and that cultivated varieties would revert to their original forms if left untended.[2]

The limitations imposed by Balfour's 'phytotheology' did not preclude a critical approach to botany. While he impressed students of the vegetable kingdom with the harmonious relation of the part to the whole that gave order to creation, at the same time he encouraged independent observation. The botanist, Balfour taught, must not be 'led away by human authority, however distinguished.' For although he might avail himself of aid supplied by eminent botanists, 'he must endeavour by personal observation to ascertain the correctness of their statements.'[3]

Balfour adopted the natural system of de Candolle, noting that the number of collected species was no longer the measure of botanical achievement. The natural system represented a dynamic scientific process of discerning plant alliances, discovering order in arrangement and function, and applying the results to classify them. Balfour urged students to take 'an enlarged and comprehensive view of the vegetation with which the earth is clothed' because it was impossible to classify scientifically without a thorough knowledge of vegetation in all parts of the world. Since plants followed principles of both order and special adaptation, botany would remain an imperfect discipline as long as vast tracts remained unexplored by botanical travellers.[4]

Such generalizations might in themselves have justified a closer look at British North America. But Balfour had a particular fascination with alpine plants and taught that their geographical distribution could best be pursued in northern regions. Even the minute lichens and mosses, he pointed out, played important roles by using 'nature's chemistry' to convert 'sterile rock' into 'a natural and luxuriant garden,' thus preparing a soil ultimately for man himself.[5] In extolling the geographical approach to botany and the botanical virtues of Scotland as a northern land, Balfour inspired at least two of his students who ended up in Canada. Once there, they found themselves well equipped to look upon the Canadian Shield as much more than an impenetrable armour of Laurentian rock.

I

The first of Balfour's students to practise botany in Canada was James Barnston. Born in 1831 at Norway House, Barnston was the eldest child of George Barnston, an officer with the Hudson's Bay Company. Barnston imbibed his father's love of natural history, especially botany. Educated first privately by his father, then at Red River, James was considered by the elder Barnston to be a 'soft boy,' sensitive and intelligent, who needed a thorough education to prepare him for the vicissitudes of life. In 1845 he went to Lachine to study with a Presbyterian minister named Simpson. In 1847 he proceeded to Scotland to study medicine at the University of Edinburgh. There, an old 'Botanical friend' of his father's introduced James to John Hutton Balfour, then dean of the medical faculty. He regularly attended Balfour's *conversaziones* and botanical excursions through the Royal Botanic Garden and was deeply influenced by his teachings.

Barnston passed his medical examinations at twenty, while still too young to practise. After touring hospitals on the continent, he graduated in 1852 with first prize in botany. He returned to Canada in October 1853 to establish a medical practice in Montreal and to collect botanical specimens there. Although European botanists had preceded him, Barnston believed that the flora of the country yet offered much to be discovered. A cholera epidemic prevented Barnston from botanizing during the summer of 1854, but he hoped to amass a 'pretty complete collection' by 1855, promising to send specimens to Balfour to illustrate the botany of his section of America.[6]

Barnston felt dismay at the dearth of natural science courses at McGill. 'Would you believe it!' he wrote to Balfour. 'In a new country like this, every one is bent on making money.' What then could be expected from the youthful minds of Canada, when no encouragement was given by their educators? To his own question Barnston, a specialist in midwifery, responded in character, 'The Natural Sciences *must be born* in every County and why should not Canada give *birth* to Geology, Zoology or Botany at this present time?' He lobbied for a chair of natural history or of botany and soon discovered that it was 'hard work to move heavy Stones.' After little initial success, he hatched an alternate plan for a botanical society and a course of popular lectures. By the spring of 1855 Barnston had set into motion the Botanical Society of Montreal, and he saw signs that a chair of botany would 'ere long be formed' at McGill.[7]

Barnston owed his success, at least in part, to his social connections in Montreal. He found support in influential acquaintances such as William Logan, 'an old Schoolmate' of George Barnston at Edinburgh. Logan, aware of James Barnston's organizational plans and successful lectures at the Mechanics' Institute, knew also that Sir William Hooker was anxious to obtain botanical specimens from Canada. He recommended Barnston as a Canadian correspondent for Kew Gardens. 'A few words of encouragement coming from this side of the water,' wrote Logan to Hooker from Edinburgh in 1856, 'has a wonderful effect in giving impulse to scientific labour on the other side. The Canadians look greatly to Europe for their standard of excellence.'[8]

Within weeks Barnston was corresponding with Hooker. 'All the knowledge we possess of Canadian plants,' he noted, 'comes thro' the discoveries and publications of British, American and French Botanists.' Barnston, who hoped to become the first indigenous professional botanist, remained confident that Canada was 'rich in Plants yet to be made known.' He founded the Botanical Society as a means to collect a complete herbarium of Canadian plants, to form a museum of the vegetable products of the country, and to publish a circular to diffuse botanical information.[9]

The Botanical Society of Montreal first met in the spring of 1856. J.W. Dawson, McGill's new principal and professor of natural history and agriculture, served as president; John George Barnston, James's younger brother, served as secretary. Other members included Thomas Sterry Hunt, A.F. Kemp, David Allen Poe, and George Shepherd, all members of the Natural History Society of Montreal. One of the new society's first actions was to initiate a transfer of Frederick Pursh's remains to Mount Royal cemetery and to erect a monument to this forgotten forerunner of their own botanical efforts. The meeting also heard papers from Dawson, Kemp, and George Barnston. Like his son, Barnston was interested in the geographical distribution of British North American plants. His lifelong experiences in the HBC territories drew his attention to this great botanical lacuna; he had gone to great lengths to obtain relevant published studies, including Torrey and Gray's handbooks and Humboldt's *Cosmos*, as soon as they became available. As an amateur, he believed that it was 'better to assume some method, however destitute of exact relation to my subject, than no method at all.'[10]

The elder Barnston's paper, published in the *Canadian Naturalist and Geologist*, remarked on the British North American distribution of the first order of flowering plants in the natural system, Ranunculaceae, a family of herbs typified by the buttercup. Barnston was intrigued both by the 'huge strides' of these species over numerous parallels of latitude and by the diversity of their forms, whose 'connecting law' he believed could be traced by the 'philosophic and scientific botanist.' Deep piety offered him personal comfort whenever obvious answers to more difficult questions were not forthcoming. 'Is it not much easier,' he asked his listeners, 'and more rational at once to suppose, that there is an Almighty Creator and wise Distributor, exercising his unfettered power and will, in all things pertaining to man's terrestrial abode?'[11]

Early in 1857 James Barnston contributed some of his own thoughts on botany to the *Canadian Naturalist*. Drawing heavily upon the teachings of his botanical mentor at Edinburgh, he highlighted botany's power to exercise the mind 'in habits of careful and accurate observation, of systematic comparisons, and of philosophical generalizations,' a power that had to be tempered by an 'honest spirit of caution.' Barnston emphasized the importance of collecting a herbarium of the district, country, and continent, for the benefit of horticulture and agriculture as well as science. Like Balfour, he romanticized the society and the field excursion as valuable social activities in which everyone possessing ordinary capabilities could partake.[12]

To this end, Barnston offered 'hints to the young botanist' on collecting, naming, and preserving plants. Even the rank beginner, he directed, should make a point of collecting and preserving 'every plant that crosses his path' and of aiming towards a complete herbarium of some district. Completeness in a limited range, he averred, was more valuable than a larger but more scattered collection; a network of companions and a system of exchanges could soon expand the district under scrutiny. A valuable Canadian desideratum, Barnston added, was information on the geographical distribution of plants in connection with the general features of the country. As a starting-point he recommended Gray's *Manual*, 'the only modern work which describes the plants that grow in this Province,' at least those 'in common with the Northern States.'[13]

But Barnston and other Canadian botanists grew increasingly aware that a number of British North American plants, Balfour's favourite alpine plants in particular, were not mentioned in Gray's *Manual*. Recognizing the contribution to be made by studying British North American plants and encouraging others to do the same, Barnston deepened his involvement in the scientific and social scene at Montreal. He drew the attention of the NHSM to Charles Smallwood's observatory, which collected related data on Canadian climate. He lobbied successfully for a chair of meteorology, to which Smallwood was appointed in 1856, at McGill. Barnston was corresponding secretary of the Montreal Horticultural Society, curator and librarian of the NHSM, and 'the most active member' of the editing committee of the *Canadian Naturalist*. He began also to catalogue A.F. Holmes's large herbarium, bequeathed to McGill in 1856.[14]

In 1857 Barnston's appointment to the new chair of botany at McGill College marked the formal separation of botany from the medical faculty. Botany became a required subject for arts students not studying chemistry in their third year at McGill. Topics included microscopy; vegetable histology, anatomy, physiology, and nutrition; taxonomy; and geographical botany. Barnston required specimens used for illustration to be 'chiefly Canadian.' In his introductory lecture, he echoed both Balfour and his own father when he enjoined arts and medical students to envision in botany 'an influence for good,' able 'to improve the quality of the mind and in some measure to regulate human action' by developing powers of observation, comparison, and judgment. He recalled his own life in the British American north-west by singling out those who would 'some day find themselves isolated in distant and little-explored regions. Far away from friends and the conversation of intelligent companions, any pursuit that can engage and occupy the mind and above all satisfy its thirst for truth by draughts from the pure and refreshing fountains of nature,' he declared, 'any such pursuit becomes a blessing and converts the desert into a paradise, one often filled with creatures yet to be named.' As 'nurseries of science' as well as of arts and literature, he added, Canadian educational institutions would become 'lasting monuments of honour and credit to the country.'[15]

These promising beginnings in Canadian botany ended abruptly before the end of Barnston's first course of lectures. 'Prostrated' by an unidentified illness, he died prematurely in May 1858. Only twenty-seven years old, Barnston left a wife and infant daughter, both of whom died within a year.

James Barnston's sudden demise after a meteoric career as a professional botanist left a gap in the development of botany in Canada. He had had too little time to build a permanent legacy. The chair of botany at McGill continued in name while Barnston's courses were covered for one session by J.W. Dawson. In 1859 botany was subsumed under natural history, and the chair ceased to exist. Dawson's personal interest in geology, zoology, and palaeobotany prevailed in later years. The Botanical Society collapsed with the loss of Barnston's organizational enthusiasm. Future botanical papers, which never compared in number to geological papers, reached the pages of the *Canadian Naturalist* through the less specialized NHSM.

Yet during his truncated career, James Barnston had managed to convey to British North American botanists some of the important concerns of mid-nineteenth-century botanical theory. His 'Catalogue of Canadian Plants,' published posthumously, earned praise first as a synopsis of the botany of the middle regions of Canada. Second, it included specimens that had since become extinct through widespread cultivation of the land. Barnston added sixty specimens of Canada trees and even more of northern or alpine plant species, especially cryptogams.[16] He had made brilliant beginnings in the reinterpretation of the flora of the Canadian wilderness as a unique and orderly environment of intellectual and practical value to future observers.

II

James Barnston embodied several important trends in Canadian botany during the 1850s. First was a growing interest in the geographical distribution of plants in British North America. J.W. Dawson suggested that the Canadian botanist should interest himself not merely in naming specimens he chanced upon, but also in the uses of native plants, the improved cultivation of plants, and the conservation of forests. Less practical aspects of phytogeography helped illustrate the geological history of the North American continent: 'Let us inquire respecting any plant,' Dawson proposed, 'what are its precise geographical limits? To what extent do these depend on climate, elevation, exposure, soil[?] What inferences may be deduced as to the centre from which it originally spread, and what as to the changes in the extent of the land and relative levels of land and sea that have occurred since its creation?' Here were 'fertile subjects of inquiry, leading to the grandest conclusions in reference to the history of life upon our planet.' More than geology, botany required a large mass of observers, and Dawson, like William Hincks, held that the theoretician relied heavily upon localized efforts: 'In truth a large proportion of the new facts added to natural science, are collected by local naturalists, whose reputation never becomes extensive, but who are yet quoted by larger workers, and receive due credit for their successful efforts.'[17]

In 1858 the *Canadian Naturalist* reprinted Richard Owen's presidential address to the BAAS on the study of plant distribution. Harking back to Buffon's Law, Owen remarked upon the differences between species and even genera of plants growing under similar conditions in Europe and in North America. From this perspective, he argued, it was important to include in botanical inventory the flora of as many localities and of places as far north as possible.[18] The obvious significance of British North America as a vast habitat of northern flora and for the geographical distribution of plants in general was not lost on its botanists. One year after his son's death George Barnston published a paper on the geographical distribution of Cruciferae throughout British North America. This family of plants, Barnston thought, held importance from a theoretical as well as a practical standpoint. Colloquially known as the cress family, Cruciferae include cabbage, cauliflower, mustard, and turnip and are native to temperate regions of the Old World. British North American Cruciferae exhibited interesting differences from their European and Asian counterparts, though not so great, thought Barnston, as to neutralize the resemblances so essential to the natural system of classification.

Barnston's interest in Cruciferae lay in their demonstrated ability to adapt to climatic variations. In British North America many Cruciferae of European origin had completely acclimatized to become 'naturalized Americans and Canadians.' 'Is not this,' wondered Barnston, 'in perfect accordance with the diffusive character of the order, as noted by botanists in those species which exist in the highest northern latitudes?' In an arresting passage he interpreted Cruciferae as forerunners to man's own northward 'diffusion' in British North America. Barnston believed that 'in those dismal regions where ice holds almost eternal empire, and where frost is arrested but for a few short weeks of the year, we still may please ourselves with discovering that wise provision is made, as far as possible under the circumstances, for the wants of man.' For him this discovery was proof that the northward advance of civilization was no mere pipe dream.

In whatever quarter of the globe man may be placed, surely by searching he may find what is best calculated to benefit him. Let him only take the trouble and time to investigate, and turn to advantage what has been so liberally – nay, often so lavishly, we may say – spread out before him, and he will not fail to discover, that an unseen hand has been long since at work to anticipate his wishes, and supply his needs.[19]

An additional consideration supported a prompt botanical inventory of British North America. The expansion of agriculture was altering the natural patterns of plant distribution by diffusing cultivated plants over wider areas. These cultivated plants were hardier and more productive, and even modified local climate by changing drainage and forest patterns. As new clearings drove native flora from many localities, records of their habitats increased in value. For this reason the herbarium of A.F. Holmes, begun during the 1820s, already held historical interest in Canada.[20]

That botany was becoming a source of Canadian historical consciousness is suggested by an 1858 letter to the *Canadian Naturalist* by W.J. Morris of Perth, likely William James Morris, brother of Alexander Morris. Morris had found onions growing wild near Lake Temiskaming and inquired whether they were indigenous to the north-west of Canada. An eloquent reply issued form George Barnston in an essay on the geographical distribution of the genus *Allium* in British North America. The onions, Barnston explained from evidence of Arctic explorers in the north-west, were not native to North America. Instead they were naturalized descendants of those once cultivated by Jesuits in the area Morris had visited. 'These floral bequests,' Barnston pointed out, 'after nearly one hundred years of neglect, have still, by the favor of nature and advantageous situation, kept their solitary hold, beautiful mementos of the pursuits and recreations of the most intelligent of the first enterprising settlers in the land.'[21]

Just as the geographical distribution of plants raised new questions about Canada's past, so also did their geological distribution. J.W. Dawson's palaeobotanical contributions during the 1850s were among the first to illustrate the nature of early vegetation on earth. Deeply influenced by Charles Lyell, Dawson saw analogies in the geographical and geological distribution of plant forms. Much as H.Y. Hind had done in his assessment of the north-west in 1857–8, Dawson applied his knowledge of fossil plants to deduce the climatic conditions in which the ancient flora grew. His specialized interest developed from investigations of the Nova Scotia coal-beds and from his analysis of the still older fossil flora of the pre-Carboniferous Devonian strata for the GSC. The result was an anomalous development in Canadian botany. Botanical morphology and anatomy in Canada, in contrast to most other countries, grew from the study of extinct plant forms, rather than vice versa. Dawson reconstructed an eerie picture of prehistoric and preglacial life that startled many as he fleshed out the GSC's estimates of the antiquity of Canada's rock formations. Somewhat whimsically, he reassured his readers in 1857, 'It is something to have a flower handed down to us from the Carboniferous period. We can now add to our picture of the coal swamps a few bright flowers, to relieve the general sombre green of ferns and pines; and are even at liberty to hope that we may discover a butterfly that flitted amongst these ancient blossoms.'[22]

Close ties between botany and geology and the growing importance of the local distribution of species in the determination of geographical distribution riveted the critical attention of some Canadian botanists on the GSC. What had been done by the survey for the advancement of zoology and botany in Canada? 'Absolutely nothing, though their parties have traversed from the heights of Gaspé to far beyond the northern limits of Lake Huron,' wrote an anonymous but 'well qualified' correspondent to the *Canadian Journal* in 1858. Although the benefits might not be as apparent as in 'the search after mines,' he insisted, a botanist or a zoologist should have been attached to the GSC from its inception. If this had been done, 'we should have had now such materials for the elucidation of the Flora and Fauna of the country as can seldom be obtained.'[23]

The GSC's annual reports on newly explored territories did include information on natural history. In 1858 W.S.M. D'Urban assisted William Logan in collecting botanical and entomological specimens in Argenteuil and Ottawa counties. Similarly, Robert Bell assisted James Richardson on the lower St Lawrence. But Logan consistently opposed the idea of appointing a botanist to the survey because occasional botanical collectors, even those funded by other sources, had nevertheless not 'proved satisfactory.' The GSC's budget, Logan held, 'barely suffice[d] for our own investigations.'[24]

Closely linked to the interest in plant distribution during the 1850s was a growing interest in northern vegetation. Amateur collectors noticed cryptogams on rocks and scanty soils typical of the more northerly reaches of the province. Encouraged by both British and American examples, members of the NHSM collected algae, mosses, and fungi, often submitting lesser-known specimens to American experts for identification. In particular, A.F. Kemp and David Allen Poe, formerly of the Botanical Society of Montreal, specialized in algae and fungi.[25]

Cryptogamic botany possessed both theoretical and practical importance. Recent research in Europe, epitomized by the work of Wilhelm Hofmeister, unlocked the reproductive processes of these seedless plants and demonstrated their affinity with flowering plant forms. Hofmeister's insight that the entire vegetable kingdom operated on a common plan was appreciated by John Lindley and transmitted through William Hincks to Canadian scientific journals. Hincks also pointed out the practical value of studying cryptogams, especially fungi, as causes of vegetable pathologies.[26]

A final trend was the call for a published flora to fill gaps left by Torrey and Gray. Canadians still looked beyond their colonial boundaries for support and recognition of their scientific achievements. Some collected for foreign botanists, especially at the Smithsonian Institution. Such requests caused a dilemma for George Barnston, who felt it more 'patriotic' to collect for British or Canadian institutions. He chided other fur traders who collected solely for the Smithsonian, but many of them valued the greater material incentives and attention accorded by the Americans for their favours.[27] The link with 'patriotism' was explained by Abbé Léon Provancher in 1858.

Ce serait méconnaître l'avenir de notre jeune patrie et lui retrancher des sources de prosperité et de grandeur, que de ne pas favoriser ce penchant pour l'étude d'une branche des connaissances humaines trop peu encouragé jusqu'à présent dans nos maisons d'éducation, même dans celles de la première classe.[28]

Provancher welcomed an increasing taste for natural science, 'tout particulière dans notre Canada,' characteristic of both English- and French-speaking professional circles. Examples outside the two major natural history societies included Auguste Delisle, a notary who began a private herbarium in 1825; Louise-Édouard Glackmeyer, a notary at Beauport and an uncle of Abbé Louis-Ovide Brunet (1826–76), who taught botany at the Petit Séminaire de Québec and at Laval University; Judge David Roy of Malbaie; Thomas Bédard, a notary at Lotbinière; J.B. Cloutier, a professor at the Laval Normal School; and the Reverend Napoléon St Cyr, who collected botanical specimens while at the seminary of Nicolet.[29]

To serve this constituency Provancher, parish priest of Saint-Joachim and primarily an entomologist, published the first Canadian botanical text, *Traité élémentaire de botanique*, for students and amateurs. He encouraged the study of botany for its utility in medicine, industry, and the arts and for pleasure, holding up popular American botany as worthy of emulation in Canada. 'Peu de pays,' Provancher thought, 'sont aussi pauvres que le Canada en fait de connaissances en Botanique.' It was time, he urged, to publish a Canadian flora. Abbé Louis-Ovide Brunet, usually judged the best French-Canadian botanist of the nineteenth century, also believed that such a flora was necessary to establish patterns of geographical distribution. Unfortunately for the progress of botany in French Canada, personal quarrels made this issue one of the few upon which Brunet and Provancher could agree.[30]

The need for a Canadian flora was recognized also by William Hincks at Toronto. By 1858 Hincks had collected at least six hundred local specimens, among them a number of naturalized European species not recorded in Gray's *Manual*. Collections were expanding so quickly that Hincks's list was obsolete when it reached the press. Unlike Brunet, Hincks did not anticipate a complete flora of Canada for many years. It seemed more realistic to aim only for a useful list of plants characteristic of a few districts. Hincks believed that 'a few journeys at a favourable season, or the opportunity of examining a few carefully formed local lists, would now settle everything excepting a small number of doubtful species.' Well over sixty years of age, he limited his botanical searches, on the old assumption that Canadian flora displayed less variety and fewer successive changes over both regions and seasons than Britain's flora did.[31]

In contrast to the active enthusiasm of more youthful botanical collectors in the Ottawa Valley and along the lower St Lawrence, Hincks still dwelled on anomalous vegetable structures. These anomalies took years to collect, but he hoped to uncover some hidden universal truth: 'the operation of some force or tendency which belongs to the being, and is constantly active, but in ordinary cases is either kept in check by other influences, or allowed to manifest itself more fully than in the special instance.' Still the philosophical idealist, Hincks visualized in the several parts of a flower 'only a difference of development, every leaf being in its origin capable of assuming any of the forms.'[32] His relatively sedentary approach to botany at Toronto, which did not yet include a botanical garden, jarred with the wider outlook of the Edinburgh tradition.

III

The second student of Balfour's who taught botany in Canada was George Lawson, appointed to Queen's College, Kingston, in 1858. A Presbyterian college, Queen's was founded in 1842 to foster natural and mechanical sciences as well as the usual ecclesiastical subjects. One chair was designated for natural history and its cognates, and another for botany and agriculture. From its inception the stability of the college was threatened by constitutional dissension and personal rivalries, all exacerbated by the lack of a strong principal. When a faculty of medicine was established in 1854, these embers were set aglow by strong new personalities with their own agendas.[33]

In 1857 the Board of Trustees initiated a search through the Colonial Committee of the Presbyterian Church in Scotland for a professor of chemistry and natural history to teach in both the medical and arts faculties. That same year they offered the position to William Lauder Lindsay (1829–80), a well-known and respected cryptogamist who had recently published a widely acclaimed text on lichens. Lauder Lindsay accepted on 23 October, then rescinded for personal reasons six days later. In February 1858 the board had better luck with a strong recommendation from John Hutton Balfour on behalf of his long-time laboratory assistant, George Lawson. Balfour assured the board of Lawson's willingness to emigrate to Canada.[34]

Lawson's immediate appointment was touted as giving Queen's 'a higher standing than before among the Schools of Scientific learning in the Province.' His distinguished academic record was recounted in local press reports, with news that Lawson would earn more than James Williamson, professor of natural philosophy, director of the astronomical observatory and brother-in-law of John A. Macdonald.[35] Lawson's move to Queen's was like a spark approaching a powder-keg.

But George Lawson (1827–95) had his own reasons for emigrating to Canada. Born at Maryton, Forfarshire, Scotland, and raised in Dundee, he had spent many summers collecting plants at a secluded cottage in Fifeshire. Apprenticed to a solicitor during the 1840s, Lawson continued botanizing, circulated a monthly botanical manuscript, and organized a local naturalists' association at Dundee.

By the time he was nineteen, Lawson was restless and dissatisfied with his future prospects as a lawyer. In 1846 he sought advice from William Jackson Hooker, he said, on behalf of 'a friend.' Feeling 'very little encouragement to stay in this Country,' Lawson's 'friend' contemplated a botanical 'stroll' in America. 'Is America,' he asked Hooker, 'a country yet likely to afford *much* that is new and interesting to the Botanist; and if one earnestly and enthusiastically devoted to the Science, & willing to undergo *any* hardships, dangers, or difficulties in its causes, were to devote his lifetime in searching out the riches of the trans-Atlantic Forests, is it likely, that his labours would be rewarded by brilliant discoveries?' Hooker, perhaps still smarting from his experience with Joseph Burke in America, appears to have been unable to offer Lawson any inducements. In 1848 Lawson enrolled instead at the University of Edinburgh to qualify as a science teacher.[36]

Lawson's organizational skills kept him employed at Edinburgh for ten years. He was deeply involved both at the university and in several local scientific institutions, often in a secretarial capacity. He served as assistant secretary and curator of the Botanical Society of Edinburgh and assistant librarian of the Royal Society of Edinburgh, and he belonged to the Caledonian Horticultural Society and the Scottish Aboricultural Society. Although he received a PH D from the University of Giessen in 1857, perhaps the single greatest influence upon Lawson's botanical career was his connection with John Hutton Balfour.

Lawson worked as a demonstrator in Balfour's botanical laboratory, where he taught James Barnston.[37]

But Lawson's professional accomplishments did not diminish the social ambitions he also harboured. He was well acquainted with James Hector, geologist to the Palliser expedition in 1857. Although he later insisted that he left Edinburgh for Kingston mainly because 'Canada presented an ample field for much useful botanical work,' especially in the cryptogams, it was nevertheless true that after ten years Lawson felt he was going nowhere at Edinburgh.[38]

Lawson's arrival in Canada with his wife and two daughters in the autumn of 1858 was hailed as 'an important acquisition not only to the University of Queen's College, but to the province generally.' One of his first public actions was to address the provincial agricultural exhibition, held that year at Kingston, casting botany's role in agriculture in a bright new light. First, he declared botany more important than soil chemistry in improving agricultural production. 'Plants are living beings like ourselves,' he told the audience. Farmers should raise new crop varieties better suited to the climate of Canada, rather than alter soil or manure compositions. A plant, in effect, was 'no mere machine acting a mechanical part.' There was still much room for well-directed experimental enquiry on the subject in Canada. Second, and more important, Lawson outlined the importance of scientific organization to the expansion of agriculture and, with it, of civilization. In earlier times, he noted, 'the noble forests that had for countless ages grown in all the wildness of nature presented a bold front, sufficient indeed to stem the tide of ordinary civilization. Like a strong enemy they would have scorned a puny hand. They called for great energy and perseverence.' But more recently, he warned, the long-term physical effects of indiscriminate deforestation were not precisely known. It was therefore unwise to trust entirely to private interest in this matter, and better for government to co-operate with science for the long-term benefit of the country.[39]

Lawson almost immediately applied his organizational skills to promote public interest in science in Canada. In 1859 he delivered a course of public lectures on the useful applications of chemistry at the Kingston Mechanics' Institute. He also set up a chemical and botanical laboratory equipped with microscopes to give students at Queen's the advantages of his own modern training, to learn by firsthand experimentation.[40] Undoubtedly Lawson's most important contribution to the organization of Canadian botany was the founding of the Botanical Society of Canada in December 1860.

During the latter months of 1860 Lawson and several colleagues agreed that Canada had too long avoided the 'obligations' to which its enormous resources bound the country. In a series of newspaper advertisements they appealed to

Canadian pride to overcome this state of relative backwardness despite the great potential of its natural resources. In Britain, they pointed out, botany was a universal pursuit. But inhabitants of countries like Canada were preoccupied 'in industrial production and trade' and less likely to pursue botany for its own sake. Botany's relations to industry were so important that 'no civilized land can allow it to fall into neglect without suffering thereby in its material interests. In England, and France, and Belgium, and Prussia, it will not be believed that a great agricultural and timber-producing country, like Canada (young as it is), is pushing on its industry in ignorance of the very science by which that industry ought to be guided.'[41]

The purpose of the proposed Botanical Society of Canada (BSC) would be to remedy this deficiency. It would encourage Canadian botanists to 'follow the examples set before them, and unite together to develop a knowledge of the Forest Flora, which is, in every sense, the richest blessing to man with which nature has gifted this great land.' Such a society would 'raise the fallen standard of botanical science among us' and direct public attention to neglected sources of industrial wealth. As a centre of communication and exchange, the BSC could facilitate the labours of Canadian botanists in their respective localities. More important, it could co-ordinate their work, giving value to their researches that no desultory observations could ever attain. The co-operation of numerous observers was the only viable means of furthering knowledge about the geographical distribution of plants.[42]

Lawson believed that amateur botanists should extend their purview to all of Canada and even to the rest of British North America. He moved beyond the localized outlook epitomized in Gilbert White's *The Natural History of Selborne*. The expanded outlook was encouraged by William Leitch, principal of Queen's and president of the BSC. 'No other science,' Leitch told the founding meeting, was more important in a new country 'in its bearings on field industry and other useful arts of life.' But just as important, the country remained as yet comparatively unexplored. The settled shores of the St Lawrence had no doubt yielded most of their botanical treasures, but there remained 'an extensive back country' where new plants might well be discovered. This possibility gave Canadian botanists the opportunity to prove Canada a productive botanical field. An astronomical society, he averred, would not enjoy the peculiar advantages of a botanical society in a country like Canada. Finally, Leitch observed that the proposed botanical society was in tune with 'a patriotic feeling rising up in Canada.' The BSC offered an opportunity 'to wipe off a reproach that has long hung over the country, by prosecuting a path of research that has been neglected.' No mere local anomaly, the embryo society was likened to a seed cast into the fertile Canadian soil. It would 'grow up into a goodly tree, spreading its branches over the length and breadth of Canada, which is yet destined to be a great country.'[43]

Leitch's expansive vision formed a prelude to Lawson's assessment of the state of botany in Canada. As secretary of the new society, he reinforced Leitch's complaint that botany in Canada was mired 'at a lower ebb than in most civilized or half civilized countries on the face of the earth.' He singled out the floras by Hooker and by Torrey and Gray not as the last word but rather as a first step towards knowledge of Canadian flora. But the implicit challenge had to be recognized before it could be taken up; instead Canada presented 'the singular anomaly of a country distinguished by its liberal patronage to science, dependent for its information respecting its native plants on the descriptions of specimens culled by early travellers.' New needs and new questions, Lawson stressed, had rendered the state of Canadian botany severely deficient.

Lawson reiterated the unique contribution the BSC could make. Winter meetings and papers would be supplemented by summer field meetings in 'distant localities' in Canada and in other parts of British North America, perhaps even in the United States. Contacts with botanists in other countries would gain access to extensive libraries and herbaria and would spur the acquisition of information on Canadian botany.

While Lawson gave a passing nod to structural and physiological botany, he emphasized that 'our position in a comparatively new country points out to us a special path of research which it will be our duty to follow.' This was the path of 'species botany,' the geographical and local distribution of the Canadian plants. 'Strewed around our path in the woods and on the shores of our lakes,' he pointed out, 'even in the midst of the City of Kingston, growing on vacant lots, and even in court yards' were interesting and useful plants, both indigenous and naturalized, to be had for the picking.

Lawson raised the BSC far above the level of a popular institution. 'We should seek,' he urged, 'rather to bring our members and the public into scientific modes of thought and expression, than to allow our Society to yield up its scientific character to suit the popular taste.' Botany, he suggested, possessed an inherent natural appeal for many social classes, for theologians, physicians, sanitary reformers, lawyers, commercial men, spinners, papermakers, farmers, gardeners, brewers, dyers, tanners, and lumbermen: 'The field is broad and the soil is rich. The extent to which we can cultivate it will depend entirely upon the number of the labourers, and the zeal and industry which they display . . . Let us lay a foundation, and persevere in the work,' he proposed, and workers would gather as they had done before in the botanical societies of other countries. Lawson reassured his audience that they were by no means alone in their botanical endeavours. 'We already have observers throughout the length and breadth of Canada, as well as in the other North American Provinces, from the Red River in the far West to the Island of Prince Edward in the East.' Many were Lawson's students, including Andrew Thomas Drummond, who collected locally and joined GSC field trips farther afield, and John Christian Schultz, who botanized at Red River in 1860.[44]

The founding meeting of the BSC attracted ninety-one members from Kingston and environs. They resolved to establish a botanical garden, to investigate the botanically rich region around Kingston and beyond, and to work towards a catalogue of Canadian plants by collecting both specimens and lists of floras of other regions. Lady members, inspired by Lawson's wife, an accomplished amateur botanist, were accorded equal privileges. By the second meeting in January 1861 the society listed 140 paying members who included local members of the Legislative Assembly like John A. Macdonald and Alexander Campbell; at the third meeting, two hundred attended. The BSC, Principal Leitch concluded, was obviously 'wanted by the country generally.'[45]

IV

The enthusiasm that greeted the founding of the Botanical Society of Canada swelled beyond Kingston and even beyond North America. Toronto's *Canadian Agriculturist* applauded the advantages to agriculture and horticulture as well as to botany that such a society with its botanical garden would confer on all the British provinces. Montreal's *Canadian Naturalist* quoted Lawson's introductory speech at length, highlighting his advocacy of forest conservation. The prestige of the BSC soared still higher in 1861, when Sir William Logan entrusted the botanical collections of the GSC to the society's care. Logan's donation, the BSC assumed, indicated his confidence in its ability to 'sustain the character of botanical science in the country.'[46]

The BSC also found acceptance abroad. It attracted more than eighty corresponding members and was mentioned in the columns of British botanical journals like the *Phytologist*. But the society took special pride in a lengthy article in the German *Bonplandia*. Named after Aimé Bonpland, Alexander von Humboldt's botanical collaborator, the journal was edited by Berthold Seemann, a renowned Hanoverian botanist who spent time in England and knew George Lawson. *Bonplandia* carried a two-and-a-half-page feature on the BSC in May 1861. Seemann's article was immediately translated and reprinted both in the society's own *Annals* and in the *British American Journal*. Ostensibly an account of the meeting to organize the Botanical Society of Canada, Canadian readers were informed, the flattering account was really 'a bird's eye sketch of Canada and of the present aspect of Canadian science from a German point of view.'[47]

Seemann described Canada in glowing terms as a place of 'well-being and progress' linked by steam to the Old World and by rail to the New; a place where new towns burst forth like mushrooms. He recognized the Canadian Institute as the germ of a national academy but welcomed, 'both as a botanist and as a German,' the rise of the BSC to the noble array of scientific institutions. As a botanist he recognized between the inhabited parts of North America and the inhospitable regions of the Arctic Circle a broad belt of land which remained to the botanist almost *terra incognita*. In Canada, therefore, a botanical society had for its operations a most extensive field, 'whereon many a (new) plant buds, blooms and withers unnamed, unknown – whereon many a species attains its northernmost limits, and awaits the hour when some savant shall record its discovery in the annals of science.' European botanists anxiously anticipated the work of the BSC and the results of its expeditions into unknown regions. Its mission, Seemann assured, interested the 'whole botanical world.' As a German, Seemann welcomed the society as the fulfilment of Frederick Pursh's plan to collect as many of the plants of Canada as possible. Pursh had died penniless in the attempt and could rest in peace now that the botanical inventory of Canada was at last to be undertaken in earnest.[48]

By mid-1861 the prestige of the BSC was such that members suffixed 'FBSC' to their names. As James Williamson reiterated at the eighth meeting in December, the society had 'struck its roots deeply into the soil, passed the period of youth, and grown up into a goodly tree, whose branches were spread far and wide.' Its 'abundant fruit' included the published *Annals* and a botanical garden, and it was expected ere long not only to add to the range of scientific knowledge but also to yield valuable economic results from experiments on plants suited to the Canadian climate. The garden, the BSC boasted, set an example for other Canadian cities, notably Montreal, and would be linked, as at Kew and Edinburgh, to a public herbarium.[49]

From its inception the BSC's garden, herbarium, and library served both practical and theoretical purposes. Farmers relied upon the society in dealing with aphids and other insect pests. New seeds were distributed, notably 'parsnip chervil [*sic*]' as a possible potato substitute. Lawson also suggested the use of seaweed as manure along the lower St Lawrence and in the eastern colonies. For manufacturers Lawson discussed plants for paper and lichens for dyes. One of his prize students, A.T. Drummond, collected lichens locally and on the Thousand Islands. He presented a paper on their economical uses and samples of dyes extracted from native lichens. In a theoretical vein, members including Drummond and Richardson and Bell of the GSC contributed papers on localized botanical collections. George Sheppard (*sic*) and J.W. Dawson of Montreal also offered studies on the geographical and geological distribution of plants in

Canada.[50] Through the efforts of the BSC, Canadian botany was finally proceeding apace.

An important stimulant for botanical studies in Canada was Sir William Hooker's announced intention to update British colonial floras in a new series. Of particular appeal was the fact that Hooker himself planned to direct the British North American volume. On the basis of an article in the London *Guardian*, Canadians anticipated that Hooker and his staff would visit Canada some time in 1860. Personal correspondence from Hooker soon disabused George Lawson of this belief, but rumours of Hooker's imminent arrival persisted even after Lawson's categorical denials at the founding meeting of the BSC in December 1860. On the contrary, Lawson told his audience, recent communications from the botanical advisers of the home government indicated that Canada must follow the example of other established British colonies and conduct its own investigations into the nature and distribution of indigenous products. This was a task, Lawson suggested, for the BSC. Much like the Geological Survey, the society could become 'instrumental in carrying out investigations of the greatest importance to the country, whether their results be viewed as intellectual achievements or as contributions to material industry.'[51]

In 1860 the secretary of state for the colonies, the Duke of Newcastle, requested that colonial governors circulate a questionnaire on natural history studied in their domains. Newcastle also urged the publication of local floras as part of a proposed twelve-volume popular account, on a 'cheap and uniform plan,' of the natural productions of all the British colonies. The series was to be published in English and 'accessible to all.' Its purpose was to provide practical information to settlers and manufacturers about the economic potential of each colony, especially its timber resources. Although the British treasury supported floras of Hong Kong and the British West Indies, its purse-strings were then pulled tight because the chancellor of the exchequer believed the other colonies could meet the expenses themselves. By 1861 the Australian and other colonial governments had accepted the challenge; where local botanists were not up to the task, British botanists were paid by individual colonies to do the work. In Canada there was no lack of enthusiasm among botanists over the project. Some, like John Macoun of Belleville, even submitted specimens to Kew Gardens. Hooker explained that Macoun was jumping the gun, for unless the Canadian legislature funded a flora, he was unable to use Macoun's specimens. He advised Macoun meanwhile to note the geographical limits in Canada of useful forest trees such as oak, pine, ash, and hickory.[52]

Hooker's intended flora, like the more fortunate geological, magnetic, and meteorological surveys, was deeply affected by the political and financial instability of the Canadian government during the final years of the Union. Newcastle's dispatch was caught up in the political transition from Edmund Walker Head to Lord Monck in 1861. Both Head and Monck approved the financing of a colonial flora, but Monck added that his office required him first to consult the governments of the other British North American colonies. Unfortunately the legislatures of the two most likely to approve grants for a flora, Nova Scotia and New Brunswick, had contracted gold fever and were preoccupied more with a geological than a botanical inventory. No reply to Monck's request was forthcoming, and in none of the colonies did the issue reach the legislature. Monck, in turn, left the British dispatch unanswered for over a year.[53]

When Hooker became restive and made private enquiries late in 1862, George Lawson explained to him that Canada had experienced, with increasing political instability and impending deadlock, yet another change of ministry. The new government suffered financial troubles and was preoccupied with the projected intercolonial railway. John A. Macdonald told Lawson that the issue had reached the Executive Council before his ministry was defeated over the militia bill in 1862, but he could not recall action having been taken. Lawson advised Hooker to apprise the government of the amount expected to be granted towards a flora. Meanwhile, Lawson pressed the matter upon 'leading public men.'[54]

Lawson complained to Hooker that in Canada it was difficult to obtain grants for such useful purposes as a botanical inventory. The observation arose from personal experience. For three years, he wrote, the BSC had been endeavouring to secure a grant to aid in a botanical exploration of the remote unsettled parts of Canada. To date the results had been negative, though the society continued its efforts. If successful, they promised, they would aid Hooker's proposed flora of British North America.[55]

An additional factor in the government's hesitation was that Canada had financed a display of domestic vegetable productions for the international exhibition in London in 1862. The government had then donated the specimens to Kew Gardens. Moreover, it was well known that a flora of Canada had also been undertaken privately, in the form of Léon Provancher's *Flore canadienne*, published at Quebec in 1862.[56]

In a final effort in February 1863, the BSC appointed a subcommittee to petition the legislature for funding of a British North American flora under Hooker's direction. Several members, including Judge Logie of Hamilton, urged the importance of the scheme both from a scientific and a commercial perspective. Logie saw the flora as 'a most effectual means of making known to Canadians, as well as to the inhabitants of European countries, the nature of the products of our rich Canadian forests which would stimulate to new branches of industry, and to the development of commercial enterprise.' This appeal to utility and national pride resembled closely that used to justify government support of contributions to international exhibitions. In case this request failed too, the BSC formed contingency plans to launch a campaign for public subscriptions.⁵⁷

In May 1863 Sir William Hooker, still awaiting a formal reply from the Canadian government, turned responsibility for the British North American flora over to his son, Joseph Dalton Hooker. In January 1864 Kew Gardens issued a formal request for information and specimens of plants 'not hitherto recorded as Canadian.' In Hooker's view the geographical distribution of plant species still remained the most critical botanical question in Canada. Except for forest trees and shrubs still desired, Hooker, annoyed by the Canadian government's neglect, abandoned hope of full Canadian participation and declared the specimens already collected at Kew 'very ample for the object in question.'⁵⁸

V

Unfortunately, by the time Hooker was ready to receive botanical contributions from Canada, George Lawson was no longer in the province to guide Canada's largest organization of amateur botanical collectors. Quarrels among the fractious faculty had persisted at Queen's College; Lawson's celebrated appointment and corresponding salary aggravated the strained relations between the trustees and the medical faculty, both of whom he served, into a seething sore. To the trustees' dismay, Lawson tried to reform the college's constitution. Then when John Stewart resigned as secretary to the medical faculty in 1861, Lawson earned his bitter enmity by succeeding him. Stewart owned the *Argus* newspaper and published in it his version of events at the college with a libellous vengeance. Lawson, he charged, was nothing but 'the ex-head gooseberry bush pruner in the experimental gardens at Edinburgh.' A new wave of 'strife and slander' divided the faculty more deeply.⁵⁹

Tensions exploded in January 1863, when new university statutes defining the authority of the trustees over professors were interpreted as a breach of faith. 'You are accustomed,' Lawson wrote to J.H. Balfour, 'to regard the severe climate as the great drawback of Canada. I can assure you however that the storm of human passion that we have had raging in Kingston ... has cast all other discomforts in the shade.' 'Public morality' had declined at Kingston, he continued, and the 'trench between College and town is so wide, that none of the Professors feel very comfortable; and, like others, I should be happy to make a change if a favourable opportunity occurred.' In January he withdrew as secretary of the medical faculty: 'One has to live a few years in a Colony,' he confided bitterly, 'to see how much evil there is in the world.' In October he submitted his resignation, effective on 1 November. A hero to the students and to some of the faculty, Lawson left the college in a state of 'riot and confusion.' Stewart had been jailed, another professor dismissed, and the principal, William Leitch, died allegedly of the strain. Lawson moved to Dalhousie University in Halifax, where he became professor of chemistry and mineralogy in November 1863. In addition, he served as secretary to the Board of Agriculture of Nova Scotia under the Tupper government, an influential position that he retained until his death in 1895.[60]

Lawson's sudden departure from Queen's left another gap in Canadian botany. He was widely acknowledged as the 'initiator and soul' of the BSC, and the threatened demise of the society was a source of private glee to some. The NHSM, while recognizing the importance of Lawson's organizational efforts, remained wary of any claim that the BSC had 'national' status or was anything more than a local institution. The NHSM aspired to recognition 'not for Montreal alone but for Canada, and [as] far as may be, for the world.' Accordingly, the *Canadian Naturalist* consistently referred to its perceived rival as the 'Kingston Botanical Society.' Nor were certain members of the NHSM willing to accept any notion that their botanical garden followed a precedent set at Queen's. Feelings of rivalry ran deeply enough to elicit attacks upon Lawson's garden in print.[61]

This rivalry was not confined to Montreal. At Toronto, William Hincks bristled when the founders of the BSC represented the science as neglected or hardly known in Canada. Hincks took this remark as a personal insult and countered that, first, there were colleges in Canada with professors expressly teaching botany. Second, the Canadian Institute was already promoting botany in Canada in its capacity as 'our national scientific association.' Third, a small network of 'zealous cultivators of Botany' had already formed around Hincks at Toronto. Although he accepted honorary membership in the BSC and wished it well, he announced his intention to submit his own work to the *Canadian Journal*.[62]

Hincks professed that Canada's condition as a new country required a united effort among 'cultivators of all the different branches of knowledge' in one institution. For this reason he believed the formation of formal scientific or literary societies to be 'an act of hostility' against the Canadian Institute, depriving it of submissions. Whatever success such distinctive scientific bodies did manage to obtain, Hincks concluded, would mean so much strength drawn away from the Canadian Institute, 'which aims at a wider usefulness, and has claims on the patronage of every man in the Country who loves and values knowledge and culture – which offers them all privileges, such as no limited body can pretend to afford.'[63]

Given his background and botanical interests, it might be expected that

Hincks did not share the wider aims of Lawson and the BSC. Hincks, unlike Lawson, did not propose to botanize north of settled Canada. With minor exceptions, he had not collected any plants not found in the northern United States; nor did he expect to do otherwise. Hincks's main concern was not to extend the list in Gray's *Manual* but merely to eliminate plants that did not belong in a purely Canadian flora.[64]

Hincks's eccentric approach to botany found reinforcement in his exclusive use of Lindley's classification not only in his own herbarium but also in the classroom. He hoped eventually to apply the system to a 'national' flora of Canada. Hincks understood the importance of uniformity in classifying botanical specimens, but his faith in Lindley led him to argue that everyone else should give up de Candolle. On the one hand he invited Abbé Louis-Ovide Brunet to co-operate towards a definitive Canadian flora, but on the other hand he dismissed Brunet's de Candollean classifications as limited in value, if only because he used the Lindleyan system at Toronto. Because Hincks identified the limits of Canadian botany with those of Canadian settlement, and because he still hoped for a flora organized along Lindleyan lines, he judged the BSC harshly both for its location and for its presumption, 'as if first promoting botanical pursuits in Canada.'[65]

William Hincks's interests diverged after 1860 to embrace not only natural history but also political economy. But he shared with other advocates of a flora of Canada the 'not unnatural feeling of a desire as a nation to provide for our own scientific wants,' instead of relying for reference upon American works. Whereas Hincks believed the main contribution of a catalogue of Canadian plants would be to eliminate confusion caused by the greater number of plants listed in Gray's *Manual*, others saw a more positive purpose. John Macoun realized, in contrast, that Gray's *Manual* failed to describe many of the plants found in Canada. Along with other members of the BSC, Macoun pinned his hopes on Hooker's proposed flora through 1864. The final blow came in 1865, when Hooker's death postponed the project indefinitely.[66] With Lawson gone and Hooker deceased, if Canadian botanists were to have their inventory of Canadian plants, they would have to produce it themselves.

13

Flower and Fruit: The Nation as Variation

Men are like plants; the goodness and flavour of the fruit proceeds from the peculiar soil and exposition in which they grow.

J.H. St John de Crevecoeur, *Letters from an American Farmer* (1782)

By the 1860s botany had joined the vanguard of inventory sciences which supported the idea of a transcontinental Canadian nation. Certain plants, their geographical distribution, and Canada's potential for cultivation were adopted as symbols and pressed into service as portents of this expansive destiny. The maple leaf, for example, had been used as an emblem in Quebec since the seventeenth century and was widely accepted by the 1830s among both French and English Canadians. Much as the Geological Survey of Canada encouraged appreciation of the Laurentian Shield as a peculiarly Canadian environmental attribute, so botany seemed to show the maple, widely distributed across Canada from Lake Superior eastward to the Atlantic seaboard, to represent Canadian unity. The maple lacked the undesirable connotation attached to the beaver, a member of the rodent family. But it posed problems later when it was shown not to grow west of Manitoba; nor could the maple, ironically, represent permanence. 'If it remains in the tree,' Sandford Fleming admitted, 'it disappears with the summer, if plucked from the tree it ... almost immediately wilts ... and perishes.'[1]

Yet in 1860 Canadians attributed great significance to the royal visit during which the Prince of Wales planted a Canadian maple in Toronto's new botanical garden. Some Canadians saw the gesture by their future sovereign as symbolic of an incipient Canadian nationhood. In recognition of the sense of nationality they deemed indispensable to the further progress of the country, Egerton Ryerson, J.H. Morris, and others organized a procession of native-born Canadians. Those born in Canada or, it was emphasized, in any other British North American colony distinguished this fact by wearing the symbolic maple leaf. Ryerson believed that the movement would 'blend the whole population of Canada in one deep, universal, unanimous feeling of devotion to the best interests of their common country.'[2]

Some Canadians took the national symbolism of their plants to ridiculous extremes. When a British newspaper complained that *Elodea canadensis*, a seaweed, had crossed the North Atlantic and caused havoc in European ports by clogging up the harbours, the indignant public response was almost silly. Two anonymous correspondents, 'VC' of Peterborough and 'Canadiensis' of Hamilton, vented spleen in the *Canadian Agriculturist*, incensed that a 'Canadian' plant could be accused of such mischief. When confronted in a rebuttal with the truth of the damages inflicted by *Elodea*, 'VC' retorted that in that case the weed could be no Canadian.[3]

Canada's vegetable products brought a positive international recognition during the same decade. Much like William Logan's achievements at the Crystal Palace, the success of agricultural and forestry displays at the London exhibition in 1862 crystallized the progress of agriculture and horticulture as a source of special pride in all the British North American colonies. Even Newfoundland, crowed the *Canadian Agriculturist*, had proven quite unlike the bleak and inhospitable country that had been supposed. Canada earned notices for exhibits of native plants and woods in the natural history department at London. Thanks to the efforts of J.B. Hurlburt, who systematically named and arranged the large Canadian collections, Canadians read in the press: 'At no previous exhibition in this or any country has so splendid and valuable a display of the products of forests and plantations been exhibited.' Science and commercial enterprise had 'gone hand in hand' to provide the plant and product information whose lack had 'rendered so much that was sent to the exhibition of 1851 completely useless.' The same spirit of critical self-appraisal measured improvement 'in all that relates to material wealth and solid progress' at annual provincial exhibitions in Canada. Natural history competitions at these exhibitions rewarded the efforts of amateur botanical collectors including Catharine Parr Traill of Peterborough, Judge Alexander Logie of Hamilton, and William Saunders and his wife of London. Supporters of a new agricultural hall in Toronto in 1863 proposed to assemble a scientific collection of agricultural plants from all parts of Canada and British North America.[4]

Canadians realized that their very future depended upon Canada's image abroad as an agriculturally promising country. The 'national' symbolism lent to certain plants was reflected also in horticulture, particularly in pomology, or fruit culture. In 1859 at the provincial exhibition held at Toronto a small group of amateur horticulturists from Niagara organized the Fruit Growers' Association of Upper Canada. Upon the death of the first president the following year, the association was reorganized at Hamilton in 1860. Its executive included Judge Alexander Logie, Dr William Craigie, J.B. Hurlburt, now also residing in Hamilton, and D.W. Beadle, an amateur naturalist from St Catharines.

As its new president, Judge Logie defined three main aims of the Fruit Growers' Association. The first was to centralize fruit culture in Canada, standardizing names of varieties to reduce confusion and fraud. The second aim was to improve indigenous fruits to withstand the rigours of the Canadian climate. The Fruit Growers' Association proposed to eliminate needless expense by identifying varieties that could not flourish in certain localities. Moreover, it would try to reverse the noted decline in productivity in older orchard regions of Canada. The third and most important aim was to increase public awareness of Canada's potential as a fruit-growing region. Logie in Canada West concurred with Abbé Léon Provancher in Canada East that fruit could successfully be cultivated much more widely in Canada. Provancher added that Canada's extremes of climate rendered foreign agricultural theories irrelevant, and he urged Canadians to develop their own unique approaches. For a hardy climate, he insisted, gave vigour to vegetation, a fact that Canadians could turn to advantage instead of shrinking from as a presumed insurmountable liability.[5]

The main interest of the Fruit Growers' Association was grape cultivation. The Viking image of North America as a 'Vinland' where grapes flourished in a natural state had gradually been obliterated by impressions of later explorers farther inland. In 1848 Louis Agassiz judged that the local grape species, although they stretched inland to the Niagara region, were commercially useless.[6] The Fruit Growers' Association hoped to alter this widespread perception.

The association understood only too well the wider implications for Canada's future of such a change in outlook. As William Hincks assured them, 'If Canada could become known as a wine producing country, the effect in creating a favorable opinion abroad of our climate and resources would be greater than that of almost any other fact that we could establish.' J.B. Hurlburt, president in 1861, reiterated that a Canada 'covered with the vine, would appear to Europeans a very different country from their present estimation of it. This one fact would be worth volumes written on our climate, and would do much to turn the tide of immigration to our shores.'

The obvious question was whether Canada could ever be covered with the vine. According to Hurlburt, the fact that Hamilton and Montreal enjoyed higher summer temperatures than most wine-growing regions, including Bordeaux, constituted strong positive evidence. Quoting Sir John Richardson and Sir George Simpson, Hurlburt added that native vines grew co-extensively with the maple, known then for its diffusiveness. He argued circularly that 'when a plant is found like the vine in Canada spontaneously springing from the soil, we have the best proof that those regions where it has established itself, are by nature adapted to it ... What then,' he continued, 'may these same vines become under the fostering hand of the skillful gardener?' Harking back to Alexander Gordon's promotion of gardening a generation earlier, Hurlburt predicted: 'That kind of culture which has educed from the sour crab apple the beautiful and delicious specimens of that fruit now before us, and done so much in improving every variety of our fruit, cannot be lost upon the wild grape, and may bring from it berries equally luscious with the European varieties and much better adapted to our climate.' He concluded optimistically that the grape could be grown profitably in British North America wherever Indian corn ripened: as far north as the Saguenay in the east and as far as 54 degrees north latitude on the Saskatchewan in the west. All that was needed to promote Canada and the territories beyond as grape-growing regions were 'adjustments,' which only botany could provide.[7]

I

Theoretical developments brewing since the 1840s spotlighted Canada's geographical position as a northern land. In a lengthy exposition on the state of botanical theory in 1853, Joseph Dalton Hooker cited the geographical distribution and variation of species as among the most challenging aspects of the science. Hooker greatly admired Edward Forbes's 1846 paper 'On the Connexion between the Existing Flora and Fauna of the British Isles, and the Geological Changes which have affected their Area,' as 'the most original and able essay' that had ever appeared on plant distribution. Forbes analysed British vegetation as comprising five characteristic floras: west and south-west Ireland; south-west England and south-east Ireland; south-east England; the British mountains; and Britain generally. He correlated each flora to a geographical region from which it had migrated: northern Spain, northern France, northern Germany, Scandinavia, or central and western Europe, during a particular geological period. Hoping to advance from British vegetation to a general theory of plant distribution, Hooker distilled four axioms from which his own work derived. First, all individuals of a species proceeded from one parent (or pair) and retained their distinctive specific characters. Second, species varied more than was generally admitted to be the case. Third, species were more widely distributed than was usually supposed. Fourth, their distribution had been effected by natural causes, although not necessarily the same causes as those to which they were now exposed.

Hooker suggested that since land masses in the northern hemisphere were far more extensive than those in the southern, careful comparative work on the floras of North America and Europe would throw light on the origins, distribution, and permanence of species. His own knowledge of plant distribution in the northern hemisphere benefited from collections by M.E. Bourgeau, botanist to the Palliser expedition in 1857, but much remained to be investigated. Even Hooker, the foremost expert on phytogeography after Forbes's untimely death in 1854, was unsure how to explain plant distribution from the axioms he had laid out.[8]

With a few more links established by the end of the decade, botanical theory became one of the sharpest cutting edges of the uniformitarian revolution in biological thought. The posthumous publication in 1858 of Forbes's work on Tertiary geological formations supported Charles Lyell's prior conviction that no geological cataclysm separated the modern from the preceding Tertiary age. For botany this meant that entirely new species of modern plants need not have been created after the Tertiary age. That same year Charles Darwin's paper to the Linnean Society entitled 'On the tendency of species to form varieties; and on the perpetuation of varieties and species by means of natural selection,' followed in 1859 by his monograph, finally forged a theoretical link between variation and distribution in plant species. Darwin had been converted to evolutionary theory by the geographic distribution of plant and animal forms. Rejecting idealist methodologies which forced nature into preconceived patterns using hypothetical land-bridges and other means to explain plant migration, Darwin propelled the biogeographical approach towards a status parallel to Lyell's uniformitarian geology.[9]

Darwin's theory of 'creation by variation,' as Hooker called it, gave botany the key to the phylogenetic relationship of plants by modifying Hooker's first and fourth axioms. Hooker had in any case retained the principle of fixed specific characters as a taxonomic tool rather than a dogmatic necessity. Whether he or other botanists explicitly accepted the mechanism of natural selection was largely irrelevant to the theory's influence on botany. The mass of detail put forward by Darwin opened new areas of botanical research, including work on the distribution of cryptogams, a large proportion of the northern vegetation that ranged so widely in British North America.[10]

In 1862 Hooker's 'Outlines of the Distribution of Arctic Plants' accepted Darwin's theory and confirmed the important place of British North America in understanding plant distribution. He postulated that the northern vegetation had migrated from Scandinavia to Asia and America during and after glaciation. He identified five circumpolar Arctic regions with distinctive floras: Europe, Asia, Western America, Eastern America, and Greenland. Closely acquainted with Darwin during the formative years of his *Origin of Species*, Hooker adopted the derivative origins of these plant species as the only logical explanation for their patterns of distribution.

Hooker's Arctic American districts (the Western from Bering Strait to the Mackenzie River, and the Eastern from there to Baffin Bay) raised interest in British North America as a northern territory. During the glacial epoch, Hooker believed, the Arctic flora had advanced southwards into what were now tropical and south temperate zones. Then during the succeeding warmer epoch, northern species in these warmer regions flourished only on mountain tops. The rest 'retreated' northwards, 'accompanied by aborigines of the countries they had invaded during their southern migration.' Struck by the clear division between the more hardy and widespread Arctic Eastern American and the more isolated Greenland types, Hooker postulated: 'The rustic denizens of Greenland, huddled upon the point of the peninsula during the long glacial cold, have never enjoyed the advantages of foreign travel; those of the adjacent continents on either side have 'seen the world', and gained much improvement and diversity thereby.' He demonstrated first that patterns of the geographical distribution of plant species could be explained; second, that existing species dated back to far more ancient times than formerly believed; and third, that these species had survived great climatic change. In North America, Hooker pointed botanical study in a northerly direction, away from the American border regions. His theory found immediate acceptance from Asa Gray at Harvard. Within five years Hooker entrusted the whole North American field of botany to Gray's care.[11]

The reception of Hooker's theory by prominent scientists in Canada was generally more reserved. At the farthest extreme and not representative of British North American botanists, J.W. Dawson abhorred Hooker's flirtation with Darwinism. The idea of unlimited variation in species, Dawson believed, ignored the permanence he saw in fossil plant forms. Dawson attempted to drive a wedge between Forbes's more innocuous glacial theory of geographical distribution and the unacceptable Darwinian corollaries utilized by Hooker. He recognized a struggle for existence not among plants but only between plants and conditions external to them.[12]

Dawson published his own geological interpretation of alpine and arctic plant distribution in 1862, before having seen Hooker's article. In his explanation, plants migrated only as seeds and thus showed little tendency to vary over time. The struggle for existence seemed to Dawson to confirm rather than to modify the characters of species. Both Hooker and Darwin believed that Dawson had intended to review *The Origin of Species* without having seen the book until Charles Lyell talked him out of it. His unbending vehemence damaged his international scientific reputation in botanical questions.[13]

But Dawson remained first and foremost a geologist, and his views by no means necessarily reflected those of botanists around him. Hooker's acceptance of Darwin's theory did not alarm Canadian botanists in the same way. One reason was that Hooker's version did not force them into an immediate stand on the issue of natural selection. If anything, it broadened their geographical horizons and encouraged them to collect plants for evidence one way or the other.[14]

George Lawson recognized the importance of Hooker's essay on arctic plants for his own understanding of their distribution in British North America. He seized the challenge of Nova Scotia botany as an opportunity to extend the work he had begun in Canada. Lawson drew a considerable distinction between the Nova Scotia flora and the continental or 'American' cast of Upper Canadian plants. In accordance with Hooker's theory, he found that northern plant forms spread farther south along the Atlantic coast than they did inland, and that Canadian swamp plants flourished instead on east-coast hillsides.[15] Lawson hesitated to pronounce judgment on the origins of plant species because he believed Darwin's case was not strong enough to be compelling. But he did encourage British North American botanists to gather specimens of heather in order to determine whether the alpine plant was indigenous to British North America (as Hooker would have it) or introduced later by settlers (as Dawson argued). Its habitat was out of the way of most ordinary collectors, but by 1865 Lawson agreed with Asa Gray that heather was indeed native to Nova Scotia and had once spread widely over North America.[16] This discovery in essence supported Hooker's theory of plant migration.

Despite Lawson's caution, at least one member of the Literary and Historical Society of Quebec, R. Sturton, was led by Hooker to accept that 'most of our species are only varieties of that variety which crossed at the north' from Eurasia to North America. In Nova Scotia Dr John Sommers, a native of Newfoundland and professor of physiology at the Halifax Medical College, agreed that the presence of common heather in Nova Scotia was evidence of prehistorical links of Nova Scotia's arctic and subarctic flora to that of northern Europe. Like Gray and Lawson, Sommers viewed heather and other rarer species as 'more probably native forms seeking refuge from extinction, than immigrants seeking establishment on a new soil.' For unlike heather, immigrant plants were found only in the vicinity of human habitation, competing with native plants for cleared land neglected by farmers.[17]

Sommers was among the few in British North America who dealt with deeper problems in species botany. He noted the difficulties of attempting to classify Compositae, typified by asters and by the Solidago family of goldenrods native to North America, because of their infinite variability, or 'plasticity.' Sommers questioned the penchant of botanists for grouping such specimens artificially into species rather than into varieties or subspecies. For Sommers it was clear that only uniformitarian forces could be responsible for both simple and complex variations in plant forms: 'It is not at all likely,' he reasoned, 'that we have had separate acts of creation for the species of two continents, or that the forces of nature had produced national distinctions of this kind: these forces are cosmopolitan, blind and rigid in their action, producing always similar results under similar conditions; light, heat, moisture, are the same everywhere in their action.'[18] Study of these dynamics, which effected the geographical distribution of plants, meshed with more localized developments in Canada to direct the attention of amateur botanists in a northerly direction.

II

When George Lawson left the province of Canada in 1863 he feared that the Botanical Society of Canada would collapse and that for lack of a competent overseer its botanical collections would default to Kew Gardens in London.[19] Although the BSC did cease to meet and to publish its *Annals*, enough of its spirit survived that Lawson's second concern was unwarranted. But the long-term result of his departure was certainly a shift in the locus of botanical activity away from Kingston.

Almost immediately the governors of Queen's College appointed a successor to the chair of chemistry and natural history. He was Robert Bell, since 1857 an assistant to the GSC. Bell's connections in government and scientific circles as well as on the college's board more than compensated for his lack of training in chemistry. He supported his application with recommendations from Sir William Logan, who had appointed Bell to the GSC as a personal favour to Bell's late father, and from J.W. Dawson, who had taught Bell at McGill. Logan and Dawson agreed that Bell had proven himself capable of collecting, naming, and classifying geological and botanical specimens and that any further deficiencies could be rectified by a term or two at the University of Edinburgh. For good measure, Bell added the compelling argument that had worked so well for Sir William Logan in 1842:

My Canadian birth, training and sympathies, together with my extensive knowledge of the province and its resources will enable me to enlist the interests of Canadian students in the study of the subjects proper to the Chair to which I aspire ... Besides the benefit which the College will reap from my thorough knowledge of the country, the extensive acquaintance which I enjoy with the Canadian people of all classes, will enable me to render the College better known to intending students.[20]

Bell received the appointment upon 'evidence of his future eminence as a lecturer and man of science,' but his primary interest had never been botany.

In some respects Lawson's move to Halifax was beneficial for Canadian botany in that it broadened the horizons and the responsibilities of Canadian botanists. Even William Hincks found himself wishing the BSC could be extended throughout British North America to take advantage of the means of communication afforded by the proposed intercolonial railway. The BSC had collected plants from southern Canada as far east as the Gaspé. Its secretary, Andrew Thomas Drummond, continued operations for a time and hoped even to expand the *Annals* to a quarterly journal. Although in the long run he was unable to sustain the society, whose international connections rested upon Lawson's reputation, Drummond did strengthen personal botanical ties throughout British North America. Abbé Louis-Ovide Brunet of Laval, for example, expressed interest in joining the BSC in December 1863. In return, he sent botanical information on the Lake St John and Saguenay regions to Drummond.[21]

A.T. Drummond was born in Kingston in 1844. His father, Andrew, had emigrated from Edinburgh a decade earlier to join his uncle on the Rideau Canal works; his mother was a cousin and adopted sister of Oliver Mowat. The elder Drummond then managed a succession of branch banks throughout Upper Canada. While living at Bytown, where Andrew Thomas received his elementary education, Andrew Drummond also served on the council of the Bytown Athenaeum. His son entered Queen's College in 1857, the youngest student to that time, and earned his bachelor of arts degree at sixteen. A student of George Lawson and a friend of Robert Bell and his brother John, another amateur botanist, Drummond excelled in geology and botany. He practised law at London, Canada West, from late 1864 to 1869, when increasing deafness forced him into commercial pursuits in Montreal. Drummond's business interests were linked to Canadian expansion into the north-west. His scientific interests meshed well with these main pursuits. Drummond's botanical collections emphasized geographical distribution, particularly among forest trees, in Canada. By 1864 he had followed Edward Forbes in dividing the vegetation of Canada into five types: Canadian, Erie, Superior, Maritime, and Alpine. He observed on a more minute level that even Upper and Lower Canada had developed unique floras, and he worked to document these differences.[22]

In addition to Brunet at Laval, Drummond corresponded with Montreal naturalists such as Elkanah Billings and the cryptogamist David Allen Poe Watt. Watt was born in Scotland in 1830 and immigrated to Montreal, in 1846. Known then merely as David Allen Poe, he entered Canada under indenture to his uncle, a shipping and exporting merchant. During the 1850s he joined the Board of Trade and helped organize the Montreal Corn Exchange. Politically an 'advanced Radical' who supported universal adult suffrage, and economically a promoter of free navigation along the St Lawrence system, Watt collected cryptogams in his spare time and became an authority on ferns and fungi.

As an amateur naturalist, Watt was driven partly by resentment of imperial botanists who overlooked native 'northern North American' flora. In an article in the *Canadian Naturalist* he rearranged a British classification of North American ferns, appending a list of his own specimens along with his views on their proper nomenclature and classification. During the 1860s he determined to produce a full catalogue of Canadian plants. Thus bringing back to Montreal the initiative to complete a published botanical inventory of Canada, Watt enlisted friends in Canada West, notably A.T. Drummond and William Saunders, as botanical collectors.[23]

By 1865 Watt had amassed a sizeable collection of local catalogues by Canadian botanists from London to the Gaspé and had begun annotating his 'Catalogue of Canadian Plants.' The task was an enormous one to undertake privately. Watt was still annoyed at the Canadian government's failure to 'cheerfully contribute' its share towards Hooker's proposed colonial flora. He chafed too at the similar failure of an obvious alternative, the appointment of a botanist to the GSC, as an unfortunate loss not only for science but also for Canada. Watt believed the survey had been kept too restricted, and he did not mince words: 'All has been devoted to the fossil, almost nothing to the living. Had Sir William been provided with means to extend his survey so as to report on the natural productions of a district as well as on its geology, the country might have been saved the thousands it has spent in making so-called colonization roads through uncolonizable territory, and in surveying lots unfitted for settlement.'[24]

In 1865 Abbé Brunet published his *Catalogue des plantes* at Laval. Brunet intended to establish exact localities for many rare Canadian plants as a step towards a fuller understanding of the geographical distribution of Canadian species. He hoped to encourage contributions towards 'l'inventaire complet de toutes nos richesses végétales,' including distinctions between indigenous, naturalized, and cultivated plants. Brunet's botanical pursuits emphasized the historical as much as the economic importance of his subject. He credited the botanical museum at Laval University with the growing popular interest in natural science he noticed among Canadians. Begun in 1862, the museum was like Kew Gardens, 'an open book to educate the public' and a permanent Canadian exhibit to apprise domestic industrialists and foreign merchants landing at Quebec of the rich resources of the country and their uses.[25]

David Watt was surprised to learn that, with all this published evidence of the botanical inventory being conducted in Canada, George Lawson still intended to assemble his own catalogue of Canadian plants in Halifax. Lawson hoped also to revive a new series of *Annals* for the BSC, which he seemed to believe still existed in 1867. He did not foresee the completion of his contemplated Canadian flora for some time: 'Every day seems to extend the list,' he lamented to Drummond, and there were 'so many men working at the same subject in ignorance of what each other are doing.' Nevertheless, 'pretty complete lists' of plants in New Brunswick, Nova Scotia, and Labrador had sharpened Lawson's botanical overview of British North America.[26]

Another ardent amateur botanist who identified specimens for Drummond and Watt was John Macoun of Belleville, who emigrated from Ireland in 1850. Macoun harboured a passion for cryptogams and named new species after their Canadian localities. He entertained Brunet as a botanical guest during the summer of 1866 and communicated regularly with George Lawson. Macoun aimed ultimately for a complete herbarium of arctic plants from all over British North America. He accordingly challenged younger botanists like Drummond to compete for greater numbers of new species and locations. He envied Drummond's opportunity, as a friend of Robert Bell, to accompany a GSC field excursion northward on the Laurentian Shield in 1866. The pill was a particularly bitter one for Macoun to swallow because he had failed years earlier to win an appointment to the survey for much the same purpose, to botanize beyond the settled limits of the province.[27] Macoun's luck was soon to improve.

III

By the mid-1860s Macoun, Drummond, and others realized that their preliminary botanical inventory of the province of Canada, like the GSC, had reached its eastern and western geographical limits. They were eager to extend their researches beyond the political boundaries of the province. For instance, botanists had catalogued all the known ferns of the Canadian and maritime colonies. Watt and others were curious to discover whether new species occurred on the north shore of Lake Superior. With Lawson working on cryptogams and arctic plants north-eastward, plenty of challenges still awaited Canadian botanists who turned their attention north-westward. Lawson too suspected that the species he studied from the northern and eastern parts of Canada would be more varied and fully developed in the HBC territories. Confederation brought these expanding conceptual horizons within the realm of possibility. After 1867 botanists endeavoured no longer to publish a limited colonial but rather a comprehensive dominion flora embracing known latitudinal ranges of plants even in the HBC territories.[28]

In 1867 A.T. Drummond followed Edward Forbes and J.D. Hooker in relating plant distribution in Canada to the country's geography and geology. By considering the influence of land-forms and constant forces of change such as light, wind, and water upon the diffusion of plants in British North America, Drummond substituted a broader and more dynamic uniformitarian conception of botany for the narrower and more static understanding held by many naturalists. Although Lawson had seen only three categories of continental, maritime, and arctic flora in British North America, and Drummond himself had previously isolated five, he now differentiated nine floral 'aspects': Canadian, Superior, Erie, lower St Lawrence, boreal, inland maritime, Upper Canadian, Lower Canadian, and local. It was not just soil constituents that determined the distribution of these various floral types but a myriad of interacting universal, continental, and local causes, which had yet to be sorted out.

Drummond considered the causes of various plants' introduction to any given locality, as well as their means of distribution and diffusion. He categorized 'habits' of naturalized species in five different types: 'incidental escapees,' 'adventive' or contiguously spreading plants, naturalized foreign plants, species both indigenous and naturalized, and native species with habits of introduced plants. Like most Canadian botanists, he did not concern himself with exact mechanisms of plant naturalization, except to wonder whether changes in species could in any sense be considered permanent. Drummond continued his phytogeographic investigations with a statistical study of the floras of Ontario and Quebec, again with a wider view which included the north-west. The patterns of distribution of many plants in the United States, he pointed out, followed a line that bent north-westward, somewhat analogously to Lorin Blodget's isotherms.[29]

While it was natural for Canadian botanists by the 1860s to set their sights upon the untrammelled flora of the Laurentian Shield and beyond, John Macoun had dreamed of it since 1850 at least. Macoun was a self-taught botanist who imbibed available works, including Humboldt's *Cosmos* and Lyell's *Principles*, supplemented by personal correspondence from prominent American and British botanists. He was deeply influenced by *Lake Superior*, the report of Louis Agassiz's expedition to the north shore in 1848. Agassiz's account offered key insights which Macoun hoped to test on his own journey to the region. Unlike A.T. Drummond, Macoun followed Agassiz in giving overriding importance to soil and moisture in determining the vegetation of a region 'as certainly as the sun rises in the east and sets in the west.' 'Of the 1300 species of plants I collect,' he confided to Drummond in this regard, 'I could tell any person the exact locality in which he might expect to find them.'[30]

When George Lawson offered financial aid for Macoun to explore lesser-known parts of Canada, Macoun joined an expedition to the Muskoka lakes and to the source of the Trent River in 1868. His chief desire, a full collection of British North American plants east of the Rockies, found support from George Barnston and David Watt, who raised subscriptions in Montreal to finance Macoun's botanical tour of the north shore of Lake Superior in the summer of 1869.[31] Macoun returned with new species for his subscribers, mainly members of the Natural History Society of Montreal, and with firm theoretical convictions.

Macoun had gone to Lake Superior aware that Agassiz classified its flora as subarctic. But six years earlier in the BSC Macoun and Lawson had noted the effects of cool bogs, swamps, and even boulders in creating local conditions suitable for northern flora in isolated stations near Kingston. Macoun applied this knowledge to refine Agassiz's account in 1869. He found that while subarctic plants did grow within a hundred yards of frigid Lake Superior, the vegetation became more temperate in character even three hundred yards back from the lake. The changing character of vegetation north of Superior convinced Macoun that he was passing into a warmer instead of a colder climatic zone. Given Agassiz's habit of comparing North America to more familiar Europe, it was natural that Macoun noticed mainly aspects of this new climatic zone that resembled his own more familiar central Canada. Coupled with Macoun's belief that cultivated plants could adapt to any soil that was clear and open and far from woods, direct observation led him to assess the open prairie of the north-west with growing enthusiasm. 'I am quite satisfied,' he assured Drummond in 1871, 'that the sub-alpine plants found around Lake Superior are an isolated colony and not connected at any point with the sub-alpine flora of the north.'[32]

Macoun was, of course, not alone in perceiving the north-west in terms of the more familiar central Canada. His promotion of a botanical inventory of all the accessible regions of British North America was part of the expansionist outlook suggested in George Barnston's earlier study of Cruciferae as well as in the writings of H.Y. Hind and others. This expansionism found particular expression in popular accounts of the Palliser and Hind expeditions. In addition to geology under the supervision of R.I. Murchison and geomagnetism under Edward Sabine, botany under W.J. Hooker was explicitly represented in the exploring party led by Capt. John Palliser in 1857. Although Hooker had instructed M.E. Bourgeau to co-operate with James Hector in noting timber trees in the north-west, the two returned with more detailed information about the wide geographical distribution of botanical species. Hector, a corresponding member of the BSC, reported on botanical distribution throughout the central part of British North America. His report reached the *Canadian Naturalist* through George Lawson in 1861. Combined with H.Y. Hind's favourable assessment, Hector's information gave Canadian agricultural expansionists the same encouragement given to their mining and manufacturing counterparts. New agricultural possibilities seemed limited for the time being to the Ottawa and Red River valleys and to the 'fertile belt' of the north-west. But farmers concerned with improvement in Perth saw the agricultural development of the Ottawa Valley as a preliminary step to Canadian north-westward expansion, towards a new agricultural frontier to supplant the old St Lawrence system. Botanical and related agricultural interests, like geological and mining interests, joined a vanguard of expansion north-westward to the Mackenzie River district and the Rocky Mountains.[33]

The expansionism exemplified by John Macoun infected another member of the BSC, John Christian Schultz (1840-96). Born in Amherstburg, Upper Canada, Schultz enrolled in medicine at Queen's and studied botany under George Lawson. Schultz took up botany for health reasons as well as professional ones. In 1860 he botanized along the old Red River trail, reporting his findings to the BSC. Lawson listed Schultz's Red River specimens in his own 'Synopsis of Canadian Ferns and Filicoid Plants' and welcomed Schultz's suggestion that Canadians revive the lucrative ginseng trade with China in the British American north-west in order to share in a similar trade centred in Minnesota.[34]

Schultz eventually moved to Red River and became the driving force of the Institute of Rupert's Land, founded in 1862. In addition to the work it could do in the limitless field for scientific inventory offered by the north-west, Schultz believed that the institute could unite political factions and even social classes upon a base of their common interest in development and progress. But, like Lawson at Queen's, he succeeded in mitigating only for a little while the clash in which he himself had participated. Within a year the institute collapsed under the weight of Schultz's opposition to the control over Rupert's Land held by the HBC, whose officers formed the majority of the institute's membership.[35]

After the new dominion acquired the company's territories, plant geography continued to transform the image of the north-west and the prospect of Canadians' immediate future there. Botany justified the expansion of Canadians north-westwards to a climate as favourable as their own. Whereas H.Y. Hind's reports had supplied mainly meteorological evidence for such favourable comparisons, by the 1870s his expansionist arguments had found amplification in new botanical evidence provided by John Macoun. Macoun crossed the continent with

Sandford Fleming's survey of a route for the Canadian Pacific Railway in 1872. Astonished to find cactus growing at 56 degrees north latitude, Macoun became convinced that vegetation in the north-west flourished over a much wider geographical range than was normal. After a second exploration with A.R.C. Selwyn of the GSC in 1875-6, Macoun pressed for a complete botanical survey of the north-west. He traced changes in vegetation from cactus to prairie sagebrush and bunch grass in the more arid regions. Believing he knew the flora of the Dominion from Atlantic to Pacific thoroughly, Macoun confidently proclaimed the north-west as 'wonderfully like the country lying northeast of the Caspian' Sea and embarked upon his well-known promotion of the territory.[36]

Even Macoun admitted to being 'confounded' by finding species of cactus, wheat, and potatoes growing 'luxuriantly in the same field' as plants of a cool climate. But his narrowing focus upon geographical distribution as a measure of unknown regions led him erroneously to rely upon 'the botanical test' as the only true criterion of agricultural potential in any given district. His heavy reliance upon vegetation as the sole indicator of soil, climate, and general suitability for agriculture revealed Macoun's inability to move conceptually beyond the traditional local approach to botany, which he had imbibed earlier. In 1878 he told Sandford Fleming that he intended to strike out on his own to popularize the potential of the north-west, independent of government authority. He crossed a narrow line from naturalist to publicist with a particular drum to beat.[37]

During the 1870s a new generation of botanists and palaeobotanists, some the offspring of the pioneers discussed above, entered the field. James Macoun literally followed in his father's footsteps by collecting arctic plants in the northwest and aiming for as complete a historical and contemporary Canadian collection as possible. George Mercer Dawson, the son of J.W. Dawson, also began his botanical and geological collections in the north-west while serving on the International Boundary Commission in 1874. Sir Joseph Dalton Hooker, writing to the elder Dawson, pinned the hopes of his own father upon these younger Canadian botanists to complete the task their fathers had begun: 'Why,' Hooker asked Dawson, 'should [George] not prepare to publish a Flora of Canada like Grays Manual? – Surely the Colonial Govt. would find the funds.'[38]

IV

No such flora was likely to appear in the near future. As had happened to Torrey and Gray in the United States decades earlier, the transcontinental expansion of the dominion overwhelmed Canadian students of plant distribution. George Lawson's goal of a thorough critical examination of Canadian plants resulted in an almost insurmountable mass of botanical materials. 'The great labour,' he wrote to A.T. Drummond in 1869, was 'to clear away the errors & rubbish w[hic]h are also immense, & w[hic]h most Canadian botanists do not seem to be aware of.' Instead of a single volume, Lawson now contemplated a series of monographs on aspects of the Canadian flora. Several such monographs appeared, each highlighting the geographical distribution of one botanical family in the dominion, but nothing approaching a complete flora ever emerged from Lawson's pen. One reason was the heavy burden and sheer diversity of professional and administrative duties he carried at Halifax. His responsibilities as secretary of agriculture for Nova Scotia, in addition to his academic duties, conflicted with Lawson's personal obsession with completeness in botanical research.[39]

Lawson also evinced an ever-widening range of interest, both theoretical and practical, in northern explorations carried out by Robert Bell and others for the GSC. Following J.D. Hooker's lead, Lawson visualized his botanical researches in terms of the variation, adaptation, and survival of plants under changing climatic conditions and over large areas of the earth's surface. To Lawson, it was more important to collocate northern American species with those of northern Europe than to discover new northern species. This comparison of species and varieties in different localities would certainly adduce much-needed evidence on one side or the other of the controversial theory of natural selection. Moreover, it would help to ascertain the potential of the northern coasts for productive cultivation and settlement. For these reasons, Lawson kept himself abreast of Robert Bell's maps and studies of arctic tree distribution.[40]

But other factors help to explain why no botanical survey of Canada was permanently institutionalized before the incorporation of natural history by the GSC in 1877. Botanists never convinced the Canadian government of their cause partly because many naturalists believed that botanists required a far smaller extent of territory than geologists did to form a general idea of any given district. Even Lawson held that inventorial botany, 'the collecting of plants, the finding of rare plants, the noting of the occurrence or absence of species in given districts, the recording of their times of leafing, flowering and fruiting,' could best be done by local residents. John Bell, Lawson's former student, agreed that for most botanists, 'the natural productions of one's own vicinity are of much greater interest than those of parts more remote.' Even after botanizing in areas as remote as the Gaspé and Newfoundland, Bell still supported the view that a limited area thoroughly botanized provided enough of an 'exploring shaft into the vegetable products of the country, to shew their nature from the highest to the lowest forms.' The thorough study of a series of districts, he concluded, was of higher scientific value than a superficial investigation of the whole province. Nor was it always easy to teach amateur collectors not to overlook local cryptogams, such as minuscule lichens clinging to bare rock and mosses that flourished on old decaying stumps in remote places.[41]

Second, no undisputed leader succeeded in the Canadian botanical community after the untimely departures of James Barnston and George Lawson. Amateurs like George Barnston who overcame the physical difficulties of collecting northern species were at pains to contact accessible experts to identify their specimens, too insecure to attempt it themselves. In 1872 a field day at Montarville sponsored by the Natural History Society of Montreal attracted a hundred persons to Boucherville Mountain. After separating into geological, zoological, and botanical groups, the zoologists formed 'a small but compact body, and looked as if they meant business,' and the geologists followed suit. But the botanists could find no leader among them to explain the relevant points of interest. This group failed even to reach the summit of the hill.[42]

Those who might have lent direction to the botanical community in Canada did not exercise their power to the fullest possible extent. At Toronto the ageing William Hincks clung to a narrower botanical vision than that which attracted younger botanists to George Lawson at Kingson. Lawson himself accomplished much after he moved to Dalhousie, but his research characterized a diligent secretary more than a creative director of Canadian botany. Lawson never broke out of the research mould cast by John Hutton Balfour and Joseph Dalton Hooker. At Quebec and Chicoutimi, Louis-Ovide Brunet and Léon Provancher dissipated the sum of their combined influence over a growing French-Canadian botanical community through their mutual enmity and inability to cooperate.[43]

The stability of botany in Canada suffered another blow in 1878. During the 1870s John Bell emerged as a leading botanical authority in Montreal. Bell had set up medical practice after studying at McGill and Queen's and interning at American army hospitals during the Civil War. Like James Barnston, he appreciated the historical value of collecting rare native species 'surrounded and imprisoned' by introduced plants, cultivated crops and 'hardy and noxious weeds.' Over the years he amassed large local collections from Kingston, the Ottawa Valley, Owen Sound, Manitoulin Island, the Gaspé, and Newfoundland. In a bizarre parallel to Barnston's fate twenty years earlier, Bell was spearheading a drive to complete the monument to Frederick Pursh begun by Barnston when he too succumbed to sudden illness and died at the age of thirty-three.[44]

One botanist who gave direction to botanical collecting in Canada after 1870 was John Macoun. Macoun's position on the Geological and Natural History

Survey after 1882 expanded both his scientific opportunities and his authority. But he was never known as a meticulous collector. Quite early in his career, Macoun was warned by Sir William Jackson Hooker to take more care in collecting, identifying, and preserving specimens, which too often reached Kew Gardens in unidentifiable clumps. Macoun never learned his lesson, and information from his enormous collections must be used with great caution. Despite his activities as a publicist for the north-west, Macoun retained his credibility as a botanist, publishing a *Catalogue of Canadian Plants* in seven parts under the auspices of the Geological and Natural History Survey beginning in 1883. But this *Catalogue*, although intended to outline the extent and distribution of the flora of the dominion, was not meant to be the long-awaited flora of Canada. Even Macoun's high ambitions stopped short of such a claim, in deference to the 'serious difficulties' that the very success of Canada's enormous expansion had imposed upon its botanists.[45]

V

More than merely arousing interest in Canada's prehistorical and historical past, the theory of geographical distribution offered a metaphor for Canada's future development. At the BAAS meeting in Montreal in 1884, Asa Gray compared the vegetation of the two sides of the Atlantic to differences in accent and intonation between British and North American speech patterns.[46] As members of the British nation, Canadians used similar analogies to explain their growing sense of cultural and even physical differentiation from inhabitants of Britain. Ideas gleaned from the theory of species distribution intertwined the Canadian loyalist tradition with a developing mythology about the country's northern location. The result was an expansive vision of Canada's destiny as a transcontinental British nation in North America. A key element in this dynamic outlook was the idea of variation in nature.

At least since the 1850s, the apparent link between biogeography and the future of British North America had interested William McDougall. Educated at Victoria College between 1836 and 1842, McDougall may have studied natural science there under Jesse Beaufort Hurlburt. McDougall was respected by his contemporaries as gifted with extraordinary intelligence, which he sharpened by a considerable range of reading and mental discipline. He exhibited a wide range of scientific interests in the pages of his periodical publications. Both the *Canadian Agriculturist* and *The North American* of Toronto included regular science columns in which McDougall wrote short articles and reprinted others from journals not readily available to his readers. Often these items addressed botanical and horticultural issues likely to interest his agricultural readers.

Recurring themes included the organization, 'social relations,' and 'instincts' of plants. McDougall lost no opportunity to promote scientific pursuits such as collecting botanical or geological specimens. He valued such activities as imbuing society with a 'higher tone of feeling' and advancing it towards 'a healthy and higher civilization.'[47]

As important as the social and cultural benefits of natural history anticipated by McDougall was the light that species botany appeared to shed on anthropological questions involving the future of mankind in North America. Before 1850 neither archaeology, ethnology, nor anthropology had fully broken from its speculative tradition, and there was little scientific basis upon which to decide controversial anthropological issues. An important open question was whether a natural law of progress guided man's physical and cultural existence. Although this question was not new, it took on new significance with the settlement of America. The presence of 'primitive' peoples required an explanation of their place in the human family depicted in the Old Testament. Theories of Indians as descendants of the lost tribes of Israel led to the discomfiting inference that they had somehow devolved from a more civilized state. It was only natural for immigrants to North America to wonder whether such an apparent devolution had any bearing on their own situation.

The first attempt to analyse the human species like any other species, in terms of its natural history, was made by Buffon. He pointed to the influence of environmental and climatic changes as causing species to 'degenerate' from their original types. But he could not explain the workings of this process of change. By the mid-nineteenth century scientific theories of the natural history of man placed the emerging disciplines of archaeology, ethnology, and anthropology at the confluence of more advanced natural and physical sciences, where they remained for decades to follow. In particular, physical and cultural anthropology drew much of their understanding of varieties in human form and culture from the theory of geographical distribution. The more highly developed botanical aspects of this theory and its relation to variation in plant forms defined a framework for the study of physical and cultural changes effected by human migration.[48]

This new framework held a particular relevance for the developing idea of a transcontinental Canadian nation. Harking back to Buffon, McDougall noted in 1854 that current theories in anthropology failed to explain why the 'human animal' seemed to 'deteriorate' physically after migrating from Europe to North America. McDougall agreed that second- and third-generation North Americans no longer looked interchangeably like their British forebears. But the long-term implications for North American society of interpreting this change as a deterioration appeared to him unduly pessimistic. Could it not be seen instead, he wondered, in the light of the botanical model, as an adjustment phase in a much longer process of acclimatization? 'Climate has great influence upon plants,' McDougall pointed out, 'frequently changing their form, and improving or deteriorating their substance. Why may it not exert an equal power over animals?' He recalled Buffon's observation that while animals introduced in America had deteriorated from the parent stock, plants had decidedly improved. From this Buffon concluded that America was 'peculiarly the continent for the vegetable, while Europe is that for the animal kingdom.' In place of this Euro-centric view, McDougall preferred the more optimistic suggestion that the historical time-scale of North America was still too short to judge physical modifications in the animal kingdom. Man might require more than two centuries to acclimatize, to 'acquire and transmit to his offspring the new and requisite physiological peculiarities' which this process entailed.[49]

This notion of human 'acclimatization' spelled danger in the view of one of the leading intellectual forces in Toronto, James Bovell (1817–80), a physician, professor of physiology and natural theology at Trinity College, and amateur anthropologist. In 1859 Bovell's *Outlines of Natural Theology for the Use of the Canadian Student* acknowledged a tendency to 'degeneracy' in the European settled in America. But as an Anglican theologian, Bovell opposed recent writings by Alfred Russell Wallace that smacked of 'Lamarckian evolutionism.' Bovell admitted that in British North America 'the magnitude and magnificence of inland seas, waving rolling plains, and accessible but far stretching mountain ranges, stir the mind to efforts of gigantic grasp, and feed it with ideas which sometimes seem too mighty to be achieved.' Yet although he accepted the theory of degeneration 'but partially,' he did not share McDougall's optimism that European man could 'acclimatize' himself in North America.

Instead he compared the development of the populations of Upper and Lower Canada in the same way that Wallace compared plant populations with and without benefit of immigration. Bovell concluded from the contrast he observed between English and French Canadians that British North American society required the constant influx of the Saxon and the Celt. To check the flow of immigration was to 'throw the onus of reproduction upon the population – no longer European, but a struggle between the European alien and his adopted father-land.' Earlier colonial societies that had been cut off from their original stock, he declared, 'gradually withered and faded, and finally died away' like 'the red man, the child of the forest,' who 'pale[d] before the intruding race.' Bovell's references to degeneration applied only to Lower Canada, where the French 'race' had been isolated for more than a century. 'If therefore, we do find small men and small horses we find all the old primitive habits of the parent stock unchanged, shewing how completely they have refrained from intermixture.' In Upper Canada, in contrast, 'tall, athletic, well formed specimens of humanity' resulted from the continued flow of British immigration.[50]

McDougall's interest in acclimatization as a clue to British North Americans' higher destiny found sympathy from another Toronto professor, Daniel Wilson (1816–92). Wilson, born and educated in Edinburgh, had published on Scottish archaeology. He held the chair of history and English literature at University College, Toronto, from 1853 until his death. In 1862 his influential study entitled *Prehistoric Man* adapted the model of natural history to compare the origins of civilization in the Old and New Worlds. Acknowledging widespread confusion over the meaning of terms such as race, stock, and family, Wilson could not draw direct anthropological analogies with natural history categories such as order, class, and species. Nevertheless, he concluded that natural history disproved the theory of European man's inevitable 'degeneracy and decrepitude' in North America, the myth that McDougall so hoped to dispel. Wilson gave scientific authority to McDougall's suspicions about human variation in declaring:

It is easy for any one familiar with the New England physiognomy to point out the Yankee in the midst of any assemblage of Englishman. He furnishes the indisputable example of a new variety of man produced within a remarkably brief period of time, by the same causes which have been at work since man was called into being, and scattered abroad to people the whole earth ... And if two centuries in New England have wrought such a change on the Englishman of the seventeenth century, what may not twenty centuries effect? or, what may be the ultimate climatic influences of Canada, the Assinaboine [sic] Territory, or Fraser's River[?][51]

The New World, he taught, provided a new atmosphere not only for the body, but also for the mind: 'We cannot overlook the silent influences of a *new* world,' he wrote, 'in making another man out of the old English stock.'[52]

The role of variation in the development of modern nations received fuller expression in Wilson's subsequent writings. In 1869 he noted that race as an element in the progress of nationalities had acquired an importance in modern times that was wholly unknown to early historians. Wilson used his anthropological studies to justify colonization as a biological struggle that inevitably entailed the disappearance of native or 'inferior' races because European colonists could turn the wilderness to better account. Unlike earlier writers who compared the Indians to native plants succumbing to the cultivated Anglo-Saxon intruder, Wilson preferred to interpret 'civilization' as a more complex process of variation. In Canada, he predicted, 'as in modern nationalities of

Europe, its ancient and prehistoric races will survive under new forms to share in the novel phases of the coming time.' He felt sure that through intermarriage, the physical variation of the race, and cultural assimilation Indians would be 'welcomed as an ethnical element in the young nation supplanting them,' much like the nationalities already merged in the Anglo-American and Canadian peoples. Native peoples, he argued, would therefore not be 'wholly excluded from a share in the advantages of such progressive civilization, or even from exercising some influence on its development.'[53] Implicit in this rationalization of the fate of native peoples was the assumption that Canadians as a 'young nation' were themselves in the process of forming a North American variation of the British nation. As an evolutionary nation, Canadians could partake in the material development of North America without renouncing through revolution the British heritage they treasured and admired.

Wilson's *Prehistoric Man* was so well received that he published a revised edition in 1865. His repudiation of the long-held assumption that the physical vigour of Saxon, Celtic, and Germanic immigrants to North America deteriorated after settlement in the New World received additional confirmation in an American report to the Statistical Society of New York in 1864. Recognizing the so-called 'American type,' distinguishable from the English, French, and German by a loss of fat in the cheeks and chest and earlier changes in the teeth, the report utilized botanical metaphors to emphasize that the North American climate had,

in the lapse of two centuries or less, considerably modified the European type, . . . a fact which we think few can deny. What that change indicates – whether it is simply a stage in the process of acclimatization, or a positive loss of vitality in the race, is a question the importance of which cannot be exaggerated. In the whole Kingdom of life no transplanting into a new soil occurs, without an apparent decline at first; but if a new and increased vigor is to be manifested, we must find the perennial root healthy and strong, though the early leaves may wither.[54]

This criterion of a 'healthy and strong perennial root' reflected a generally narrowing line of distinction between laws governing plant distribution and those governing human geographical distribution. J.D. Hooker's work on the geographical distribution of arctic plants in the light of Darwin's theory of creation by variation invited the application of this new approach to physical and cultural variations in the human species as well. Hooker and others attempted with some success to explain national affinities between unusual and widely separated cultures like the mountain-dwelling Basques and the northern Finns as analogues of similarly distributed arctic plant species.[55]

VI

A Canadian corollary of such speculations about the development of northern nations was the popular myth of the north. Emphasis upon the north in the idea of a transcontinental Canadian nation as a British variation in North America derived from the expanding conception which was as natural to botanists as it was to geologists. The myth of the north is most often associated with members of the Canada First movement. J.C. Schultz pursued geographic botany as a student of George Lawson, but the influence of both men as carriers of scientific ideas and speculations reached farther afield. Schultz shared his interest in both botany and anthropology with his friend Charles Mair, a young poet from Perth and one of the original Canada Firsters.[56]

Mair (1838-1927) grew up near the lumber camps of the Ottawa Valley and developed a romantic attraction to the seemingly endless pine forest. He studied medicine at Queen's before and after Lawson's tenure there but caught the attention of Lawson and the BSC through his poem 'The Pines,' read to the society in 1862. As a tribute to his 'truly Canadian production,' the BSC named Mair an honorary member. Through his acquaintance with Schultz and H.Y. Hind, with whom he corresponded during the 1860s, Mair combined his interest in nature with a vision of the British American north-west as an attractive wilderness that held the key to Canada's future.[57]

Mair was observing in the House of Commons in 1867 when William McDougall, then member of parliament for North Lanark and minister of public works in John A. Macdonald's first administration, steered through his resolutions that Canada acquire the vast agricultural lands of the HBC. In 1868 Mair gave up his medical studies to work for McDougall, researching the history of the north-west in preparation for McDougall's negotiation of the transfer. The two men shared not only an expansionist outlook but also a view of the north as the birthplace of a hardy new Canadian spirit.[58]

The myth of the north had deep roots in Western thought. The idea of the dominance of hardy northern over weaker southern forms reached back in the British heritage to Francis Bacon and derived authority among Victorians from Charles Lyell's *Principles of Geology* during the 1830s. It received further scientific approbation in J.D. Hooker's essay on the geographical distribution of arctic plants. Both Mair and McDougall were predisposed to accept its premises by personal experiences during their formative years. As a youth Mair became convinced that 'real Canadians' were forged by life near the timber-line. For McDougall the connection was more eccentric. All his life he retained an indelible impression of the rebellion of 1837 in Toronto, when as a youth he had witnessed the flight north along Yonge Street from the disturbance at Montgomery's Tavern. He attributed special emotional significance to this scene as the microcosm of a great 'migration to the north,' which he recalled vividly even many years later as part of the natural pursuit of liberty and survival.[59]

The theme of the north has been studied as 'one of the most arresting themes in the emerging nationalist creed of Canada First.' Articulated most fully by Robert Grant Haliburton in a lecture to the Montreal Literary Club in 1869, it embodied, in Haliburton's view, the most unique characteristic of the new Canadian dominion, namely that it was 'a Northern country inhabited by the descendants of Northern races.'[60] The conceptual links of this northern theme to the idea of variation and the theory of geographical distribution were explicit.

Haliburton (1831–1901), a native of Halifax, was a leading figure alongside George Lawson in both the Nova Scotian Institute of Natural Science and the movement for Nova Scotian agricultural improvement. He witnessed repeated reports from Lawson on the distribution of northern flora. Furthermore he pursued researches in ethnology and belonged to the Royal Society of Northern Antiquaries of Copenhagen.

Haliburton's outlook had been tempered by that of his well-known father, a writer whose varied interests included science and the improved farming advocated by 'Agricola' a generation earlier. Thomas Chandler Haliburton (1796–1865) had years before been struck by the bitter fact that a colonist who visited the mother country to transact business on behalf of his country 'is utterly nowhere. He is neither Scotch, Irish, [nor] British; ... a native [n]or a foreigner, an American [n]or an Englishman ... neither fish nor flesh ... He has no nationality at all – he is nobody.'[61]

R.G. Haliburton and his contemporaries drew from their knowledge of science to clarify their predicament as British North Americans and also to seek a means towards its improvement. In *The Men of the North*, Haliburton reassessed this predicament as a blessing in disguise. First, he pointed out, 'we are sprung from a common race'; second, like Daniel Wilson he argued that such hardy roots could only be strengthened by their obvious destiny in North America: 'As British colonists we may well be proud of the name of Englishmen.' But the British people were themselves 'a fusion of many northern elements which are here again meeting and mingling, and blending together to form a new nationality.' Haliburton urged that Canadians, like hardy new varieties of a plant species, must diffuse themselves, 'in our national aspirations take a wider range, and adopt a broader basis which will comprise at once the Celtic, the Teutonic, and the Scandinavian elements, and embrace the Celt, the

Norman French, the Saxon and the Swede, all of which are noble sources of national life.' He asked rhetorically: 'Is the northern land which we have chosen, a congenial home for the growth of a free and a dominant race? What is the stock from which we are sprung? . . . Is it climate that produces varieties in our race, or must we adopt the views . . . that the striking diversities . . . of nations, must have existed *ab initio*?' He, for one, chose the former option:

Why should a strange chance have planted the dominant families of mankind in northern latitudes? Climate, it is true, cannot in the lapse of a few centuries produce any very marked physical change, though even in one or two generations, its effects are sometimes visible. But if we can allow forty, fifty, or a hundred centuries for the effect of climatic influences, we may bring ourselves to believe that even the woolly heads and the black skin of the negro may have been the result of a tropical sun.[62]

Haliburton and other members of Canada First perceived the 'wild tribes of the west' as subjects for ethnological research, in much the same way as Daniel Wilson did. His writings struck a chord with Mair, convinced too of Canada's northern destiny. Mair assured Schultz that once a road was built to Red River, 'I think you shall see such an inroad of hardy young Canadians as will astonish the natives.'[63]

Just as plants near the border resembled American flora, and the hardier, more typically 'Canadian' varieties were diffused farther north, Haliburton predicted that hardy and virtuous variations of the Canadian people would ultimately occur in the north-west: 'Our corn fields, rich though they are, cannot compare with the fertile prairies of the West, and our long winters are a drain on the profits of business,' he conceded, 'but may not our snow and frost give us what is of more value than gold or silver, a healthy, hardy, virtuous, dominant race?' Haliburton urged Canadians to recognize that they were not Britons, but 'descendants of Britons,' 'the sons of the New World.' He hoped to name the Canadian north-west, and perhaps one day all of Canada, 'Norland.'[64]

Botany followed its own patterns of development during the nineteenth century. More than the other Victorian inventory sciences, botany, with its emphasis upon the geographical distribution of plants in Canada, relied upon voluntary contributions of amateur collectors. With the appointment of D.P. Penhallow, an American trained at Boston College, as professor of botany at McGill in 1885, Canadian universities began to train professionals in anatomical, physiological, and palaeontological botany. As a result, George Lawson feared a decline in the species botany he still deemed paramount to the progress of the science in Canada.[65] To some extent the slack was taken up by schoolteachers and their pupils, who collected plant specimens on field trips throughout the country. But what was still lacking, he regretted, was 'some bond of union' to forge groups of local botanists into 'an army of explorers pervading the whole extent of the Dominion.'

Lawson hoped to alleviate this problem by organizing local groups around a central botanical depot; otherwise, he felt, Canada would risk losing forever information highly important to larger botanical questions.[66] At a meeting of the Royal Society of Canada in 1891, Lawson proposed to establish the loosely organized Botanical Club of Canada, modelled on its predecessor at Edinburgh. But like the Botanical Society of Canada before it, the Botanical Club could not hope to attain the recognition enjoyed by a public institution like the Geological Survey of Canada or the Toronto observatory. The failure of the Province of Canada to support the publication of a flora before 1867 probably precluded any such institutionalization of a botanical survey after Confederation, when problems far more immediate pressed public funds beyond their limit. The rapid transcontinental expansion of the dominion by 1871 brought an embarrassment of botanical riches that would take decades to process.

Botany experienced its uniformitarian revolution almost a generation after Charles Lyell completed this task in geology. The far-reaching biological implications which culminated in Darwin's theory of 'creation by variation' took time to assimilate. Many amateur naturalists felt intimidated by the Latin nomenclature which continued to dominate botany while the discipline broadened its scope as a science.[67] Some who did wish to expand their botanical knowledge, like Catharine Parr Traill and George Barnston, complained that it was often difficult in the backwoods to gain access to the most recent botanical literature.

But despite this more amorphous picture, Victorian botany was no less important than other inventory sciences to Canadian intellectual development. Botany enhanced Canadians' awareness of their own historical past. By the 1860s theoretical advances made possible by Darwin's *The Origin of the Species* concluded the long search for a key to the relationships of plant forms. The result was a more historical understanding of present-day varieties and even species as descendants of prior plant species. J.D. Hooker's explanation of the geographical distribution of plants over the earth's surface designated British North America as a transitional zone with prehistorical ties to European forms. The emphasis in Victorian botany upon the geographical distribution of plants expanded conceptual horizons of Canadian botanists both eastwards to the lower provinces and north-westwards to the HBC territories.

Yet, more than merely a concern with Canada's prehistorical past, the theory of plant distribution outlined an alluring vision of Canada's future. As an important impetus to the formulation of the myth of the north, the theory offered a model for explaining the development of a Canadian nation in terms analogous to the inevitable variation of plant forms diffused from a common centre over different geographical regions. Victorian botany thus made unique and complex contributions to the idea of a transcontinental Canadian nation. The theory of the geographical distribution of northern plant forms, which adopted Darwin's theory of the role of variation in the proliferation of species, was recast to legitimize Canada's place both in the British Empire and on the North American continent. In essence, botany offered the compelling view of an incipient Canadian or British North American nation not as a contradiction in terms or as a political, economic, and cultural impossibility but rather as an inevitable development to be welcomed as part of the natural order of things.

While these ideas found their best-known expression in the Canada First movement, others shared them. William McDougall's questions about the ability of European man to acclimatize himself in North American, answered in the affirmative by Daniel Wilson's anthropological studies, illuminate the scientistic foundations of McDougall's Clear Grit ideology. More than merely a shadow of American frontier democrats, the Clear Grit leader claimed that British Canadians were not just transplanted Britons doomed to permanent deterioration in their colonialism. Instead they were new Britons in North America, variations no longer identical to the original British type. As such, British Canadians had to be permitted to 'naturalize' British institutions in the manner best suited to their own needs.

McDougall's ideas as a Clear Grit drew from just such an understanding of Canadians as a British variety modified by life in North America. He shared this line of thinking with Charles Clarke, co-author with McDougall of the Clear Grit platform of 1851 and an amateur botanist and ornithologist. In essence, the Clear Grits proposed to emulate the successful acclimatization of an older British variety, the United States, to the North American environment. The first plank of their platform, elective institutions and an elective governor, was premised on the dual nature of that developing binomial creature, the 'British Canadian.' A British subject born or resident in Canada, they declared, was as much a British subject as one born in Britain and was quite as able to bestow all proper care upon imperial interests. 'A man who has lived in the Canadian wilderness, battled with its difficulties, and become practically conversant with the necessities of the country' was a better judge of Canadian affairs than 'a recent importation': he would be 'Canadian in heart and feeling.'[68]

The scientist basis of McDougall's vision of Canada helps to explain some of his controversial political decisions. As commissioner of Crown lands in 1862, he oversaw the acquisition of a large portion of Manitoulin Island from its Indian inhabitants. McDougall was responsible for the reversal of the traditional government policy of protecting these tribes from the encroachment of white settlement by granting them land north-west of Canadian settlement. In 1869 he became lieutenant-governor of Rupert's Land and the Northwest Territories, and his attempt to proclaim Canada's acquisition of the HBC lands despite the resistance led by Louis Riel ended in fiasco. It appeared to some that wherever McDougall had anything to do with Indians, he 'created discord and strife, where all had been peace before.' His decisions, widely considered hasty and ill-judged from a political perspective, make sense in the light of his anthropological understanding of Canadians' destiny as a hardy northern nation, a corollary of the botanical theory of geographical distribution. Further inquiry into the intellectual roots of the Clear Grit platform will perhaps be able to transcend the traditional dichotomy between frontierism and metropolitanism to portray William McDougall more coherently than the capricious political enigma he is usually taken to be.[69]

The tendency to visualize the Canadian as a variety of the British nation was not exclusive to Clear Grits and Canada Firsters. The generation that attained the age of majority during the 1840s was the first among whom many were natives of British North America. People as diverse as William McDougall, J.C. Schultz, Egerton Ryerson, and others began to question the place of native-born Canadians in their own society. All shared an educational background which included strong faith in the power of science to aid in this task as well as a fundamental desire to nurture this incipient Canadian variety of the British nation to the best of their abilities.

In a classic case of the use of analogy, Canadians used a scientific theory for their own ideological purposes, to create the *idea* of a commonality among Canadians as a transcontinental nation rather than to describe one already in existence. The model of geographical distribution, with its implications for the evolution of a Canadian nation forged by its northward expansion, offered an attractive interpretation of Canada's role in North America. Certain selected features of the model offered grounds for a Canadian nation; but other aspects were conveniently suppressed, including the innumerable centuries that even plant, let alone human, varieties required to transform their specific identities.

Like Darwin's theory of 'creation by variation,' the idea of a Canadian nation was foremost a historical development in the minds of its supporters. The idea of such a Canadian nation, like any other natural group,

> is steadily fixed, though not precisely limited; it is given, though not circumscribed; it is determined, not by what it strictly excludes, but by what it eminently includes; by an example, not by a precept.[70]

The ultimate proof of its existence lay in the believing.

Conclusion

> Our mental business is carried on much in the same way as the business of the State: a great deal of work is done by agents who are not acknowledged. In a piece of machinery, too, I believe there is often a small unnoticeable wheel which has a great deal to do with the motion of the large obvious ones.
>
> George Eliot, *Adam Bede* (1859)

Science did more to define the hopes and dreams of Victorian Canadians than even they perhaps realized. The basis of science as it was practised in Canada during the Victorian age was inventory, systematic surveys of the land and its resources with the ultimate goal of assessing its material potential. Inventory science grew from the amateur naturalist tradition so popular in British culture, adding to it the dynamic of an entrepreneurial outlook. The result was intellectual as much as economic and social, the transformation of conceptions of what Canada – and Canadians – were to be.

It is natural to assume, from a conventionally compartmentalized perspective, that science could have little to do with the idea of creating a nation. 'What,' asks a professor of chemistry, 'could be less national than "science", or conversely less scientific than "nationalism"?'[1] But an integrative historical approach to the culture of Victorian Canada shows the reality to be more complex. Both inventory and nation-building were important organizational processes which Canadians believed would arm them to meet the challenges of industrialization and modernization. Victorian inventory science defined premises upon which a certain Canadian nation could be built, and which gave rise to 'national' policies designed to safeguard that existence. New information offered raw material for developing conceptions; without it, the idea of a nation in British North America could no doubt still have emerged, but it might not have been the same as that informed by Victorian science.

Nor could it be the same as that created by the American Revolution a century before. The United States was planted squarely within the Newtonian tradition, upon an intellectual foundation that held a relatively static and mechanistic view of the world. It seemed 'self-evident' for the American Founding Fathers to define their new nation as subject to natural law, and their rights as 'inalienable.'[2] Such claims were no longer 'natural' by the time British North Americans debated Confederation. Their outlook was more historical, conscious of their involvement in a process of organic change over time. The notions of variation and of evolution in nature, for which Darwin's theory was not the source but the synthesis of a tradition, had eroded the certainties of a generation that had witnessed the evolution of the American constitution away from its decentralized origins; this generation constructed the British North America Act accordingly.

Four main inventory sciences (geology, terrestrial magnetism, meteorology, and botany) contributed to the idea of a transcontinental nation in Victorian Canada. Each did so in its own way, some quite directly and others more subtly. Geology promoted new perceptions of the potential wealth of British North America by pointing out mineral deposits in regions beyond the pale of familiarity. The mineral-rich Laurentian Shield, once thought to be a waste land, earned new appreciation. Even it its negative aspects, geology exerted an enormous impact upon Canada. The painful conclusion that no workable coal deposits existed in the province of Canada meant that it could not attain industrial self-sufficiency within the confines of Victorian technological development. As a result, Victorian central Canadians cast covetous eyes eastwards to known Carboniferous deposits in the maritime provinces, and hopeful eyes north-westwards to the limitless possibilities of the unknown. They enjoyed international recognition in the achievements of the Geological Survey of Canada under the direction of Sir William Logan. Participation in international exhibitions added unaccustomed prestige and a new sense of the value of the land they inhabited. Logan's richly deserved reputation made him a national hero whose glory he invited fellow Canadians to share. The basis of Logan's geology was the Lyellian approach to nature, which emphasized the importance of the historical process and thereby drew Canada – and Canadians – into the historical mainstream.

Terrestrial magnetism and meteorology focused international attention on Canada as a land whose northern location offered valuable information. J.H. Lefroy and his associates integrated British North America into larger imperial and North American scientific networks. More importantly, their work encouraged co-operation among British North Americans. Canadians supported the Toronto Magnetic and Meteorological Observatory fully convinced that their new responsibility would be seen as an upward step on the scale of civilization. Meteorology furthermore encouraged the view of the British North American north-west as a land fit not only for habitation but even for cultivation. These scientific speculations did much to transform Canadians' attitudes towards the HBC lands; from seeing the region as an unknown waste land, they came to accept it as a necessary extension of their own territory, without which they believed Canada could not reach its full potential. H.Y. Hind's emulation of Francis Parkman's *The Oregon Trail* in the guise of a scientific report on the potential of the north-west convinced Canadians that beyond Lake Superior's foreboding shores lay a 'fertile belt' inviting the relentless north-westward march of peoples.

Botany, the least centralized of the inventory sciences, contributed less tangibly to Canadian thought. But it stood at the vanguard of biological sciences then emerging from the older natural history tradition. To trace botany's impact upon Canadians' self-perceptions is to begin to comprehend the complexities of an age of great intellectual transition. In particular the theory of geographical distribution gave them a deep sense of their prehistorical and historical past and of their own potential as a hardy new northern variation of the British nation. Through the application of the theory to anthropology, botany informed Canadians with concepts of variation that seemed controversial only when explicitly associated with Darwin's theory of evolution by natural selection. In its tamer form, the theory of geographical distribution provided a rationale for the future to which many Canadians aspired. As members of the British nation, yet modified by life in North America, they visualized an evolution towards nationhood that remained loyal to the British heritage while welcoming as natural a changing relationship with the mother country. Botany contributed an early attempt at a 'national' scientific society, exposing fundamental disagreements over who represented truly 'Canadian' interests.

The idea of scientific inventory touched other areas of natural history as well as botany. Entomology, for example, found legitimacy in the practical needs that arose during times of insect infestation. The Entomological Society of Canada, founded at Toronto in 1863, soon sprouted branches in Quebec and London, and it remains Canada's oldest continuing 'national' specialized scientific society. Entomology reflected many of the same intellectual qualities that characterized the four major Victorian inventory sciences, perhaps not surprisingly, since its practitioners were many of the same people. The influence of William Hincks upon members from Canada West helps explain resemblances between entomology and the botany he practised at Toronto: the tendency, for example, to downplay field-work beyond the limits of settlement. Nevertheless, members like Charles Bethune made efforts to correlate insects collected in central Canada with those from the maritime provinces and to underline their similarities. And much as in geology and botany, Canadian exhibitions of insect collections abroad transformed Canada's hyperborean image to one of a temperate land with seasons resembling the tropics.[3]

Each of the inventory sciences found a relatively enthusiastic reception that crossed political party and even provincial boundary lines. Seen by the professional and entrepreneurial classes as a key to material progress, science revealed some of the undisputed and therefore often hidden assumptions of the age. Faith in science provided an important common denominator for right- and left-wing, urban and rural, business and labouring, and sometimes even English and French Canadians. Natural history was one of the few cultural interests that united inhabitants of all the British North American colonies in a common pursuit during the nineteenth century.

Yet supporters of each of the inventory sciences experienced difficulty attaining the degree of recognition and financial support they desired. Limited financial resources in a struggling colony made rivals of the sciences themselves as they fought for public funds. The degree of aid varied not only with the prospects of material gain but also with the economic cycle and with the level of organization involved. In practice, this meant that although Sir William Logan did not have an easy time, the GSC fared better than the Toronto observatory and better still than the more informal studies of plant distribution. Sir William Logan campaigned tirelessly to keep the accomplishments of the GSC before the public; his survey suffered confusion and frustration when confronting an infant and sometimes arbitrary civil service. And while his contributions to Canadian mining and industrial ventures were widely recognized, he never acceded to farmers' repeated requests for local rock and soil analyses.[4]

In the physical sciences, Sir Henry Lefroy fought hard to convince Canadians to accept the torch of imperial science, which he passed to them in the form of the Toronto Magnetic and Meteorological Observatory. His ideas for a Canadian and even British North American meteorological network caught fire only slowly, and not before he left Canada. In botany, neither James Barnston nor George Lawson put down roots deep enough to initiate a botanical survey as a public institution. As an organized inventory science, botany was a late bloom on the Canadian scene; it never attained a position of relative strength before Confederation drastically expanded the field to be surveyed.

A major influence in the development of all the inventory sciences in Victorian Canada was their connection with science in Great Britain and the United States, forming a cultural North Atlantic triangle. In geology Logan initially maintained a close relationship with Sir Henry De la Beche of the Geological Survey of Great Britain. But the torpid pace of overseas communication turned his attention to James Hall, the New York State geologist at Albany. In geomagnetism and meteorology, J.H. Lefroy did not always see eye to eye with his superior in faraway Woolwich, and he too looked for support and recognition to his colleagues in the United States. In botany, the early interest in British North America was centred at Kew Gardens, but Sir Joseph Dalton Hooker willingly abdicated 'responsibility' for co-ordinating the study of the geographical distribution of North American plants to Asa Gray at Harvard. The dominant trend was for the focus of Canadian science to shift away from Britain and closer to home, notwithstanding Sir J.W. Dawson's rhetorical proposal in later decades for an imperial geological union.[5]

In addition to its colonial scientific relationship with Britain, Canada enjoyed special intellectual ties with Scotland. From the Edinburgh education of Sir William Logan and earlier geologists to the training by John Hutton Balfour of young botanists who migrated to Canada, inventory science carried implicitly the old Scottish vision of a new northern nation in America. This vision joined the newer practical approach to the problem of economic backwardness to an older sense of Scotland's prophetic destiny. It transferred to British North America the goal of establishing this new nation instead of trading stations or other short-term means of exploitation.[6] And not only trained scientists but also many Scottish immigrants to Canada imbibed the usefulness of science through the Scottish educational system during the late eighteenth and early nineteenth centuries. In an age when science formed an important part of popular culture and was easily accessible to educated persons, many Canadians shared the view that science and its concomitant, technology, made a transcontinental nation seem viable and even inevitable.

This relationship was clearly a two-way street; Victorian science nurtured the idea of a transcontinental Canadian nation, but the reverse was also true. The fact that Sir William Logan was a native Canadian helped to loosen the Canadian government's purse strings, and his efforts did not disappoint his sponsors. The contribution of Canada's inventory sciences to knowledge in the nineteenth century was not inconsiderable, yet they could do no more than scratch the surface in tasks of such enormous scale that Canadians are still groping for a detailed 'national inventory.'[7] Nor could the sciences impose unanimity on all rival cultural, regional, and social visions of the country. But the assumptions and values they carried were so deeply absorbed by our industrial society that a century later they are seldom recognized. Until recently most prominent Canadian historians continued to promote 'nation-building' interpretations in tones not unlike those of the Victorians; in particular the Metropolitan and Laurentian schools of Harold Innis and Donald Creighton, with their prominent geographical determinism, were heirs of the ideas presented here.[8]

Inventory sciences contributed substantially to the modern intellectual framework within which Canada was 'invented' and even re-invented by historians of the 'nationalist' school. These sciences shaped Canadians' expectations about the potential and the limitations of the country they inhabited as well as their own position within it. 'Invention,' wrote Mary Shelley, 'consists in the capacity of seizing on the capabilities of a subject and in the power of moulding and fashioning ideas suggested to it.' It 'can give form to dark, shapeless substances,' but it 'cannot bring into being the substance itself.'[9] Victorian science helped form a continuum for a transcontinental Canadian nation, but it remained for Canadians themselves to fill it as they saw fit.

Notes

The following abbreviations are used in the notes.

AAAS	American Association for the Advancement of Science
BAAS	British Association for the Advancement of Science
BAC	*British American Cultivator*
BAJ	*British American Journal of Medical and Physical Science*
BSC	Botanical Society of Canada
CAg	*Canadian Agriculturist*
CHR	*Canadian Historical Review*
CJ	*Canadian Journal of Industry, Science and Art*
CN and G	*Canadian Naturalist and Geologist*
DCB	*Dictionary of Canadian Biography*
DHE	*Documentary History of Education in Upper Canada*
DP	J.W. Dawson Papers
FO	Foreign Office
GSC	Geological Survey of Canada
HBC	Hudson's Bay Company
HOP	*Historical and Other Papers and Documents Illustrative of the Educational System of Ontario, 1853–1868*
HSSM	Historical and Scientific Society of Manitoba
JEUC	*Journal of Education for Upper Canada*
JHB	*Journal of the History of Biology*
LHSQ	Literary and Historical Society of Quebec
LP	Sir William Logan Papers
MUA	McGill University Archives
NAm	*The North American*
NHSM	Natural History Society of Montreal

NSINS   Nova Scotian Institute of Natural Science
OHS   Ontario Historical Society
PAC   Public Archives, Canada
PAM   Public Archives of Manitoba
PAO   Public Archives, Ontario
PRO, MO Public Record Office, Metrological Office
RASC   Royal Astronomical Society of Canada
RSC   Royal Society of Canada

INTRODUCTION

1 Marcia Kline, *Beyond the Land Itself: Views of Nature in Canada and the United States* (Cambridge, Mass., 1970); Gaile McGregor, *The Wacousta Syndrome: Explorations in the Canadian Langscape* (Toronto 1985); for a broader view see George Altmeyer, 'Three Ideas of Nature in Canada, 1893–1914,' *Journal of Canadian Studies* 11 (Aug. 1976): 21–36.
2 Robert A. Stafford, 'Geological Surveys, Mineral Discoveries, and British Expansion, 1835–71,' *Journal of Imperial and Commonwealth History* 12/3 (May 1984): 23
3 J.D. Bernal, *Science in History*, 4 vols. (London 1969), II: *The Scientific and Industrial Revolutions*, 515; John Herman Randall, Jr, *The Making of the Modern Mind*, revised ed. (Cambridge, Mass., 1940), 394
4 Nathan Reingold, 'Cleveland Abbe at Pulkova: Theory and Practice in the Nineteenth-Century Physical Sciences,' *Archives Internationales d'Histoire des Sciences* 17 (1964): 143; Daniel J. Boorstin, *The Discoverers: A History of Man's Search to Know His World and Himself* (New York 1985), 276, 640, 651
5 M.J. Cullen, *The Statistical Movement in Early Victorian Britain* (New York 1975), 10–11; Rosalind Mitchison, *Agricultural Sir John: The Life of Sir John Sinclair of Ulbister 1754–1835* (London 1962), 120–1; Thomas Pennant, *A Tour in Scotland 1769*, 2nd ed. (London 1772), questionnaire reprinted 302–13; David Daiches, Peter Jones, and Jean Jones, eds., *A Hotbed of Genius: The Scottish Enlightenment 1730–1790* (Edinburgh 1986)
6 McGill University Library, NHSM, 'First Report of the Indian Committee,' 26 May 1828. Cf. Pennant's questionnaire.
7 Morris Berman, *Social Change and Scientific Organization: The Royal Institution, 1799–1844* (Ithaca, NY, 1978), 40–1, 108–9
8 Étienne Parent, 'L'industrie considerée comme moyen de conserver la nationalité canadienne-française,' speech to the Institut Canadien (1846), reprinted in J. Huston, comp., *Le repertoire national*, IV (Montreal 1893), 10, 14
9 Gunnar Eriksson, *Kartläggarna: Naturvetenskapens tillväxt och tillampningar i det industriella genombrottets Sverige 1870–1914* [Laying down the map: the growth

and application of science in Sweden in the early industrial era, 1870–1914] (Umeå 1978), 143
10 *Parliamentary Debates on the Subject of the Confederation of the British North American Provinces* (Quebec 1865) [*Confederation Debates*], speeches by John A. Macdonald, 26; George Brown, 99; J.S. Sanborn, 118; T.D. McGee, 128; W. McCrea, 169; D. Christie, 213; J.G. Currie, 274; H.L. Langevin, 383; A. Morris, 443–5; C. Alleyn, 670–1; F.H. Chambers, 771; J.L. Biggar, 886–7; W. Shanly, 901; H.J. Cameron, 968; Taché, 343
11 Ibid., Harwood, 825–6; Cauchon, 561. Anthony W. Rasporich, 'Positivism and Scientism in the Canadian Confederation Debates,' in L.A. Knafla, M.S. Staum, T.H.E. Travers, eds., *Science, Technology and Culture in Historical Perspective* (Calgary 1976), 206–34
12 CN and G 3/5 (Oct. 1858): 392–5. Internal and external evidence points to J.W. Dawson, principal of McGill College and president of the NHSM, as the author. See Morris quoting Dawson's strikingly similar remarks in *Confederation Debates*, 443.
13 Thomas D'Arcy McGee, 'A Further Plea for British American Nationality,' *British American Magazine* 1 (Oct. 1863): 563
14 *Confederation Debates*, Harwood, 832; see also McGee, 141; Campbell, 175; Morris, 443–4
15 Ibid., McCrea, 169
16 Ibid., Scoble, 908
17 Ibid., Cartier, 60

CHAPTER 1 Exposing the Strata

1 Walter B. Hendrickson, 'Nineteenth-Century State Geological Surveys: Early Government Support of Science,' in Nathan Reingold, ed., *Science in America since 1820* (New York 1976), 135; Ernest Greeley, 'The Search for Salt in Upper Canada,' OHS, *Papers and Records* 26 (1930): 406–31. The official history of the GSC is Morris Zaslow, *Reading the Rocks* (Ottawa 1974).
2 See David Elliston Allen, *The Naturalist in Britain: A Social History* (London 1976), chaps. 1–2; also Morris Berman, *Social Change and Scientific Organization: The Royal Institution, 1799–1844* (London 1978). For Victorian views of Bacon see A.D. Orange, 'The Idols of the Theatre: The British Association and Its Early Critics,' *Annals of Science* 32 (1975): 277–94; and Jack Morrell and Arnold Thackray, *Gentlemen of Science: Early Years of the British Association for the Advancement of Science* (Oxford 1981).
3 Roy Porter, *The Making of Geology: Earth Science in Britain 1660–1815* (Cambridge 1977), 152, 171–3
4 Ibid., 132; Porter, 'The Industrial Revolution and the Rise of the Science of Geol-

ogy,' in M. Teich and R. Young, eds., *Changing Perspectives in the History of Science* (London 1973), 320-43; D.H. Hall, *History of the Earth Sciences during the Scientific and Industrial Revolutions* (Amsterdam 1976), 206; J.D. Bernal, *Science in History* (London 1954), 360

5 Robert A. Stafford, 'Geological Surveys, Mineral Discoveries, and British Expansion, 1835-71,' *Journal of Imperial and Commonwealth History* 12/3 (May 1984): 7; see also Horace B. Woodward, *The History of the Geological Society of London* (London 1907).

6 J.D. Overton, 'A Theory of Exploration,' *Journal of the History of Geography* 7/1 (1981): 53-8; Geoffrey Blainey, 'A Theory of Mineral Discovery: Australia in the Nineteenth Century,' *Economic History Review* 23 (1970): 307

7 Hew Strachan, 'The Early Victorian Army and the Nineteenth-Century Revolution in Government,' *English Historical Review* 95/376 (July 1980): 782-809; Peter Burroughs, 'The Ordnance Department and Colonial Defence, 1821-1855,' *Journal of Imperial and Commonwealth History* 10/2 (Jan. 1982): 125-49; George Raudzens, *The British Ordnance Department and Canada's Canals 1815-1855* (Waterloo, Ont., 1979), 5-139. J.W. Bain, 'Surveys of a Water Route between Lake Simcoe and the Ottawa River by the Royal Engineers, 1819-1827,' *Ontario History* 50/1 (Winter, 1958): 15-27; Duke of Wellington, 'Minute of Report of Surveys,' 10 Jan. 1827, in F.B. Murray, ed., *Muskoka and Haliburton 1615-1875*, Champlain Society (Toronto 1963), 47-8. Kingston *Chronicle and Gazette*, 28 Dec. 1836

8 Robert England, 'Disbanded and Discharged Soldiers in Canada Prior to 1914,' CHR 27/1 (Mar. 1946): 8. Jean S. McGill, *A Pioneer History of the County of Lanark* (Toronto 1968), 58-9; William Canniff, *The Medical Profession in Upper Canada, 1783-1850* (Toronto 1894), 119, 123. B.J. Harrington, *Life of Sir William E. Logan* (Montreal 1883), 133

9 Hamnet P. Hill, 'The Bytown *Gazette*: A Pioneer Newspaper,' OHS, *Papers and Records* 27 (1931): 420-2; C.C.J. Bond, 'Alexander James Christie, Bytown Pioneer, His Life and Times 1787-1843,' *Ontario History* 56 (Mar. 1964): 16-36. Bytown *Gazette*, 2 and 15 Nov. 1836; Canniff, *Medical Profession in Upper Canada*, 652; Harrington, *Life of Sir William E. Logan*, 288-9. Those in the more fertile southern regions of the provinces were interested in geology from an agricultural perspective; *Western Herald and Farmers' Magazine*, 3 Jan. 1838.

10 DCB XI, A.W. Rasporich, 72

11 J.J. Bigsby, *Notes on the Geography and Geology of Lake Huron* (London 1824); 'On the Utility and Design of the Science of Geology,' *Canadian Review and Literary and Historical Journal* 2/2 (Dec. 1824); introduction to Bigsby, *Localities of Canadian Minerals*, comp. LHSQ (Quebec 1827), iii

12 *Canadian Review and Magazine* No. 4 (Feb. 1826): 319

13 Ibid., 320-1, 332-3

279 Notes to pages 19-24

14 André Lefèbvre, La 'Montreal Gazette' et le nationalisme canadien (1835-1842) (Montreal 1970), x-xi; Canadian Review and Magazine No. 4 (Feb. 1826): 249-51, 319
15 C.G. Karr, 'The Two Sides of John Galt,' Ontario History 59 (1967): 94; C.G. Karr, The Canada Land Company (Ottawa 1974); Richard Brown, The Coal Fields and Coal Trade of the Island of Cape Breton (London 1871), 74-5. PAO, Canada Company Papers, Minutes of the Court of Directors, Series A-2, Vol. 2, 13 Mar. 1827, 75; for Logan see J.D. Borthwick, History and Biographical Gazetteer of Montreal to the Year 1892 (Montreal 1892), 55.
16 Canada Company Papers, Series A-2, Vol. 2, Minutes of the Court of Directors, First Report of the Committee of Correspondence, 11 Sept. 1826, 'Instructions to the Warden of the Forests'
17 W.H. Graham, The Tiger of Canada West (Toronto 1962), 33-4, 38; W.J. Rattray, The Scot in British North America, II (Toronto 1881), 445, 447
18 William Dunlop, 'Report of the Warden of the Forests' (1827 and 1828), reprinted in PAC, Report on Canadian Archives, 1898, comp. Douglas Brymner (Ottawa 1899), 9, 13; Porter, Making of Geology, 174-5
19 PAC, CO42, G Series, Vol. 65, 385, Murray to Colborne, 12 Dec. 1829, Despatch No. 50; Graham, Tiger of Canada West, 48; Dunlop, 'Report,' 13-14, 16
20 Dunlop, 'Report,' 16; Canada Company Papers, Series A-6-2, Vol. 2, Correspondence with the Commissioners, 71, Ellice to Commissioners, 12 June 1829
21 G.P. Ure, The Hand-Book of Toronto (Toronto 1858), 185; W. Dunlop, An Address to the York Mechanics' Institution March 1832 (York 1833), 3-14; Graham, Tiger of Canada West, 110
22 York Commercial Directory, Street Guide, and Register, 1833-4 (York 1834), 136; Paul M. Romney, 'A Man Out of Place: The Life of Charles Fothergill; Naturalist, Businessman, Journalist, Politician, 1782-1840,' PHD thesis, University of Toronto, 1981, 546-9; William Dunlop, 'A Paper on Peat Mosses,' The Canadian Literary Magazine 2/1 (May 1833): 98-100
23 Upper Canada, House of Assembly, Journals, 1832-3, 52-3
24 R.W. James, John Rae Political Economist, I (Toronto 1965), chaps. 2-3; Journals, 1831-2, 100
25 Journals, 1831-2, 103; 1826-7, App. K, 'Copy of the Arrangements made between His Majesty's Government and Canada Company.' Canada Company Papers, Series A-6-3, Vol. 1, Jones to Court of Directors, 23 July 1834; Series A-6-3, Vol. 2, Correspondence with the Commissioners, 12 Feb. 1829, 13; Series A-6-2, Vol. 2, 362, Court of Directors to Jones, 19 Dec. 1833. Dunlop, 'Report,' 10; Graham, Tiger of Canada West, 53
26 Canada Company Papers, Series A-6-3, Vol. 1, Jones to Court, 23 July 1834; Toronto Patriot and Farmer's Monitor, 19 and 20 Nov. 1833. Italics added

27 Toronto *Patriot*, 13 Dec. 1833
28 Dunlop, 'Report,' 10. *Journals*, 1833-4, 59-60, 63
29 See Elizabeth Hulse, *A Dictionary of Toronto Printers, Publishers, Booksellers and the Allied Trades 1798-1900* (Toronto 1982), 77. Raymond Card, 'The Daltons and *The Patriot*,' CHR 16/2 (June 1935): 176-8; Ian R. Dalton, 'The Kingston Brewery of Thomas Dalton,' *Historic Kingston* No. 26 (Mar. 1978): 38-50. M.L. Magill, 'James Morton of Kingston-Brewer,' *Historic Kingston* No. 21 (Mar. 1973): 28-36. Toronto *Patriot*, 5 June 1832; 31 Oct. 1834; 14 Mar. 1834
30 Toronto *Patriot*, 11 Feb. 1834; PAC, Upper Canada State Papers, G1, Vol. 71, Spring Rice to Colborne, 29 July 1834, Despatch No. 14, 244; PAC, CO42, Q Series, Vol. 383, Part 2, Colborne to Spring Rice, 15 Nov. 1834, Despatch No. 69, 222. The orders are reprinted in Murray, *Muskoka and Haliburton*, 72-3.
31 R.H. Bonnycastle, *The Canadas in 1841* (London 1841), 58
32 Ibid., 19; Thomas Roy, 'On the Internal Communications of Upper Canada,' in *The Correspondent and Advocate*, 5 Oct. 1836. Roy surveyed canal and railway routes in Upper Canada, and was interested in geology. In 1837 he explained landforms around Lake Ontario in a paper to the Geological Society of London. Roy accompanied Charles Lyell on his visit to the Toronto area in 1842; see chap. 2 and Charles Lyell, *Travels in North America*, II (New York 1845), 85-6.
33 *Journals*, 1836, 92; Toronto *Patriot*, 15 Mar. 1836; PAC, Colborne Papers, Colborne to Goderich, 30 Apr. 1833, 102. Toronto *Constitution*, 19 July 1837
34 Captain Dunlop is thus not to be confused with Dr Dunlop; cf. Zaslow, *Reading the Rocks*, 17. Henry J. Morgan, *Sketches of Celebrated Canadians* (Quebec 1862), 363-4; Robina and Kathleen Macfarlane Lizars, *In the Days of the Canada Company* (Toronto 1896), 180; Thelma Coleman, *The Canada Company* (Stratford, Ont., 1978), 103; Graham, *Tiger of Canada West*, 145; I. Annette Stewart, 'Robert Graham Dunlop: A Huron County Anti-Compact Constitutionalist,' MA thesis, University of Toronto, 1947, 89
35 *Journals*, 1836, Vol. 3, App. 126, 'Report of the Committee on Geological Surveys,' 1 Mar. 1836
36 Kingston *Chronicle and Gazette*, 1 Oct. 1836
37 Ibid., 19 Oct. 1836. This theme recurred in Baddeley's writings; see 'A Geological Sketch of the Most South-Eastern Portion of Lower Canada,' LHSQ, *Transactions* 3 (1833): 272-3, where he also suggested a geological survey.
38 *Journals*, 1836-7, 15, 27, 58, 171; Toronto *Royal Standard*, 19 Nov. 1836
39 Kingston *Chronicle and Gazette*, 30 Nov. 1836
40 Ibid., 24 Dec. 1836, 12 Aug. 1837
41 Sir Francis Bond Head, *Rough Notes Taken during Some Rapid Journeys across the Pampas and among the Andes* (London 1826), v; 'Cornish Miners in America,' in Bond Head, *Descriptive Essays Contributed to the Quarterly Review*, I (London 1857),

18–19, 21. Sydney W. Jackman, *Galloping Head: The Life of the Right Honourable Sir Francis Bond Head* (London 1958), chap. 3

42 S.F. Wise, 'John Macaulay: Tory for All Seasons,' in Gerald Tulchinsky, ed., *To Preserve and Defend: Essays on Kingston in the Nineteenth Century* (Montreal 1976), 185–302; H.P. Gundy, 'Publishing and Bookselling in Kingston since 1810,' *Historic Kingston* No. 10 (Jan. 1962): 25–7

CHAPTER 2 Montreal Masonry

1 *Montreal Gazette*, 22 Nov. and 22 Dec. 1836; *Quebec Gazette*, 9 Jan. 1837
2 NHSM, *Annual Report* (Montreal 1828)
3 NHSM, *Annual Report* (Montreal 1830, 1834, 1836)
4 *Montreal Gazette*, 7 June 1836
5 Ibid., 13 June 1837
6 Stanley Brice Frost, *McGill University*, I (Montreal 1980), 128. PAC, MG24 A27, Vol. 22, Durham Papers, Sec. v, Vol. 1, Holmes to Sir J. Doratt, 20 July 1838, 590
7 McGill University Library, NHSM, 'Report of the Indian Committee,' 1828. André Lefèbvre, *La 'Montreal Gazette' et le nationalisme canadien (1835–1842)* (Montreal 1970), x; Robert W.S. Mackay, comp., *The Montreal Directory* (Montreal 1843). Andrew may have edited the *Gazette* for a time; see J.D. Borthwick, *History of Montreal* (Montreal 1897), 146. Brondgeest was a director of the Bank of British North America and had a personal interest in geology and mining.
8 John Charles Dent, *The Story of the Upper Canadian Rebellion* (Toronto 1885), I: 249, 279–80; II: 147, 238. Colin Read, *The Rising in Western Upper Canada 1837–38: The Duncombe Revolt and After* (Toronto 1982). Robina and Kathleen Macfarlane Lizars, *In the Days of the Canada Company* (Toronto 1896), 216; W.H. Graham, *The Tiger of Canada West* (Toronto 1962), 174–81
9 *Le Canadien*, 15 Dec. 1841; 10 Jan. and 8 Apr. 1842
10 Fernand Ouellet, *Economic and Social History of Quebec 1760–1850*, Carleton Library No. 120 (Ottawa 1980), 384–6, 388. Philip Goldring, 'Province and Nation: Problems of Imperial Rule in Lower Canada, 1820 to 1841,' *Journal of Imperial and Comonwealth History* 9/1 (Oct. 1980): 50–1. *Kingston British Whig*, 15 Dec. 1836 and 28 Feb. 1837; *Kingston Chronicle and Gazette*, 28 Jan. and 11 Feb. 1837
11 Morris Berman, *Social Change and Scientific Organization* (London 1978), 189
12 Stuart J. Reid, ed., *Life and Letters of the First Earl of Durham* (London 1906), I: 25–49; Leonard Cooper, *Radical Jack: The Life of John George Lambton* (London 1959), 24; Chester New, *Lord Durham: A Biography* (Oxford 1929), 4–8, 13–14
13 David Spring, 'The Earls of Durham and the Great Northern Coal Field, 1830–1881,' CHR 33/3 (Sept. 1952): 237; 'The English Landed Estate in the Age of Coal and Iron: 1830–1880,' *The Journal of Economic History* 11/1 (Winter 1951): 6

14 Spring, 'Earls of Durham,' 239; Roy Porter, *The Making of Geology* (Cambridge 1977), 133-4
15 Cooper, *Radical Jack*, 49; Berman, *Social Change*, 37-9
16 A.D. Orange, 'The Idols of the Theatre: The British Association and Its Early Critics,' *Annals of Science* 32 (1975): 290. Montreal *Gazette*, 29 May, 30 June, 10 July 1838
17 Durham Papers, Sec. 5, Vol. 1, William Sheppard to Doratt, 5 July 1838
18 Ibid., Holmes to Doratt; Montreal *Gazette*, 30 June 1838
19 Montreal *Gazette*, 26 July 1838
20 *Lord Durham's Report on the Affairs of British North America*, ed. Sir Charles Lucas (Oxford 1912), II: 12-13
21 Ibid., 13
22 Ibid., I: 15; II: 316-17. Donald Creighton, *The Empire of the St. Lawrence* (Toronto 1956; revised ed. 1970), 326; Albert Faucher, *Histoire économique et unité canadienne* (Montreal 1970), 32-8. Durham Papers, Vol. 46, Durham to Melbourne, 17 July 1838. Gerald M. Craig, *Upper Canada: The Formative Years 1784-1841*, Canadian Centenary Series (Toronto 1963), 268-72
23 Faucher, *Histoire économique*, 52-3
24 Newton Bosworth, ed., *Hochelaga Depicta: The Early History and Present State of the City and Island of Montreal* (Montreal 1839), 216, 220. Bosworth was a Baptist pastor in Montreal.
25 Montreal *Gazette*, 3 Jan. 1839, 4 and 9 Jan. 1840; topics included 'On the existence of Coal Fields in the District of Montreal or Three Rivers; on the most probable localities of such fields, and the modes of search; and also on the existence of Gypsum, fine Marbles, or metals in the said Districts' and 'On the mineralogy of the district of Montreal.'
26 The citizens of Saint John requested a popular course in mineralogy during Christmas recess, 1841; Montreal *Gazette*, 31 Mar. 1840
27 Quebec *Gazette*, 4 Mar. 1840; Toronto *British Colonist*, 24 Mar. 1841; Montreal *Gazette*, 10 Nov. 1840
28 Montreal *Gazette*, 29 Oct. 1840
29 Lizars, *In the Days of the Canada Company*, 180; I. Annette Stewart, 'The 1841 Election of Dr. William Dunlop as Member of Parliament for Huron County,' *Ontario History* 39 (1947): 51-62. Montreal *Gazette*, 6 Apr., 28 May 1841
30 *Debates of the Legislative Assembly of United Canada*, Vol. 1, 1841, 217, 220, 278, 527, 540, 908, 910
31 Montreal *Gazette*, 14 Sept., 22 July 1841; *Debates*, Vol. 1, 1841, 424
32 Montreal *Gazette*, 13 Sept. 1841
33 Ibid.
34 Ibid.; Toronto *British Colonist*, 15 Sept., 3 Feb. 1841. The prospectus was carried by

most newspapers for several months during 1841; Rae's book was never published.
35 Bagot was friendly with Michael Faraday, whose advice he sought in selecting a professor of chemistry for the newly founded University College at Toronto; see PAC, Bagot Papers, Vol. 4, 710–11, 737–9, 742, 238–43.
36 Montreal *Gazette*, 9 Nov. 1841
37 Ibid.
38 National Museum of Wales, Cardiff, De la Beche Papers, Logan to De la Beche, 1 Aug. 1840; B.J. Harrington, *Life of Sir William E. Logan* (Montreal 1883), 31
39 Sir Charles Lyell, *Principles of Geology, Being an Attempt to Explain the Former Changes of the Earth's Surface by References to Causes Now in Operation*, 3 vols. (London 1830–1, reprint ed. New York 1969), I: 1; see the critical introduction by Martin Rudwick, xiii, xxv. Porter, *Making of Geology*, 129; and Stephen Jay Gould, *Time's Arrow, Time's Cycle: Myth and Metaphor in the Discovery of Geological Time*, The Jerusalem-Harvard Lectures (Cambridge, Mass., and London 1987). For Hutton's place in the Scottish Enlightenment see David Daiches, Peter Jones, and Jean Jones, eds., *A Hotbed of Genius: The Scottish Enlightenment 1730–1790* (Edinburgh 1986), esp. 116–36.
40 Montreal *Gazette*, 4 Nov. 1841
41 Metropolitan Toronto Reference Library, Baldwin Room, W.E. Logan, Journal, Vol. 3, 14 Aug. 1841. Logan was also collecting specimens for the Museum of Economic Geology of the Swansea Institution, of which he was honorary secretary and curator for geology; De la Beche Papers, Logan to De la Beche, 5 Oct. 1940, 3 Dec. 1841. MUA, LP, Logan to James Logan, 6 Sept. and 16 Aug. 1841
42 Logan Journal, Vol. 3, 31 Aug. and 4 Sept. 1841; LP, Logan to James Logan, 6 Sept. 1841; De la Beche Papers, Logan to De la Beche, 19 Oct. 1841. Perhaps he recalled De la Beche's remark in 1839 concerning a job-seeking cousin of Logan's, that 'if he got connected with government as a scientific man there might be no knowing how far he might advance himself'; Institute of Geological Sciences, London, GSM 1/289, Director General's Letterbooks, De la Beche to Logan, 28 June 1839.
43 MUA, DP, Lyell to Dawson, 2 May 1843, 2 Feb. 1843, 30 May 1854; see also Charles Lyell, *Travels in North America in the Years 1841–2*, 2 vols. (New York 1845) and E.A. Collard, 'Lyell and Dawson: A Centenary,' *Dalhousie Review* 22 (1942–3): 133–44.
44 De la Beche Papers, Logan to De la Beche, 3 and 11 Dec. 1841; LP, George Murray to Anthony Murray, 26 Nov. 1841; Anthony Murray to Logan, 30 Nov. 1841; Alexander Murray to Logan, 30 Jan. 1842
45 Gerald J.J. Tulchinsky, *The River Barons: Montreal Businessmen and the Growth of Industry and Transportation 1837–53* (Toronto 1977), shows the close interconnections among the Montreal business community during this period; Harrington, *Life of Sir William E. Logan*, 78. James Logan served with the Molsons and with

J.S. McCord in the Constitutional Association of Montreal in 1838; he was a director of the Bank of Montreal and had served on the committee of management of the City Gas Company chaired by John Molson.

46 LP, 12 July 1839. A.J.B. Milborne, *Freemasonry in the Province of Quebec 1759-1959* (n.p. 1960), 66. J.H. Graham, *Outlines of the History of Freemasonry in the Province of Quebec* (Montreal 1892), 168ff; the Montreal Freemasons included J.S. McCord, Frederick Griffin, and William Badgley. Frances A. Yates, *The Rosicrucian Enlightenment* (London 1972), 219 and chaps. IX, XV; Margaret C. Jacob, *The Radical Enlightenment: Pantheists, Freemasons and Republicans* (London 1981), chap. V; W.J. Hughan, 'Old British Lodges,' in Osborn Sheppard, comp. and ed., *A Concise History of Freemasonry in Canada* (Hamilton 1930), 5, argues that this connection with modern science was retained in Canadian Freemasonry during the nineteenth century. Reid, *Life and Letters of the First Earl of Durham*, I: 348; J. Ross Robertson, *The History of Freemasonry in Canada*, II (Toronto 1900), 156

47 McGill had called on William Logan shortly after his arrival in Montreal; see Logan Journal, Vol. 3, 1 Sept. 1840. PAC, RG7 G20, Vol. 10, Civil Secretary's Correspondence, 3 Feb. 1842, 1232; 2 Feb. 1842, 1202; Bagot Papers, Vol. 9, Bagot to Stanley, 18 Feb. 1842, Despatch No. 35; PAC, RG7 G17C, Governor General's Office, Letterbooks, Bagot to Holmes, 19 Feb. 1842. Montreal *Gazette*, 24 Feb. 1842

48 McCullough was Logan's physician; see Logan Journal, Vol. 3, 21 Aug. 1840. PAC, RG7 G20, Vol. 19, Civil Secretary's Correspondence, McCulloch to Murdoch, 6 Jan. 1842, 1018-1/2

49 LP, Holmes to Logan, 3 Jan. 1842; Logan to James Logan, 6 Sept. 1841

50 NHSM, Minutes of the Council, 31 July 1840. PAC, RG7 G20, Vol. 11, 1329-59, Holmes to Bagot, 22 Mar. 1842

51 Ibid. These were the negotiations to which Benjamin Holmes referred in the assembly debates shortly before Sydenham's death.

52 RG7 G17C, Vol. 6, Governor General's Office, Letterbooks, Murdoch to Holmes, 29 Mar. 1842, 167-8. Testimonials reprinted in Harrington, *Life of Sir William E. Logan*, 126-32. Prior to recommending Logan, Sedgwick asked Murchison, 'Can you help me with any good clinching facts to enable me to answer [the Secretary of State's] letter point blank?'; Archives of the Geological Society of London, M/SII/192, Sedgwick to Murchison, 1842 (photocopy). Public Record Office, London, CO326/281, letter to Logan, 15 Apr. 1842

53 Montreal *Gazette*, 16 and 20 May, 22 June 1842; Kingston *Chronicle and Gazette*, 4 June 1842; Toronto *British Colonist*, 28 May 1842. Holmes retained other executive positions such as the presidency of the Montreal Library.

54 De la Beche Papers, Logan to De la Beche, 4 Aug. 1842. Logan knew about the letter on 3 Feb. 1842.

55 LP, copy, 19 July 1842. Walter B. Hendrickson, 'Nineteenth-Century State Geological Surveys: Early Government Support of Science,' in Nathan Reingold, ed., *Science in America since 1820* (New York 1976), 134, 140; see also Stanley M. Guralnick, 'Geology and Religion before Darwin: The Case of Edward Hitchcock, Theologian and Geologist (1793-1864), ibid., 131-45. W.F. Cannon, 'History in Depth: The Early Victorian Period,' *History of Science* 3 (1964): 33; reprinted in S.F. Cannon, *Science in Culture: The Early Victorian Period* (New York 1978), 225-62
56 David S. Landes, *The Unbound Prometheus* (Cambridge 1972), 54
57 Bytown *Gazette*, 2 Nov. 1843

CHAPTER 3 Logan's Geological Inventory

1 William Norris, 'Canadian Nationality: A Present-Day Plea,' *Rose-Belford's Canadian Monthly and National Review* 4 (Feb. 1880): 118. Norris, a Toronto lawyer, was a member of the Canadian National Association, a political offshoot of the Canada First movement, during the 1870s.
2 Morris Zaslow, *Reading the Rocks* (Ottawa 1974), 38, 41, 71. See also B.J. Harrington, *Life of Sir William E. Logan* (Montreal 1883); Robert Bell, *Sir William E. Logan and the Geological Survey of Canada* (Ottawa 1907); Alexander Murray, 'Anecdotes of the Life of Sir W.E. Logan,' manuscript in LP.
3 LP, Murray manuscript, 9; Logan to De la Beche, 11 Nov. 1844
4 Ibid., Logan to De la Beche, 12 May 1845; reprinted in Harrington, *Life of Sir William E. Logan*, 234-5
5 Reprinted in Harrington, *Life of Sir William E. Logan*, 235-6
6 Ibid., chap. 1 and p. 53; LP, Logan to De La Beche, 24 Apr. 1843
7 Quoted in Harrington, *Life of Sir William E. Logan*, 50-1
8 Institute of Geological Sciences, Ordnance Geological Survey of Great Britain, Correspondence of the Director General (1836-44), GSM1/289, De la Beche to Logan, 5 Jan. 1839. GSC, 'Report of Progress for the Year 1843 [R of P 1843],' in Province of Canada, House of Assembly, *Journals*, 1844-5, App. W; LP, Logan to De la Beche, 24 Aug. 1843; for the problem of nomenclature and its relation to palaeontology, lithology, and locality, see Patsy A. Gerstner, 'Henry Darwin Rogers and William Barton Rogers on the Nomenclature of the American Paleozoic Rocks,' in Cecil J. Schneer, ed., *Two Hundred Years of Geology in America* (Hanover, NH, 1979), 176-7.
9 Archibald Geikie, *The Founders of Geology*, reprint ed. (New York 1962), chap. 13; Geikie, *Life of Sir Roderick I. Murchison* (London 1875), I; chaps. 9-10; J.W. Clark and T. McK. Hughes, *The Life and Letters of the Reverend Adam Sedgwick*, 2 vols.

(Cambridge 1890); and Martin J.S. Rudwick, *The Meaning of Fossils*, revised ed. (New York 1976), chap. IV; Rudwick, *The Great Devonian Controversy* (Chicago and London 1985); James A. Secord, *Controversy in Victorian Geology: The Cambrian-Silurian Dispute* (Princeton 1986)

10 LP, W.E. Logan, 'Remarks on the Mode of Proceeding to Make a Geological Survey of the Province,' Sept. 1842

11 W.E. Logan, 'On the Packing of Ice in the River St. Lawrence,' read to the Geological Society of London, 15 June 1842, in CN *and* G 3/2 (Mar. 1858): 117-20; reprinted in Harrington, *Life of Sir William E. Logan*, 84-100

12 Logan, 'Remarks'

13 C. Gordon Winder, 'Logan and South Wales,' Geological Association of Canada, *Proceedings* 17 (Sept. 1966): 103-24; D. Trevor Williams, *The Economic Development of Swansea and of the Swansea District to 1921* (Swansea 1940), chaps. 2 and 3; D.B. Barton, *A History of Copper Mining in Cornwall* (Truro 1961); Sir Ronald Prain, *Copper: The Anatomy of an Industry* (London 1975). Logan, 'Preliminary Report,' 6 Dec. 1842, in House of Assembly, *Journals*, 1844-5, App. W; questionnaires do not appear to have survived.

14 GSC, R of P 1843

15 Ibid.

16 Logan, 'Preliminary Report,' 1842; LP, Logan to De la Beche, 31 May 1843, and Logan to J.W. Dawson, 4 June 1843

17 GSC, R of P 1843; LP, Logan to De la Beche, 20 Apr. 1844

18 MUA, DP, Logan to Dawson, 4 June 1843, and Charles Lyell to Dawson, 2 May 1843

19 GSC, R of P 1843

20 Province of Canada, *Statutes*, 1843, 7 Vict., c. 45; Montreal *Gazette*, 23 Jan. 1845

21 GSC, R of P 1843

22 Ibid.

23 GSC, R of P 1846, and Logan's 'Remarks on the Supposed Mining Region of Lake Superior,' 24 Mar. 1846, in *Journals*, 1847, App. AAA; PAO, Crown Lands Dept., RG1, Series G-1, Mining Lands Branch, Vol. 3, Mining Patents Issued, 1846-75, 3. Michigan, Geological Survey, *Geological Reports of Douglass Houghton First State Geologist of Michigan 1837-1845*, ed. G.N. Fuller (Lansing 1928), 59, 544, 554

24 Logan to Murray, 7 Mar. 1844, quoted in Harrington, *Life of Sir William E. Logan*, 180-1

25 LP, Logan to De la Beche, 11 Nov. 1844; 20 Apr. 1844; 11 Nov. 1844

26 Murray manuscript; GSC, R of P 1844, in *Journals*, 1846, App. GGG

27 LP, Logan to De la Beche, 12 May 1845; GSC, R of P 1844; LP, Logan to De la Beche, 20 Apr. 1844. Logan's early obsession with copper is evident in Metropolitan Toronto Reference Library, Baldwin Room, Logan Journal, Vol. 1, especially the entries for 20 June and 27 July 1834 in Spain.

287 Notes to pages 62–8

28 Ibid., 12 May 1845
29 GSC, R of P 1844, Murray's Report
30 Quoted in Bell, *Sir William E. Logan*. For the turtle incident, see Logan Journal, Vol. 3, 14 Aug. 1841.
31 Ibid.
32 LP, Logan to De la Beche, 12 May 1845; Montreal *Gazette*, 21, 23, and 24 Jan. 1845.
33 *Statutes*, 1844–5, 113–14
34 Ibid.; LP, Logan to De la Beche, 27 Dec. 1845; R of P 1844
35 Montreal *Gazette*, 30 Jan. 1845; Toronto *British Colonist*, 4 Apr. 1845. Logan understood the importance to his career of good press in Britain; see LP, Logan to De la Beche, 12 May 1845.
36 Toronto *British Colonist*, 5 Dec. and 6 May 1845
37 Ibid., 28 Nov. 1845; Montreal *Gazette*, 17 Nov. 1845
38 Ibid.; Toronto *British Colonist*, 25 Nov. and 8 Aug. 1845.
39 Mines Register, III: *The Copper Handbook* (Houghton, Mich., 1903), 65; Copper Development Association, *Copper Through the Ages*, revised ed. ([England] 1955), 40–2, 53
40 Allison Butts, *Copper* (New York 1954), 11; J. Percy, *Traité complet de métallurgie*, I (Paris 1864), clxxix; A. Snowdon Piggott, *The Chemistry and Metallurgy of Copper* (Philadelphia 1858), see table VII, 386.
41 R of P 1845–6, in *Journals*, 1847, App. C; LP, Logan to De la Beche, 12 May 1845
42 *Journals*, 1846; and 1847, App. AAA; *Statutes*, 1847; PAO, Crown Lands Dept., RGI, Series G-1, Mining Lands Branch, Vol. 1: Mining Privileges Lakes Superior and Huron 1845–68, and Vol. 2: Mining Location Tickets, Lakes Superior and Huron, 1845–50
43 *Journals*, 1847, App. AAA; *Journals*, 10 Dec. 1845; Montreal *Gazette*, 15 May 1846
44 Toronto *British Colonist*, 14 and 28 Apr. 1846. Prince had had mining interests on Lake Superior since 1841, after which his American partners withdrew because he was unable to obtain a charter of incorporation from the Canadian government; see Province of Canada, *Debates of the Legislative Assembly*, 1841, 484, 'Petition of Lewis Davenport and others.' The anti-American tinge of Prince's rhetoric was used to advantage by others with mining interests; George Moffatt of the Montreal Mining Company claimed to have severed previous ties with American capitalists because of the Oregon boundary dispute in 1846; see Montreal *Gazette*, 15 May 1846.
45 *Journals*, 1847, App. AAA; PAO, Crown Lands Dept., Vol. 1, 23 Mar. and 7 July 1846. See also Logan Journal, Vol. 4, 9 July 1846.
46 *Journals*, 1847, App. AAA, Civil Secretary to James Logan et al., 4 Apr. 1846; Montreal *Gazette*, 26 May 1847; 15 May 1846
47 Montreal *Gazette*, 20 Nov. 1846. PAO, Crown Lands Dept., RG 1, Series G-1, Mining

288 Notes to pages 68–71

Lands Branch, Vol. 2, 'List of Grants made to the Montreal Mining Company,' and 'Mining Location Ticket, petition of 19 Sept. 1845'; *Statutes*, 1847, 10-11 Vict., c. 68

48 PAO, Crown Lands Dept., RG 1, Series G-1, Mining Lands Branch, Vol. 3. Logan's notion of the government as 'landlord' was in keeping with British tradition in mineral rights and closely paralleled his experience in Swansea, where 'landlords retained their mineral rights when selling any property'; see Williams, *Economic Development of Swansea*, 65.

49 PAO, Crown Lands Dept., Vol. 7, Miscellaneous Papers 1845-81, Logan to D.B. Papineau, 15 Mar. 1847; Montreal *Gazette*, 2 July 1847. For Logan's interest in locating workable copper deposits on this side of the lakes see his Journal, Vol. 4, 4 Aug. 1846.

50 *Journals*, 1846, App. WW, Count de Rottermund to Provincial Secretary, 17 Apr. 1846

51 Toronto *Globe*, 16 Dec. 1846; Montreal *Gazette*, 20 Nov. 1846. For Logan's opinion see his Journal, Vol. 4, 20 Aug. 1846.

52 Montreal *Gazette*, 27 Sept., 27 Oct., 20 Nov. 1846

53 Toronto *Globe*, 16 Dec. 1846

54 Ibid.; Ontario, Legislative Assembly, *Report of the Royal Commission on the Mineral Resources of Ontario* (1889–90); J. Castell Hopkins, ed., *Canada: An Encyclopedia of the Country*, III (Toronto 1898), 441; Adam Shortt and Arthur G. Doughty, eds., *Canada and Its Provinces* (Toronto 1914) XVIII: 619. Thomas W. Gibson, *Mining in Ontario* (Toronto 1937), 99, offers other practical reasons. Gerald J.J. Tulchinsky, *The River Barons* (Toronto 1977), argues that Montreal businessmen were averse to taking financial risks. Zaslow, *Reading the Rocks*, 52; Montreal Mining Company, *Report of the Directors* (Montreal 1852), 6

55 E.E. Rich, *Hudson's Bay Company 1670–1870*, III: *1821–1870* (Toronto 1960), 501. J.S. Galbraith, *The Hudson's Bay Company as an Imperial Factor 1821–1869* (Toronto 1957), 40–2; and J.S. Galbraith, *The Little Emperor* (Toronto 1976), 175–7. Beckles Willson, *The Great Company* (Toronto 1899), 461: in 1847 the company offered to undertake colonization of its territories; about the same time it requested a collector of customs at Sault Ste Marie to help control the area. *Journals*, 1847, App. AAA; Crown Lands Dept., Vol. 2; *Statutes*, 1847, 10-11 Vict., c. 72 and c. 73

56 *Statutes*, 1847, 10-11 Vict., c. 69. Miles Macdonell, his uncle, had been governor of Assiniboia; E.E. Rich, ed., *The Publications of the Hudson's Bay Record Society*, III: Minutes of Council, Northern Department of Rupert's Land 1821–1831 (London 1940), 394–5, 411, 445; W.J. Rattray, *The Scot in British North America*, IV (Toronto 1884), 1182–3

57 Piggott, *Chemistry and Metallurgy of Copper*, 254; *Toronto Globe*, 22 Nov. 1849, 7 Feb.

1850; Macdonell's later testimony on the Hudson's Bay Company in *Journals*, 1857, App. 17; Rattray, *Scot in British North America*, 1184; Galbraith, *Imperial Factor*, 40–4; *Journals*, 1851, App. V; John Weiler, *Michipicoten*, Ontario Ministry of Natural Resources, Historical Sites Branch, Research Report 3 (Toronto 1973), 41–3; Dianne Newell, *Technology on the Frontier* (Vancouver 1986)

58 Gilbert N. Tucker, *The Canadian Commercial Revolution 1845–1851*, reprint ed. (Toronto 1964) 28–31. PAC, CO42, Vol. 498 (1842), mfm reel B-379, 134–6; *Journals*, 1846, App. O. One writer noted, 'Lakes Superior and Huron, must be united, the products of the mineral region secured[;] this achieved, we shall possess a more extensive shore, a more productive area, than any nation on the globe'; *Barker's Canadian Magazine* quoted in Montreal *Gazette*, 27 Oct. 1846.

59 Montreal Mining Company, *Quarterly Report of the Directors* (Montreal 1852), 7; PAC, MG28 III 22, British North American Mining Company, 1846–1910, Minute Book, 14 Dec. 1847 and 9 Jan. 1849; James Logan was a director of this company. BAJ 3, 6 Oct. 1847, 154; *Journals*, 1847, App. QQ

60 *Journals*, 1848, petition of J.T. Brondgeest; also 1851 and 1853; Toronto *Globe*, 21 Oct. 1853; Francis Hincks, *Reminiscences* (Montreal 1884), 348–9

61 Toronto *Globe*, 28 July 1847. J.M.S. Careless, *Brown of 'The Globe'*, I (Toronto 1959), 230

62 Toronto *Globe*, 12 Nov. 1850

63 MacNab was a trustee of the Montreal Mining Company; *Statutes*, 1847, 1643, 'An Act to Incorporate the Montreal Mining Company,' 10-11 Vict., c. 68

64 Macdonell, 'Observations upon the Construction of a Railroad from Lake Superior to the Pacific,' in *Journals*, 1851, App. UU

65 For Macdonell see Doug Owram, *Promise of Eden: The Canadian Expansionist Movement and the Idea of the West 1856–1900* (Toronto 1980).

66 Concerned farmers in Frontenac County were told that free trade altered the filial relationship with the mother country, that 'the existing connection would at least be materially weakened'; Montreal *Gazette*, 11 Mar. 1846.

67 Ibid.; J.T. Brondgeest, 'On Mines and Mining, especially referring to the Mining Districts on the Great Lakes of North America,' *Simmonds's Colonial Magazine and Foreign Miscellany* 13 (1848): 404. GSC, R of P 1846–7, in *Journals*, 1847, App. C

68 BAJ 3/6 (Oct. 1847): 144, 154, agreed with Logan that the 'coal problem' would be alleviated if electricity could soon be applied to smelt ores at Lake Superior. Montreal *Gazette*, 21 Jan. 1846 and 15 Jan. 1847

69 *Montreal Transcript and Commercial Advertiser*, 13 Oct. 1849; Montreal *Gazette*, 6 May 1850; British American League, *Minutes of the Proceedings of the Second Convention of Delegates* (Toronto 1849), 9.

70 British American League, *Minutes*, xvi

71 *Journals*, 1852–3, 3 Sept. and 11 Oct. 1852, 100, 279. The members were C.F. Fournier

(L'Islet), John Prince (Essex), Robert Christie (Gaspé), and Marc Pascal de Sales La Terrière (Saguenay); see the report in ibid., App. zzz.
72 Toronto *Globe*, 17 Mar. 1849; BAJ 5/1 (May 1949): 15–16
73 LP, Murray to Logan, 18 Feb. 1850; Logan to James Logan, 13 Feb. 1850.
74 J.E. Hodgetts, *Pioneer Public Service* (Toronto 1955), 186; J.P. Merritt, *Biography of the Hon. W.H. Merritt* (St Catharines 1875), 372–86; *Journals*, 1851, App. T. LP, Logan to James Logan, 13 Feb. 1850. Toronto *Globe*, 16 July 1850; *Statutes*, 1850, 13 Vict., c. 12

CHAPTER 4 'Grandeur and Historical Renown,' 1851–1856

1 GSC, 'Report of Progress for the Year 1850-1' [R of P 1850-1], in Province of Canada, House of Assembly, *Journals*, 1852, App. O; Audrey Short, 'Canada Exhibited, 1851–1867,' CHR 48/4 (Dec. 1967): 353–64
2 Richard D. Altick, *The Shows of London* (London 1978), 457; William Whewell, 'The General Bearing of the Great Exhibition on the Progress of Arts and Science,' in *Lectures on the Results of the Great Exhibition of 1851, delivered before the Society of Arts, Manufacturers, and Commerce* (London 1852), 13
3 London, Exhibition of 1851, *Official Descriptive and Illustrative Catalogue of the Great Exhibition of the Works of Industry of All Nations, 1851*, Part I (London n.d.), 23–4. Whewell, 'General Bearing of the Great Exhibition,' 6
4 GSC, R of P 1850-1
5 Montreal *Gazette*, 18 Feb. and 27 Mar. 1850
6 *Journals*, 1850, App. L, Report of the Select Committee on the Industrial Exhibition
7 Ibid.; the other sections were Machinery; Manufacturers; and Sculpture, Models, and the Plastic Arts. Montreal *Gazette*, 18 July, 13 Aug., and 24 Oct. 1850; *Official Catalogue*, II (London 1851), 957. GSC, R of P 1848-9, in *Journals*, 1850, App. V
8 *Official Catalogue*, II: 957. For an alternative view on organization at the exhibition, see Gottfried Semper, *Wissenschaft, Industrie und Kunst*, ed. H.M. Wingler (Mainz 1966); Semper's 'Wissenschaft, Industrie und Kunst: Vorschläge zur Anregung nationalen Kunstgefühles' (1851), trans. in Elizabeth Holt, ed., *The Art of All Nations 1850–1873: The Emerging Role of Exhibitions and Critics* (Garden City 1981); Wolfgang Hermann, *Gottfried Semper im Exil* (Basel 1978), 50–1. Semper, an architect in exile after manning barricades in 1848, was hired in London in 1851 to arrange several exhibits, including Canada's. From the Canadian exhibit he developed his 'principle of architectonically decorative order.'
9 *Official Catalogue*, II: 957–8; Nova Scotia exhibitors claimed that the province was 'capable of supplying the whole British Empire with steel and charcoal iron, equal to the best foreign articles, and at greatly reduced prices,' 970. GSC, R of P 1850-1.

10 Kenneth Warren, *The British Iron and Steel Sheet Industry since 1840* (London 1970), 18-19; Harry Scrivenor, *History of the Iron Trade* (London 1854), vi, 269, 283
11 Scrivenor, *History of the Iron Trade*, 315; Allan Birch, *The Economic History of the British Iron and Steel Industry 1784-1879* (London 1967), 215-16; A.S. Hewitt, *On the Statistics and Geography of the Production of Iron* (New York 1856), 13. Ludwig Beck, *Die Geschichte des Eisens in Technischer und Kulturgeschichtlicher Beziehung*, 4: 1801-1860 (Brunswick 1899), 661, 143-4
12 Toronto *Globe*, 8 and 17 May, 17 July 1851; *Journals*, 1851, 2; Montreal *Gazette*, 28 May 1851. *Quebec Morning Chronicle*, 10 June 1851
13 Montreal *Gazette*, 14 June 1851; *Journals*, 1851, 'First Report of the Commissioners,' 7 Aug. 1851, App. KKK. *Quebec Morning Chronicle*, 10 June 1851
14 Montreal *Gazette*, 14 June 1851
15 Toronto *Globe*, 17 July 1851
16 Ibid., 24 July 1851; Montreal *Gazette*, 9 June 1851; *A Few Words Upon Canada, and Her Productions in the Great Exhibition* (London 1851), 5-6; CJ carried articles on this theme in Sept., Nov., and Dec. 1852. GSC, R of P 1850-1
17 GSC, R of P 1850-1
18 [H.Y. Hind], 'Provincial Agricultural Show,' CJ 1/3 (Oct. 1852). Provincial agricultural shows in Canada were peripatetic exhibitions; Toronto *Globe*, 4 Aug. 1853, 29 Sept. 1854. Toronto *British Colonist*, 12 Sept. 1853
19 Montreal *Gazette*, 18 Apr. and 13 Oct. 1853; LP, 1207/11, Box 2, Folder 5, Nov. 1853
20 Sandford Fleming, 'The Valley of the Nottawasaga,' CJ 2/10 (May 1853): 205. Toronto *Globe*, 9 Dec. 1853
21 Toronto *British Colonist*, 11 Nov. 1853; cf. Samuel Strickland, *Twenty-Seven Years in Canada West* (London 1853), 10
22 DP, Logan to Dawson, 909A/11, ref. 2, 10 Jan. 1853; copy also in LP
23 *Journals*, 1854-5, 123. Herbert R. Balls, 'John Langton and the Canadian Audit Office,' CHR 21/2 (June 1940):150-76. W.A. Langton, 'The Letters of John Langton about Canadian Politics, 1855-56,' CHR 5/4 (Dec. 1924): 341; Part 5/II, CHR 6/1 (Mar. 1925): 47
24 Reprinted in CJ 2/8 (March 1854), Supplement, 201-4
25 Jean S. Magill, *A Pioneer History of the County of Lanark* (Toronto 1968), 28, 161, 212. *Journals*, 1854, App. L. Logan repaid the favour when Hall's own survey was reviewed in 1855; New York State Library (Albany), James Hall Papers, Folder 197, Hall to Logan, [15?] Mar. [1855]; Logan to Hall, Folder 824, [17?] and 20 Mar. 1855.
26 *Journals*, 1854, App. L
27 Ibid.
28 J.L. Gourlay, *History of the Ottawa Valley* (n.p. 1896), 124-6. *The Packet and Weekly Commercial Gazette*, 28 Dec. 1850

29 *The Packet and Weekly Commercial Gazette*, 28 Dec. 1850; Part II, 11 Jan. 1851. *The Ottawa Citizen*, 12 July 1851
30 *Ottawa Citizen*, 2, 9, 16, 30 Oct.; 6, 13, 27 Nov.; 11, 18 Dec. 1852; 16 Aug. 1854
31 *Ottawa Citizen*, 2 July 1853; *Journals*, 1846, App. WW; Morris Zaslow, *Reading the Rocks* (Ottawa 1974), 46-7; Henry J. Morgan, *Sketches of Celebrated Canadians* (Quebec 1862), 767-8; *Journals*, 1854, App. L
32 *Journals*, 1854, App. L; *Le moniteur canadien*, 29 Mar. 1855, in Metropolitan Toronto Reference Library, Baldwin Room, W.E. Logan Scrapbook
33 *Journals*, 1854, App. L
34 Ibid.
35 GSC, R of P 1853-6, in *Journals*, 1857, App. 52; J.G.C. Anderson, 'The Concept of Precambrian Geology and the Recognition of Precambrian Rocks in Scotland and Ireland,' in W.O. Kupsch and W.A.S. Sarjeant, eds., *History of Concepts in Precambrian Geology*, Geological Association of Canada Special Paper no. 19 (n.p. 1979), 3-4. *Journals*, 1854, App. L
36 *Journals*, 1854, App. L. D.E. Allen, *The Naturalist in Britain* (London 1976), 87. Sir Archibald Geikie, *Memoir of Sir Andrew Crombie Ramsay* (London 1895), 204-5
37 *Journals*, 1854, App. L
38 *Journals*, 1854-5, 785, 1292
39 James Gibson, 'Sir Edmund Walker Head and Technology in the Victorian Age,' paper delivered at University College Symposium on 'Canadian Nationhood in the Nineteenth Century,' University of Toronto, 19 Jan. 1981. See also D.G.G. Kerr, *Sir Edmund Head: A Scholarly Governor* (Toronto 1954), 16, 208, chap. 7
40 Toronto *Globe*, 5 Mar. 1855; LP, 'Report of the Executive Committee of the Provincial Commission appointed to Ensure a Fitting Representation of the Industry and Resources of Canada at the World's Exhibition to be held in Paris in the Year 1855,' Nov. 1854
41 LP, 909A/11, ref. 3, Logan to Dawson, 17 Feb. 1855; copy in DP. LP, Dawson to Logan, 6 Mar. 1855
42 London *Times*, 7 Sept. 1855, in Logan Scrapbook. LP, 1207/16, Logan to Francis Hincks, 11 May 1855
43 *Times*, 7 Sept. 1855
44 *Journals*, 1856, App. 46, J.C. Taché, 'Canada at the Universal Exhibition of 1855,' and 'Observations on the Exhibition'
45 Ibid., Thomas Sterry Hunt, ed., Introduction to 'A Sketch of the Geology of Canada.' *The Canadian Statesman*, 6 Sept. [1855?], Logan Scrapbook; actual copies of this issue appear no longer to exist. The editor was Rev. John Climie, a Liberal and a former Congregationalist minister who acquired the paper in August 1855. His son William sometimes edited the paper as well; see R.G. Hamlyn et al., *Bowmanville: A Retrospect* (Bowmanville 1958), 61, 78. The editor took a swipe at

J.C. Taché's rumoured failure to carry his share of the burden at Paris, leaving Logan to do most of the work. He blamed all French Canadians for their 'habitant' mentality, compared with what he termed the ability of Upper Canadians to think 'large thoughts.'

46 LP, Lyon Playfair to Logan, 1207/11, ref. 44, 3 Nov.; 1207/76, 7 Nov.; 1207/77, 10 Nov. 1855; 1207/11, ref. 78, Robert Ellice, Whitehall, to Logan, 15 Nov. 1855; 1207/11, ref. 45, unsigned note to Logan, 19 Nov. 1855. Imperial College of Science and Technology, Lyon Playfair Papers, Logan to Playfair, 20 Nov. 1855

47 LP, Simpson to Logan, 11 Mar. 1856. Canadian Institute, 'Address to Sir William E. Logan,' *CJ* n.s. I/4 (July 1856); quoted in B.J. Harrington, *Life of Sir William E. Logan* (Montreal, 1883), 313

48 Canadian Institute, 'Address to Sir William E. Logan'; LP, E.J. Chapman to Logan, 22 Mar. 1856; also Logan Scrapbook. The only available copy of the *Ottawa Citizen* article is in the Logan Scrapbook, dated '1859' by hand; internal evidence, however, suggests that the article was published in 1856. Sandford Fleming, 'The Canadian Geological Survey and Its Director,' *CJ* n.s. I/3 (May 1856). *Journals*, 1856, 211; the act for renewal was 19-20 Vict., c. 13.

CHAPTER 5 'Permanence,' 1857-1869

1 Quoted in B.J. Harrington, *Life of Sir William E. Logan* (Montreal 1883), 331-2
2 J.W. Dawson, 'Geological Survey of Canada-Report of Progress for 1857,' CN *and* G, 4/1 (Feb. 1859); E.J. Chapman, 'Review of Geological Survey of Canada, Report of Progress 1857,' *CJ* n.s. IV/22 (July 1859); 'Geological Survey in Great Britain and Her Dependencies,' CN *and* G 3/4 (Aug. 1858), Art. 25. Dawson, 'Things to be Observed in Canada, and especially in Montreal and Its Vicinity,' CN *and* G 3/1 (Feb. 1858), Art. 1; 'Report of the Geological Survey of Canada 1853-55,' CN *and* G 3/1 (Feb. 1858), Art. 5
3 Canada Legislative Assembly, *Journals*, 1857, App. 5. Not among de Rottermund's valid theories were those on the role of electricity in the formation of metallic ores.
4 Toronto *Globe*, 1 Feb. 1855; LP, Hector Langevin to Logan, 14 Feb 1859. Canadian geologists rejected de Rottermund's coal, putting the government in a temporary quandary: LP, 1207/10, Russell to Logan, 23 Feb. 1855; 1207/9, Chapman to Logan, 4 Mar. 1855; 1207/2, Chapman to Logan, 10 Mar. 1855; 1207/11, ref. 137, Chapman to Logan, 9 Apr. 1855; E.J. Chapman, 'Review of (Count de Rottermund)-Report on the Exploration of Lakes Superior and Huron,' *CJ* n.s. I/5 (Sept. 1856); LP, Dawson to Logan, 1 June 1858; Dawson, 'Coal in Canada: The Bowmanville Discovery,' CN *and* G 3/3 (June 1858), Art. 23; 'To Our Reviewers,' ibid. 3/5 (Oct. 1858)

5 Alexander Morris, *Nova Britannia* [1858] (Toronto 1884), 37-9; and his *Canada and Her Resources* (Montreal 1855), 58. J.C. Taché, *Des Provinces de l'Amérique du Nord et d'une Union fédérale* (Quebec 1858), 42-6, 222-3, 238-40. Joseph E. Cauchon, *Étude sur l'Union projétée des Provinces Britanniques de l'Amérique du Nord* (Quebec 1858), 5, 22; Cauchon later changed his mind.
6 Toronto *Globe*, 1 Feb. 1855
7 LP, Simpson to Logan, 11 Mar. 1856. When renewal of the HBC charter came before a select committee, Simpson denied knowledge of valuable minerals in its territories; *Journals*, 1858, App. 3. Draper was president of the Canadian Institute in 1857; see his address in CJ n.s. 2/8 (March 1857). LP, A.R. Roche to Logan, 15 May 1857; Murray to Logan, 8 June 1857
8 LP, H.Y. Hind to P. Vankoughnet, 3 July 1857; Murray to Logan, 8 June 1857; Hind to Logan, 3 July 1857. Logan to Hind, Private, 7 July 1857. W.L. Morton, *Henry Youle Hind 1823-1908* (Toronto 1980), chap. 2, esp. 26
9 *Journals*, 1858, App. 3; Great Britain, Parliament, House of Commons, Select Committee on the Hudson's Bay Company, *Report* (London 1857); House of Commons, *Report*, questions no. 255, 398-400; nos. 3092-3109, 4300; no. 4464; 6090. The question of the real interests of the HBC is complex: R.S. Young, 'Minerals and the Fur Trade,' *Alberta History* 24/3 (Summer 1976): 20-4; R.G. Ironside and S.A. Hamiton, 'Historical Geography of Coal Mining in the Edmonton District,' *Alberta Historical Review* 20/3 (Summer 1972): 6-16; cf. Howard N. Eavenson, *The First Century and a Quarter of American Coal Industry* (Pittsburgh 1942), 362.
10 W.O. Kupsch, 'Métis and Proud,' *Geoscience Canada* IV/3 (Sept. 1977): 146-8; A.K. Isbister, 'On the Geology of the Hudson's Bay Territories, and of Portions of the Arctic and North-Western Regions of America, with a Coloured Map,' *Quarterly Journal of the Geological Society of London* 11 (1855): 497-520
11 *Journals*, 1857, 5 and 19 Mar. Doug Owram, *Promise of Eden* (Toronto 1980), 46-7
12 Dawson, 'Report of the Geological Survey of Canada 1853-55,' 37-8. C.W. Cooper, *Frontenac, Lennox & Addington: An Essay* ([Kingston 1856]; reprint ed., Ottawa 1980) 80-1; Elkanah Billings, 'On the Iron Ores of Canada and the cost at which they may be worked,' CN and G 2/1 (March 1857): 20-8.
13 Eric W. Morse, *Fur Trade Canoe Routes of Canada: Then and Now*, second ed. (Toronto 1979), chap. V. F.B. Murray, 'Agricultural Settlement on the Canadian Shield,' in Ontario Historical Society, *Profiles of a Province* (Toronto 1967), 180-1. Dawson, 'Geological Survey of Canada, Reports of Progress for the Years 1853-58,' CN and G 3/2 (March 1858): 82
14 Alexander Morris, 'The Hudson's Bay and Pacific Territories [1858-9],' in *Nova Britannia*, 52-3, 59
15 *Journals*, 1854, App. L. H.Y. Hind, 'Preliminary Address,' CJ I/1 (Jan. 1856): 2
16 A.N. Rennie, 'Annual Report of the Natural History Society of Montreal,' CN and

G 2/3 (July 1857): 228. Some British luminaries were disappointed when Canada did not send a steamer for them. Imperial College, A.C. Ramsay Papers, Logan to Ramsay, 7 Aug. 1857. W.E. Logan, 'On the Division of the Azoic Rocks of Canada into Huronian and Laurentian,' CJ II/12 (Nov. 1857): 439-40; 'On the Probable Subdivision of the Laurentian Rocks of Canada,' ibid. III (Jan. 1858): 1. R.H. Dott, Jr, 'The Geosyncline-First Major Geological Concept "Made in America",' in Cecil J. Schneer, ed., *Two Hundred Years of Geology in America* (Hanover, NH, 1979), 239-64

17 'Eleventh Meeting of the AAAS,' CN and G 2/4 (Sept. 1857): 241. 'Annual Report of the Natural History Society of Montreal, 18 May 1858,' ibid. 3/3 (June 1858): 229. Dawson, 'Farther Gleanings from the Meeting of the American Association in Montreal,' ibid. 2/5 (Nov. 1857): 356-9. While a student at Edinburgh Dawson was dismayed to be treated as a colonial; not long before this meeting, he had lost out in the competition for a chair of natural history there, again ostensibly because he was a colonial. 'Inauguration of the New Building of the Natural History Society,' ibid. 4/2 (April 1859): 143-6; 'Annual Meeting,' ibid. 2/4 (Aug. 1865): 303

18 GSC, R of P 1858 (Montreal 1859), 65-6. LP, James Logan to Logan, 23 May 1862

19 O.D. Skelton, *The Life and Times of Sir Alexander Tilloch Galt* (Toronto 1920), chap. 10. Toronto *Globe*, 18 May 1861; Montreal *Gazette*, 10 and 18 May 1861. *Journals*, 1861, 7 May; 1862, 3 June. LP, Logan to James Logan, 31 Mar., 1862; James Logan to Logan, 30 May, 20 June 1862. PAC, Robert Bell Papers, I, Elizabeth Notman Bell to Robert Bell, 27 May 1863

20 GSC, *Report of Progress from Its Commencement to 1863 [Geology of Canada]* (Montreal 1863), iii; University of Edinburgh Library, Charles Lyell Papers, Dawson to Lyell, 13 Nov. 1860; MUA, Robert Bell Papers, Logan to Bell, 4 July 1863. Dawson, 'Geological Survey of Canada-Report of Progress from Its Commencement to 1863,' CN and G n.s. 1/1 (1864): 65-7. [E.J. Chapman?], 'Review of Geological Survey of Canada, Report of Progress,' CJ 9/50 (March 1864): 207-10

21 Dawson, '1863,' 66. Montreal *Gazette*, 25 Sept. 1863. Quebec *Gazette*, 28 Sept. 1863, 10 June 1864; Harrington, *Life of Sir William E. Logan*, 352

22 *Journals*, 1864; 27-28 Vict , c. 8. Dawson, 'Geological Survey of Canada, Report of Progress for 1857,' CN and G 4/1 (Feb. 1859): 66

23 Lyell Papers, Dawson to Lyell, [7?] Mar. 1856; Nov. 1859; 20 Mar. 1858; 28 Jan. 1859; 26 Apr. and 11 June 1860; 15 May 1861. *Geology of Canada*, 48-9. GSC, R of P 1863-6, 14-17; T.C. Weston, *Reminiscences Among the Rocks* (Toronto 1899), 26-7, 93-4

24 GSC, R of P 1866-9, 15-16; Lyell Papers, Dawson to Lyell, 1 Mar. 1864. MUA, Robert Bell Papers, Logan to Bell, 15 Dec. 1864; LP, Logan to Dawson, 909A/11, ref. 4, 13 Oct. 1864. CN and G n.s. 2/2 (1865). DP, Logan to Dawson, 18 May 1865; Lyell to Dawson, 6 Dec. 1865. C.F. O'Brien, '*Eozoon Canadense* "The Dawn Animal of

296 Notes to pages 104–7

Canada",' *Isis* 61 (1970): 206–23; N.A. Rupke, 'Bathybius Haeckelii and the Psychology of Scientific Discovery,' *Studies in the History and Philosophy of Science* 7/1 (1976): 53–62; Stephen Jay Gould, *The Panda's Thumb: More Reflections in Natural History* (New York 1980), chap. 23: 'Bathybius and Eozoon'
25 Quoted in J.M. Harrison and E. Hall, 'William Edmond Logan,' Geological Association of Canada, *Proceedings* 15 (Nov. 1963): 41
26 *The Nation* 1/7, 14 May 1874; 1/8, 21 May 1874; 2/51, 24 Dec. 1875
27 'Proceedings,' CN *and* G 8/5 (May 1877): 301
28 Skelton, *Life and Times of Sir Alexander Tilloch Galt*, 352. Since the 1840s mining speculation a growing lobby of 'western' interests had pressed the government to reduce duties to match the Gaspé region. In 1849 Francis Hincks promised that 'if the Gaspé Fisheries were again to incur certain Dutiable privileges, the Mining and Lumber interests would not be forgotten'; British North American Mining Company, Minute Books, 49. Throne Speech, quoted in Quebec *Gazette*, 19 Feb. 1864. Harrington, *Life of Sir William E. Logan*, 378ff.
29 *Geology of Canada*, 529; there were exceptions: Weston wrote that during the 1880s, 'coal was still popularly anticipated' in the area around Quebec, 214; W.M. Whitelaw, *The Maritimes and Canada before Confederation* (Toronto 1934), 273
30 'On the Geology of the Country between Lake Superior and the Pacific Ocean,' CN *and* G 6/4 (Aug. 1861): 330
31 William Stanley Jevons, *The Coal Question: An Inquiry Concerning the Progress of the Nation, and the Probable Exhaustion of Our Coal Mines* (London 1865), 349. PAC, GSC, Director's Letterbooks, Logan to William McDougall. 25 Aug. 1866. PAC, Molson Company Archive, Correspondence, Letterbooks, Vol. 14, Thomas Molson to Lowe, Gibb, and Co., 28 May 1859, 18; Inventory and Balance Sheet, Vol. 183. R.G. Haliburton, *The Coal Trade of the New Dominion* (Halifax 1868), delivered to the NSINS, 1866–7, 5
32 Memorial University, Centre for Newfoundland Studies, Robert Bell Papers, 'Newfoundland Coal–Extract from "Lectures, Literary and Biographical" by Rev. M. Harvey' (1864); Copy of an 'Agreement between D. Ross of Montreal, Canada, R. Bell of Kingston, Canada, and A. Shea of St. John's Newfoundland, 6 April 1867.' PAC, Robert Bell Papers, Bell to Ambrose Shea, 4 Apr. 1866; Shea to Bell, Private, 19 Apr. 1866; Vol. 2, J.H. Warren to Bell, 21 Dec. 1869. James Hiller, 'Confederation Defeated: The Newfoundland Election of 1869,' in Hiller and Peter Neary, eds., *Newfoundland in the Nineteenth and Twentieth Centuries* (Toronto 1980), 67–94
33 PAC, Robert Bell Papers, Richard Berford to Bell, Nov. 1865; T.C. Lombard to Bell, 26 Dec. 1865; A. Campbell to Bell, Private, 16 Dec. 1865; Lombard to Campbell, 18 Dec. 1865; Bell to Campbell, 16 Dec. 1865; Campbell to Bell, Private, 8 Mar. 1866; G.H. Frost to Bell, 24 Jan. 1866; 'Report on the Oil Region of Gaspé,' 3 Feb. 1866;

'Memo. of Conversation with Hon. Mr. Campbell, 16 May 1866,' in which Bell was to 'go to see the priest & pretend to be a devote [sic]-son of Mother Church' in order to secure Indian consent for lands on Manitoulin; A.T. Drummond to Bell, 30 Jan. 1867. Those interested in petroleum included Thomas Sterry Hunt, Sandford Fleming, T.C. Keefer, Charles Robb, and Thomas Macfarlane; CN and G and CJ 1861-3, passim. LP, Bell to Logan, 18 Dec. 1865; MUA, Robert Bell Papers, Logan to Bell, 17 Feb. 1866. For Logan's theory of the anticlinal origins of petroleum, see *Geology of Canada*, 521. GSC, Director's Letterbooks, Logan to Bell, 9 and 16 June 1866; H.G. Vennor warned Bell that Logan suspected a conflict of interest, PAC, Robert Bell Papers, Vennor to Bell, Private, 14 Dec. 1867.

34 Dawson, 'Notes on the Meeting of the BAAS at Birmingham, 1865,' *CN and G* 2/6 (Dec. 1865): 415; E.A. Collard, 'Lyell and Dawson: A Centenary,' *Dalhousie Review* 22 (1942-3): 139. *The Nation* 2/11, 19 Mar. 1875, 124

35 LP, Robert Barlow to Logan, 1862; DP, Logan to Dawson, 2211/21, ref. 7, 28 Jan. 1867; Bell to Logan, 27 Nov. 1864. Director's Letterbooks, Logan to Richards, 24 Aug. 1866, 173; Logan to Crown Lands Dept., 16 June 1866, 130

36 Letterbooks, Logan to Carter, 21 May 1866, 115. LP, Murray to Logan, 23 Apr. 1864; Edward Morris to Logan, 16 Oct. 1866; Logan to Morris, 5 Nov. 1866. Weston, *Reminiscences Among the Rocks*, 86.

37 DP, Richard Brown to Dawson, 18 Nov. 1845; S. Cunard to Dawson, 28 Nov., 8 Dec. 1845; Brown to Dawson, 9 Dec. 1845; Dawson to G.R. Young 1846; Brown to Dawson, 10 Mar. 1846; Cunard to Dawson, 19 Feb. 1846; Young to Dawson, Apr. 1848; Lyell Papers, Dawson to Lyell, 4 July 1854. Lyell Papers, Dawson to Lyell, 10 Oct. 1868; LP, Dawson to Logan, 18 Mar. 1858; 10 Oct. 1861. Nova Scotia, House of Assembly, *Debates and Proceedings*, 18 Apr. 1864, 229. Public Archives, Nova Scotia, RG7, Vol. 50, Honeyman to Provincial Secretary, 19 Apr. 1864; *Debates and Proceedings*, 1865, App. 17; Director's Letterbooks, A.R.C. Selwyn to[?], 23 Dec. 1870; DCB X: 361-2, XI: 420-1

38 LP, Hind to Logan, 31 May 1864

39 W.O. Kupsch, 'Boundary of the Canadian Shield,' in Kupsch and W.A.S. Sarjeant, eds., *History of Concepts in Precambrian Geology*, Geological Association of Canada Special Paper no. 19 (n.p. 1979), 125. Director's Letterbooks, Logan to Murchison, 29 Nov. 1867, 286; Logan to Bell, 25 Jan. 1867, 233

40 Director's Letterbooks, Logan to Ramsay, 2 Dec. 1868, 405

41 Ibid., Selwyn to Logan, 11 Apr. 1870, 143. Weston, *Reminiscences Among the Rocks*, 80. A.H. Lang, 'Sir William Logan and the Economic Development of Canada,' *Canadian Public Administration* (1969): 551-65; C.G. Winder, 'Sir William E. Logan (1798-1875)-Founder of Canadian Geology,' Geological Association of Canada, *Proceedings* 24/2 (Nov. 1972): 39-41

42 DP, 'On Some Characteristics of the British American Mind,' Aug. 1868

43 Richard Brown, *The Coal Fields and Coal Trade of the Island of Cape Breton* (London 1871), 82, 93; S.A. Saunders, 'The Coal Problem in Canada,' Royal Bank of Canada, *Essays on Canadian Economic Problems*, II (Montreal 1929), 18–21; Rand Matheson, 'The Coal Problem in Canada,' in ibid., 33; A.A. den Otter, 'Sir Alexander Tilloch Galt, the Canadian Government and Alberta's Coal,' Canadian Historical Association, *Historical Papers* (1973): 21–42
44 *The Nation* II/19, 14 May 1875, 'Nationality,' 224. On Lyell's legacy see II/11, 19 Mar. 1875, 124
45 Geoffrey Best, *Honour among Men and Nations*, the 1981 Joanne Goodman Lectures (Toronto 1982), 44
46 W.J. Rattray, *The Scot in British North America*, II (Toronto 1881), 933
47 E. Jerome Dyer, *The Routes and Mineral Resources of North Western Canada* (London 1898), 3–4

CHAPTER 6 The Spirit of the Method

1 'The President's Annual Address to the Canadian Institute,' Toronto *British Colonist*, 1 Feb. 1853
2 Susan Faye Cannon, 'Humboldtian Science,' in Cannon, *Science in Culture: The Early Victorian Period* (New York 1978), chap. 4. Klaus Hammacher, ed., *Universalismus und Wissenschaft im Werk und Wirken der Brüder Humboldt* (Frankfurt 1976); Nathan Reingold, 'Alexander Dallas Bache: Science and Technology in the American Idiom,' *Technology and Culture* 2 (1970): 163–77; Walter E. Gross, 'The American Philosophical Society and the Growth of Meteorology in the United States, 1835–1850,' *Annals of Science* 29/4 (Dec. 1972): 321–38
3 John Cawood, 'Terrestrial Magnetism and the Development of International Collaboration in the Early 19th Century,' *Annals of Science* 34 (1977): 551–87; R. Glenn Madill, 'The Magnetic North,' *The Beaver*, Outfit 266/2 (Sept. 1935): 12–15, 66. William Whewell, *History of the Inductive Sciences*, III (London 1847); W.E. Gross, 'The American Philosophical Society and the Growth of Science in the United States, 1835–1850,' PHD thesis, University of Pennsylvania, 1970, 23–6
4 Dorothy Stimson, *Scientists and Amateurs: A History of the Royal Society* (New York 1948), 12, 138; A.K. Khrgian, *Meteorology: A Historical Survey*, I, 2nd ed., trans. and ed., K.P. Pogosyan (Jerusalem 1970), 70–1; Howard Daniel, *One Hundred Years of International Co-operation in Meteorology (1873–1973)* (Geneva 1973), 2; Marcus Benjamin, 'Meteorology,' in G.B. Goode, ed., *The Smithsonian Institution 1846–1896* (Washington 1897), 648
5 W.E. Knowles Middleton, *A History of the Theories of Rain and Other Forms of Precipitation* (New York 1965), chaps. 6–8; Cannon, *Science in Culture*, 97. Howard Frisinger, *The History of Meteorology to 1800* (New York 1977), 140; 'Early Theories

299 Notes to pages 118-21

on the Cause of Thunder and Lightning,' American Meteorological Society *Bulletin* 46/12 (Dec. 1965): 785-93. Maxime Bôcher, 'The Meteorological Labors of Dove, Redfield and Espy,' American Meteorological Society *Bulletin* 46/8 (Aug. 1965): 448-52; J.E. McDonald, 'James Espy and the Beginnings of Cloud Dynamics,' American Meteorological Society *Bulletin* 44 (1963): 634-41

6 Edward Sabine, 'Contributions to Terrestrial Magnetism,' Royal Society of London, *Philosophical Transactions* 136, Part III, 237-55. One of the best explications of nineteenth-century geomagnetism and magnetic instruments remains Balfour Stewart's 'Meteorology' in the *Encyclopedia Britannica*, 9th ed. (London 1890), XVI: 159-84; see also 'Magnetism,' XV: 238-9.

7 A.F. Kemp, 'Review of *The Life, Travels, and Books of Alexander von Humboldt*,' CN and G 6/3 (June 1860): 209. Cawood, 'Terrestrial Magnetism,' 552; *Science in Culture*, 104

8 G. Waldo Dunnington, *Carl Friedrich Gauss: Titan of Science* (New York 1955), chap. 13. Jack Morrell and Arnold Thackray, *Gentlemen of Science* (Oxford 1981), 354-70, 517-27; John Cawood, 'The Magnetic Crusade: Science and Politics in Early Victorian Britain,' *Isis* 70/254 (1979): 493-518; Whewell, *History of the Inductive Sciences*, 68-70, 75. Reingold, 'Alexander Dallas Bache,' 173; American Philosophical Society, *Proceedings* 1/7 (Aug. 1839): 104, 111; 1/8 (Sept. 1839): 116-17; 1/9 (Dec. 1839): 148, 151-2

9 BAAS, *Annual Report*, 1839, xxi, 3-4; 1841, 42. Dunnington, *Carl Friedrich Gauss*, 157; Edward Sabine, *Observations at the Magnetic and Meteorological Observatory at Toronto in Canada* (London 1845), I (1840-2), II. Vernon Harcourt, Presidential Address, BAAS, *Annual Report* (1839), 3-4

10 PRO, MO, Meteorology, Historical Records, Part VI (Letters to Sabine from Canada), no. 13, Riddell to Sabine, 5 Mar. 1840. Mfm at PAC.

11 Upper Canada, House of Assembly, *Journals*, 1833-4, 38, 42. J.G. Hodgins, ed., DHE II: 144-5

12 PAC, Upper Canada State Papers, Q Series, Vol. 388, Part 1, 7, Public Officers and Miscellaneous, Secretary of the Admiralty to Hay, 7 July, 14 July 1835; CO42, G Series, G74, Glenelg to Colborne, no. 29, 29 July 1835, 206; G76, Glenelg to Head, no. 27, 29 Feb. 1836, 296, refers to Colborne's no. 70 of 23 Nov. 1835; G86, Glenelg to Arthur, no. 105, 29 June 1838, 171; James Fitzgibbon to Macaulay, 8 Oct. 1838

13 Fred Kaplan, ' "The Mesmeric Mania": The Early Victorians and Animal Magnetism,' *Journal of the History of Ideas* 35/4 (Oct.-Dec. 1974): 691-702. *Le Canadien* 8/53 (5 Sept.), 8/58 (17 Sept.), and 8/66 (5 Oct.) 1838; 'Les Séances de Magnetisme,' *Le Bulletin des Recherches Historiques* 48/10 (Oct. 1942): 317; I am indebted to John H. Noble for these two sources. *Quebec Mercury*, 12 Mar. 1840. 'Electro-Psychology or Mesmerism,' Toronto *British Colonist*, 16 Apr. 1852. Dickens, *American Notes and Pictures from Italy* (London 1893), 178 (first published 1842); Fred

Kaplan, *Dickens and Mesmerism* (Princeton 1975), chap. 1. For the link between mesmerism and radical politics, see Robert Darnton, *Mesmerism and the End of the Enlightenment in France* (Cambridge, Mass., 1968).
14 A. Burnett Lowe, 'Canada's First Weathermen,' *The Beaver*, Outfit 292 (Summer 1961): 4-7; and M.K. Thomas, 'A Century of Canadian Meteorology,' Canadian Atmospheric Environment Service, *Annual Report of Operations* (1971-2): 1-20. I am grateful to Mr Thomas of the Atmospheric Environment Service for a copy of this issue. F.J. Toole, 'The Scientific Tradition,' in A.G. Bailey, ed., *The University of New Brunswick Memorial Volume* (Fredericton 1950), 69-71; A. Foster Baird, 'The History of Engineering at the University of New Brunswick,' ibid., 26. BAAS, *Annual Report* (1839), 170; (1840), 149
15 James H. Morrison, 'Soldiers, Storms and Seasons; Weather Watching in 19th-century Halifax,' *Nova Scotia Historical Quarterly* 10/3-4 (Sept.-Dec. 1980): 224-5. Marjorie Whitelaw, ed., *The Dalhousie Journals*, I (n.p. 1978), 7, 10-11. W.F. Ganong, ed., 'The Journal of Captain William Owen, R.N.,' *Collections of the New Brunswick Historical Society* 2/4 (1899): 24-5
16 William Kelly, 'Abstract of the Meteorological Journal,' LHSQ *Transactions* 3 (1837): 46-65, delivered in 1832; 'On the Effect of Clearing and Cultivation on Climate,' 3 (1837): 309. R.H. Bonnycastle, 'An Account of Some Meteorological Phenomena oberved in Canada,' 1 (1824-9): 47-51
17 Upper Canada, House of Assembly, *Journals*, 1833-4, report of Select Committee quoted in Hodgins, DHE, II: 144. T.A. Reed, 'The Observatory at Toronto, 1840-1908,' *Canadian Geographical Journal* 15/6 (Dec. 1957): 235. Edmund W. Gilbert, *British Pioneers in Geography* (London 1972), chap. 4
18 McGill University Library, NHSM, Minutes of the Council (1833-40), 21 Apr. 1837; Montreal *Gazette*, 19 Jan., 25 Apr., and 13 June 1837; 2 Aug. 1838; 3 Jan. 1839
19 Gross, 'American Philosophical Society,' 327. Bache founded the first American magnetic observatory at Girard College, Philadelphia, in 1840. PRO, MO, Part VI, Riddell to Sabine, no. 1, 23 Sept. 1839; no. 12, 4 Feb. 1840
20 PRO, MO, Part VI, Riddell to Sabine, no. 3, 5 Oct. 1839; no. 4, 14 Oct. 1839; no. 5, Bayfield to Beaufort, 18 Oct. 1839. Riddell to Deputy Adjutant General, 14 Nov. 1839, reprinted in A.D. Thiessen, 'The Founding of the Toronto Magnetic Observatory and the Canadian Meteorological Service,' RASC, *Journal* 34 (1940): 319
21 PRO, MO, Part VI, no. 6, Riddell to Sabine, 9 Nov. 1839; Riddell to Sabine, 25 Aug. 1840, in Thiessen, 'Founding of the Toronto Magnetic Observatory,' 337
22 Sabine, 'Contributions to Terrestrial Magnetism,' 238-9. PRO, MO, Part VI, no. 13, Riddell to Sabine, 5 Mar. 1840; no. 14, 1 May 1840; no. 16, 25 June 1840; no. 23, 15 Nov. 1840
23 PRO, MO, Part VI, no. 15, 26 May 1840; 25 June 1840
24 PRO, MO, Part IV (Sabine to Canada), no. 10, Riddell to Younghusband, 23 Dec. 1841

25 Sabine to Lefroy, 25 Nov. 1840, quoted in J.H. Lefroy, *Diary of a Magnetic Survey* (London 1883), vi–vii. Sabine to Lefroy, 10 Apr. 1839, quoted in A.D. Thiessen, 'Founding of the Toronto Magnetic Observatory,' RASC, *Journal* 36/2 (Feb. 1942): 62–3
26 PRO, MO, Part VI, no. 7, Lefroy to Sabine, n.d., from Gravesend; no. 10, 8 Sept. 1842
27 Ibid., no. 11, 25 Oct. 1842. BAAS, *Report* (1842), 4–5; (1843), 59. Sabine, *Observations*, I: 19
28 PRO, MO, Part VI, Lefroy to Sabine, no. 12, 21 Nov. 1842; no. 13, 5 Dec. 1842
29 G.F.G. Stanley, ed., Introduction to Lefroy, *In Search of the Magnetic North* (Toronto 1955); the instruments supplied to Lefroy are listed in Sabine, 'Contributions,' 240. Lefroy, *Autobiography*, ed. Lady Lefroy (London [1896]), 66. Lefroy to George Simpson, 25 Sept. 1842, in A.D. Thiessen, 'Founding of the Toronto Magnetic Observatory,' RASC, *Journal* 35/4 (Apr. 1941): 146; disturbances were to be measured simultaneously with Younghusband at Toronto and Bache at Philadelphia: see PRO, MO, Part V, Sabine to Younghusband, 16 Oct. 1843, 33–4.
30 PRO, MO, Part IV, Sabine to A. Barclay, 5 Dec. 1844, 81–2. The implication was that Lefroy was lost; he later wrote that his dip circle had been accidentally broken and he stopped at Fort Garry for repairs; *Autobiography*, 72–3.
31 PRO, MO, Part IV, Sabine to Dr John Locke, Cincinnati, 22 Nov. 1843, 39; Part VI, Lefroy to Sabine, no. 26, 24 July 1843; Part V, Sabine to Lefroy, 29 Mar. 1844, 40
32 Lefroy, *Autobiography*, 83–7. PRO, MO, Part VI, no. 40, Lefroy to Sabine, 30 Aug. and 5 Sept. 1844
33 PRO, MO, Part VI, John Richardson to Sabine, no. 9, 5 July 1842. Lefroy, *Diary*, 141, 151
34 PRO, MO, Part IV, Sabine to Barclay, 5 Dec. 1844, 82; Part VI, no. 40, Lefroy to Sabine, 30 Aug. and 5 Sept. 1844. BAAS, *Annual Report* (1844), 147
35 PRO, MO, Part V, Sabine to Lefroy, 17 Dec. 1844, p. 91; Part IV, Sabine to Barclay, 5 Dec. 1844, 81
36 Ibid., Part VI, Lefroy to Sabine, 21 Nov. 1842. Simpson to Lefroy, 25 Apr. 1843, in Thiessen, 'Founding of the Toronto Observatory,' RASC, *Journal* 35/4 (April 1941): 147
37 PRO, MO, Part VI, Lefroy to Sabine, no. 47, 16–20 Jan. 1845; no. 16, 27 Mar. 1844; Richardson to Sabine, no. 9, 5 July 1842; Lefroy to Sabine, no. 16, 15 Feb. 1843; no. 40, 30 Aug. and 5 Sept. 1844
38 Ibid., Part V, Sabine to Lefroy, 31 Jan. 1845, 94–5; Part VI, Lefroy to Sabine, no. 51, 10 Feb. 1845; no. 53, 14 Feb. 1845; Part V, Sabine to Lefroy, 14 Feb. 1845, 97–99
39 Ibid., Part V, Sabine to Lefroy, 1 Jan. 1845, 93; Part V, Sabine to Lefroy, 20 Feb. 1845, 101–3; Part VI, Lefroy to Sabine, no. 63, 18 Apr. 1845; no. 67, 1 May 1845; no. 68, Lefroy to Royal Society of London, 10 May 1845; Part V, Sabine to Lefroy, 2 July 1845, 123–5; Part VI, no. 70, Lefroy to Sabine, 23 May 1845

CHAPTER 7 Mutual Attractions, 1845-1850

1 J.H. Lefroy, *In Search of the Magnetic North*, ed. G.F.G. Stanley (Toronto 1955), 70. Mary Larrat Smith, *Young Mr. Smith in Upper Canada* (Toronto 1980), 22
2 Lefroy, *In Search*, 156, 159-60; Smith, *Young Mr. Smith*, 22, 113-14, 192. Lefroy, *Autobiography*, ed. Lady Lefroy (London [1896]), 104
3 Lefroy, *Autobiography*, 63, 106; Lefroy, *In Search*, 160. PRO, MO, Part VI, no. 94, Lefroy to Sabine (Private), 22 Nov. 1845. Lefroy felt that for disciplinary reasons the officers of the magnetic service ought to be married men too; no. 88, 6 Oct. 1845.
4 J.G. Hodgins, ed., DHE, I (Toronto 1894), 42
5 John Graves Simcoe to Joseph Banks, quoted in Hodgins, DHE, I: 11, 55, 109-10, 132; Upper Canada, House of Assembly, *Journals*, 1805, 1806, quoted in ibid., 51-6
6 See esp. Terry Cook, 'John Beverley Robinson and the Conservative Blueprint for the Upper Canadian Community,' *Ontario History* 64/2 (June 1972): 79-94. Lefroy founded the first 'Toronto Book-Club,' which included Strachan, Robinson, G.W. Allan, and J.B. Macaulay, all Family Compact members; see W.S. Wallace, ed., *The Royal Canadian Institute Centenary Volume* (Toronto 1949), 124, n. 2. Charles Walker Robinson, *Life of Sir John Beverley Robinson* (London 1904), 368
7 PRO, MO, Part VI, no. 40, Lefroy to Sabine, 30 Aug. and 5 Sept. 1844; no. 194, Lefroy to Sabine, 2 Nov. 1848. Lefroy, *In Search*, 156-8
8 Edward Sabine, *Observations Made at the Magnetic and Meteorological Observatory at Toronto in Canada*, I (1840-2) (London 1845); PRO, MO, Part VI, no. 65, Lefroy to Sabine, 21 Apr. 1845. BAAS *Annual Report* (1845), xxi-xxii
9 BAAS, *Annual Report* (1845), xxxiv-xxxv; (1844), 146-7. PRO, MO, Part IV, Sabine to Sir John Harvey, 22 July 1844, 71, and 13 Aug. 1845, 102; to Herschel, 11 Apr. 1845, 97; 18 July 1846, 115; to Bayfield, 22 Dec. 1846, 123-4; to Capt. Smythe (Halifax), 27 Nov. 1848, 181-2; to Sir William Colebrooke, 3 Feb. 1847, 130. Sabine, 'Scheme of Observations at Newfoundland,' [1846], 116-7
10 PRO, MO, Part V, Sabine to Lefroy, 1 July 1845, 121-3
11 Hodgins, DHE, V: 207. Copies to the Governor General and to the LHSQ; see PRO, MO, Part V, Sabine to Lefroy, 1 Apr. 1845, 106; Part VI, no. 76, Lefroy to Sabine, 24 July 1845
12 PRO, MO, Part V, Sabine to Lefroy, 18 Aug. 1845, 141
13 R.D. Gidney, 'The Rev. Robert Murray: Ontario's First Superintendent of Schools,' *Ontario History* 63/4 (Dec. 1971): 191-204. Francis Hincks, *Reminiscences* (Montreal 1884), 85-6
14 PRO, MO, Part V, Sabine to Lefroy, 18 Aug. 1845, 141; Part VI, no. 85, Lefroy to Sabine,

303 Notes to pages 136-140

11 Sept. 1845; no. 97, Lefroy to Sabine, 8 Dec. 1845
15 PRO, MO, Part VI, no. 73, Lefroy to Sabine, 25 June 1845
16 W.I.B. Beveridge, *Influenza: the Last Great Plague* (London 1977), 1-2, 28-9. Lefroy, *Autobiography*, 114; PRO, MO, Part VI, no. 156, Lefroy to Sabine, 24 Sept. 1847. Charles E. Rosenberg, *The Cholera Years: The United States in 1832, 1849 and 1866* (Chicago 1962), 164-5; Geoffrey Bilson, *A Darkened House: Cholera in Nineteenth-century Canada* (Toronto 1980), 4, 151. In 1849 John Snow demonstrated waterborne origins of cholera; see Edmund W. Gilbert, *British Pioneers in Geography* (London 1972), 81.
17 BAJ 1/1 (Apr. 1845): 30; 1/3 (June 1845): 86; 2/4 (Aug. 1846): 89. William Craigie, 'Meteorological Observations at Hamilton,' CN and G (Feb. 1854): 187; JEUC (Aug. 1863)
18 Schönbein's report translated in BAAS, *Report* (1845), 92-102. R.J. Morris, *Cholera 1832: The Social Response to an Epidemic* (London 1976), 190-1. 'On Influenza and Ozone,' BAJ 5/3 (July 1849): 99-100; 'The Ozone Hypothesis of the Production of Cholera,' ibid. 5/6 (Oct. 1849): 163
19 Smallwood, 'Contributions to Meteorology,' BAJ 6/11 (Mar. 1851): 489; 'On Ozone,' CN and G 2/5 (Nov. 1857): 321-8; 'The Observatory at St. Martin's, Isle Jésus, Canada East,' CJ n.s. 3/16 (July 1858); 'On Ozone,' CN and G 4/3 (June 1859): 169-72, 5/5 (Oct. 1859): 343-4, 4/6 (Dec. 1859): 408-10. See also 'Report of the Sub-Committee ... to Examine the Meteorological Observatory of Charles Smallwood,' CJ n.s. 1/4 (July 1856): 409. A.D. Thiessen, 'Her Majesty's Magnetical and Meteorological Observatory, Toronto,' RASC, *Journal* 39/9 (Nov. 1945): 367. PRO, MO, Part VI, Lefroy to Sabine, no. 210, 4 Oct. 1849
20 BAJ 5/12 (Apr. 1850); Montreal *Gazette*, 29 Sept., 13 and 14 Oct. 1846, 25 Jan. 1847; Toronto *Globe*, 30 Jan. 1847. Lefroy had been deeply moved while witnessing a spectacular auroral display in the high north in 1843; see *Diary*, p. 106. PRO, MO, Part VI, Lefroy to Sabine, no. 205, 7 June 1849; Part VI, no. 192, Lefroy, 'Instructions for the Observation of Aurora,' 11 Oct. 1848; no. 178, Lefroy to Sabine, 9 May 1848; no. 200, Lefroy to Sabine, 2 Dec. 1848. Lefroy, 'Auroral Display of Nov. 17,' BAJ 4/9 (Jan. 1849): 231-2. In 1859 the magnetic interference of an extraordinary aurora forced telegraph operators in Quebec, New Orleans, and England to shut off electrical power. It induced an electrical current sufficient to continue operations; JEUC (Sept. 1859): 132.
21 BAJ 2 (1846-7): 50-2; 3/1 (May 1847): 63
22 Ibid., 3/1 (May 1847): 66. PRO, MO, Part VI, no. 153, Lefroy to Sabine, 20 July 1847
23 Edward Sabine, 'On the Meteorology of Toronto in Canada,' in BAAS, *Annual Report* (1844), 42. LP, 1207/11, Box 2, Folder 3, 1846-7.
24 McCord, 'Observations on Meteorology,' BAJ 3/9 (Jan. 1848): 228. *A Few Words upon Canada, and Her Productions in the Great Exhibition*, (London 1851,), 11-12

25 PRO, MO, Part VI, no. 130, Lefroy to Sir John Richardson, 8 Feb. 1846; no. 61, Lefroy to Sabine, 27 Mar. 1844. Richardson was testing the agricultural potential of the north-western regions. Lefroy's findings contrasted the widespread belief that 'from the great breadth of the American continent towards the North Pole, a vast surface is overspread by snow and ice, which almost bids defience to the summer heat'; see Abraham Gesner, *New Brunswick* (n.p. 1847), 224. Lefroy, *Autobiography*, 78. Wheat, barley, and oats were long known to be cultivable at Cumberland House; David Thompson, 'Mean Temperature of Cumberland House and Bedford House, Hudson's Bay Territory, 1789-90, 1795-96,' BAJ 4/11 (Mar. 1849): 302.

26 Montreal *Gazette*, 19 May 1842; 'The Barometer and the Tempest,' ibid., 31 Oct. 1844; Toronto *British Colonist*, 1 Aug. 1845. J.P. Espy, *The Philosophy of Storms* (Boston 1841); Maxime Bôcher, 'The Meteorological Labors of Dove, Redfield and Espy,' American Meteorological Society, *Bulletin* 46/8 (Aug. 1965): 448-52; Gisela Kutzbach, *The Thermal Theory of Cyclones* (Boston 1979), 27; Bruce Sinclair, 'Gustavus A. Hyde, Professor Espy's Volunteers, and the Development of Systematic Weather Observation,' American Meteorological Society, *Bulletin* 46/12 (Dec. 1965): 779-84. PRO, MO, Part V, Sabine to Younghusband, 14 Oct. 1844, 73-5. Sabine, 'Meteorology of Toronto,' 43, 55

27 PRO, MO, Part IV, Sabine to Airy, 29 Nov. 1848, 183-7; Part VI, no. 153, Lefroy to Sabine, 20 July 1847; no. 137, 4 Jan. 1847. BAAS, *Annual Report* (1848), 99

28 PRO, MO, Part IV, Sabine to Marquis Northampton, 18 May 1847, 134-7; Sabine to Count de Rosen, 22 Apr. [1848], 164-6; Sabine to Bache, 3 Oct. 1845, 104. Loomis to Sabine, 28 Feb. 1845, quoted in BAAS, *Annual Report* (1845), 21. W.E. Gross, 'The American Philosophical Society and the Growth of Science in the United States: 1835-1850,' PHD dissertation, University of Pennsylvania, 1970, 82-3

29 Gross, ibid., 78. PRO, MO, Part IV, Sabine to Joseph Henry, 11 Sept. 1847, 138-41; Part V, Sabine to Lefroy, 26 Oct. 1849, 217; Part VI, no. 214, Lefroy to Sabine, 15 Dec. 1849. Smithsonian Institution, Annual Report of the Secretary (1849) in *Annual Report* (Washington 1850), 15

30 PRO, MO, Part VI, no. 216, 11 Mar. 1850; no. 218, 3 May 1850

31 Lefroy to Sabine, 16 May 1850, in A.D. Thiessen, 'Her Majesty's Observatory, Toronto,' RASC, *Journal* 39/10 (Dec. 1945): 396

32 PRO, MO, Part IV, Sabine to Henry, 17 May 1850, 212-14

33 Lefroy to Younghusband, 14 June 1850, in Thiessen, 'Her Majesty's Observatory, Toronto,' RASC, *Journal* 39/8 (Oct. 1945): 319; Lefroy to Sabine, 6 June 1850, 399; Henry to Lefroy, 8 June 1850, 39/10 (Dec. 1945): 401; see also Henry to Lefroy, 14 June 1850, 401-2; Bond to Lefroy, 23 July 1850, 402-3. PRO, MO, Part V, Sabine to Lefroy, 21 June 1850, 227. 'Preliminary Report on the Observations of the Aurora Borealis,' BAJ n.s. 6/1 (May 1850): 72-8

34 PRO, CO42/568, Secretary of the AAAS to Abbott Laurence, 29 Aug. 1850, 226-8;

Abbott Laurence to Palmerston, 17 Sept. 1850, 225; FO to Merivale, 23 Sept.1850, 220; Blackwood to Merivale, 24 Sept. 1850: 'I scarcely know how this should be dealt with. Perhaps the Astronomer Royal should be consulted'; Blackwood to Airy, 10 Oct. 1850; Sabine to Earl of Rosse, 22 Oct. 1850; Blackwood to Trevelyan, 6 Nov. 1850, 223. PRO, FO5/524, Palmerston to Merivale, 23 Sept. 1850; FO to Merivale, 19 Nov. 1850, 141; FO to Merivale, 19 Nov. 1850, 301. PRO, MO, Part IV, Sabine to Loomis, 30 Sept. 1850, 225; Sabine to Henry, 26 Dec. 1850, 228-30; Part V, Sabine to Lefroy, 22 Oct. 1850, 246

35 Lefroy to Sabine, 5 Sept. 1850, in Thiessen, 'Her Majesty's Observatory, Toronto,' RASC, *Journal* 40/7 (Sept. 1946): 258
36 Hodgins, DHE, IX: 23
37 Lefroy to Sabine, 5 Sept. 1850, in Thiessen, RASC, *Journal* 40/7 (Sept. 1946): 258

CHAPTER 8 Science as a Cultural Adhesive, 1850–1853

1 In 1880 Lefroy suggested to Sandford Fleming that time zones be added to future maps of Canada, 'thus familiarizing the popular mind with the relations of Time and Longitude.' PAC, Sandford Fleming Papers, Vol. 28, Folder 199, 7 Feb. 1880
2 PRO, MO, Part VI, no. 146, Lefroy to Sabine, 29 Apr. 1847; no. 148, 24 May 1847; no. 157, 9 Oct. 1847
3 Egerton Ryerson, *The Story of My Life*, ed. J.G. Hodgins (Toronto 1883), 35–6; 'A Lecture on the Social Advancement of Canada,' JEUC (Dec. 1849): 184; 'Importance of the Insignificant' (Nov. 1848): 336; 'Importance of Teaching Children to Observe' (June 1850): 87
4 C.B. Sissons, *Egerton Ryerson: His Life and Letters*, II (Toronto 1847), 230, 345. John Langton, *Early Days in Upper Canada*, ed. W.A. Langton (Toronto 1926), 254, 354. PAO, RG2, (Education Papers), Series C-6-A, Register of Incoming General Correspondence, Vol. E (1850), Secretary of the Province to Ryerson, 30 July 1850, 395
5 'Prospectus,' JEUC (Jan. 1848): 3. United Church Archives, Egerton Ryerson, Inaugural Address on the Nature and Advantages of an English and Liberal Education (Toronto 1842), 22
6 United Church Archives, John George Hodgins Papers, Box 3, File 97, 'Sketch of the Rev. Dr. Ryerson,' 5. Robin S. Harris, 'Egerton Ryerson,' in R.L. McDougall, ed., *Our Living Tradition*, series 3 (Toronto 1959), 254. John Wesley, *The Journal*, ed. Nehemiah Curnock, IV (London n.d.), entry for 11 Dec. 1758, 295; *A Survey of the Wisdom of God in the Creation*, 2nd ed., I (Bristol 1770), iii, 13 and II: 110; 1770 edition, I: 10. Wesley, *A Compendium of Natural Philosophy, being a Survey of the Wisdom of God in the Creation*, ed. Robert Mudie (London 1836), III: 282–3
7 C.B. Sissons, *A History of Victoria University* (Toronto 1952), 47; A.B. McKillop, *A Disciplined Intelligence: Critical Inquiry and Canadian Thought in the Victorian Era*

(Montreal 1979), 18; Goldwin S. French, 'Egerton Ryerson and the Methodist Model for Upper Canada,' in Neil McDonald and Alf Chaiton, eds., *Egerton Ryerson and His Times* (Toronto 1978), 45–58

8 'Alexander von Humboldt,' JEUC (Nov. 1849): 174. Ryerson, *Story of My Life*, 356, 358
9 JEUC (May 1850): 69; (Jan. 1866): 22. Ryerson, 'The Importance of Education to an Agricultural People,' JEUC (Sept. 1848): 267
10 Ryerson, *Story of My Life*, 262
11 J.G. Hodgins, ed., HOP (Toronto 1911), I: 245
12 PAO, RG, Series B, Vol. 1, Letterbook A, 1846–52, Ryerson to John Rintoul, 3 Mar. 1847, 19. 'The Chief Superintendent's Annual Report for 1847,' in Hodgins, HOP, V: 45. Sissons, *Life and Letters*, II: 154, 220
13 Toronto *British Colonist*, 9 Mar., 23 Apr., and 15 June 1847. PRO, MO, Part VI, no. 147, Lefroy to Sabine, 21 May 1847
14 Sissons, *Life and Letters*, II: 154; Toronto *British Colonist*, 9 Nov. 1847. H.Y. Hind, 'Agricultural Education in Upper Canada,' JEUC (Apr., June, July, Aug. 1848), esp. (Aug. 1848): 225
15 Hind, *Lectures on Agricultural Chemistry* (Toronto 1850), 2nd ed. (Toronto 1851), 28
16 H.Y. Hind, *A Comparative View of the Climate of Western Canada, considered in relation to its influence upon agriculture* (Toronto 1851,), iv, 16, 25. W.L. Morton, *Henry Youle Hind 1823–1908* (Toronto 1980), 18
17 Hind, *Comparative View*, vi. [Review of] 'H.Y. Hind, A Comparative View of the Climate of Western Canada,' JEUC (Sept. 1851): 137. 'Agricultural Chemistry–Mr. Hind's Two Lectures,' Letter to the Editor, Toronto *British Colonist*, 24 Dec. 1850
18 PAO, RG2, Series B, Vol. 1, Ryerson to Robinson, 4 Apr. 1849, 66–7; R.D. Gidney, 'Centralization and Education: The Origins of an Ontario Tradition,' *Journal of Canadian Studies* 7/4 (Nov. 1972): 38. *Niagara Mail*, quoted in JEUC (Jan. 1850); 'Normal and Model School,' Toronto *Patriot*, 9 Mar. 1850
19 *Dictionary of National Biography*, IX: 892
20 Sissons, *Life and Letters*, II: 388, 449. Toronto *British Colonist*, 19 July 1850. Province of Canada, House of Assembly, *Journals* (1850), 106, 155, 231. 'Chief Superintendent's Annual Report for 1849,' in Hodgins, HOP, V: 79
21 'On Some of the Collateral Advantages which may be derived from a well-organized system of Public Schools,' JEUC (Mar. 1851): 42–3 and (Sept. 1851): 129–30; see also 'Literary and Scientific Items from Various Sources,' ibid. (Oct. 1850): 158. Peter Perry (York East) in Canada, Legislative Assembly, 1850, *Debates*, gen. ed. Elizabeth Gibbs (Montreal 1982), 1228
22 Sir David Brewster, 'Report respecting the two series of Hourly Meteorological Observations kept in Scotland,' BAAS, *Annual Report* (1839), 27. John Drew, *Practical Meteorology* (London 1855), 237

23 Toronto *British Colonist*, 11 Dec. 1839. Bernard Semmel, *The Methodist Revolution* (New York 1973), 171; Ryerson, *Story of My Life*, 26–9. PRO, MO, Part VI, Lefroy to Sabine, 26 Sept. 1849. 'Collateral Advantages,' JEUC (Sept. 1851): 130
24 Semmel, *Methodist Revolution*, chap. VII, esp. 171. French, 'Ryerson and the Methodist Model,' 52. H.Y. Hind, 'Popular Science,' JEUC (Nov. 1850): 172–3. *Journals* (1844–5), 261; the House appointed a select committee on the expediency of establishing public chairs for the teaching of practical sciences to the mechanical classes of the cities of Montreal and Quebec. Joseph Cauchon, who moved the formation of the committee, had written *Notions élémentaires de physique, avec planches á l'usage des maisons d'éducation* (Quebec 1841).
25 John S. Moir, *Church and State in Canada West* (Toronto 1959), xii. 'The Chief Superintendent's Annual Report for 1848,' in Hodgins, HOP, V: 75, 117. 'Canadian Patriotism, the Lever of Canadian Greatness,' JEUC (Mar. 1850): 40–1. 'Nationality–Another of the Wants of Canada,' reprinted from the *Norfolk Messenger* in JEUC (May 1852): 75
26 'Ceremony of Opening the New Buildings of the Normal and Model School for Upper Canada,' JEUC (Dec. 1852): 179
27 W.S. Wallace, ed., *The Royal Canadian Institute Centenary Volume* (Toronto 1949), 131, 136
28 Sandford Fleming, 'The Early Days of the Canadian Institute,' Canadian Institute, *Transactions* 6 (1898–9): 13, 20–1. JEUC (Jan. 1849): 11; the reference was to the Royal Society of London.
29 Fleming, 'Early Days of the Canadian Institute,' Part II, Canadian Institute, *Transactions* 6 (1898–9): 658; Charles Walker Robinson, *Life of Sir John Beverley Robinson* (London 1904), 369. Don W. Thomson, *Men and Meridians* I (Ottawa 1966), 217
30 Fleming, 16
31 Lefroy, 'Remarks on Thermometric Registers,' CJ (Sept. 1852): 29–31; (Nov. 1852): 75–7. 'Extracts from the First Report of the Secretary of the Board of Registration and Statistics on the Census of the Canadas for 1851–52,' CJ (Oct. 1853): 92–7
32 Joseph Henry quoted in BAAS, *Annual Report* (1851), 320–5. PRO, MO, Part VI, no. 220. Lefroy to Bruce, 7 Apr. 1851; CO42/572, dispatch no. 57, Elgin to Colonial Secretary, 7 Apr. 1851; Elgin to Grey, 11 Apr. 1851, 338
33 PRO, MO, Part V, Sabine to Lefroy, 24 Dec. 1852, 280. 'The President's Annual Address to the Canadian Institute,' Toronto *British Colonist*, 1 Feb. 1853. Edward Sabine, *Observations of the Magnetical and Meteorological Observatory at Toronto*, II (1843–5) (London 1853), III (1846–8) (London 1857); PRO, WO44/514, Lefroy to Provincial Government, 19 Nov. 1852
34 PRO, WO44/514, Sabine to Ordnance Board, 30 Nov. 1852; Sabine to Deputy Adjutant-General Royal Artillery, 11 Dec. 1852; Lefroy to Bruce, 10 Jan. 1853. PRO, MO,

Sabine to Lefroy, 11 Dec. 1852, 279. J.G. Hodgins, ed., DHE (Toronto 1894), X: 1
35 Toronto *British Colonist*, 11 Jan. 1853
36 PRO, WO1/566, Memorial of NHSM to Elgin, 7 Feb. 1853, 40–9
37 CJ 1/7 (Feb. 1853): 146
38 Ibid. 1/6 (Jan. 1853): 121
39 PRO, WO1/566, Elgin to Newcastle, 24 Feb. 1853; WO44/514, Extract from a report of a committee of the Hon. the Executive Council on Matters of State, 7 Feb. 1853; dispatch no. 10, Elgin to Newcastle, 10 Feb. 1853; A.N. Morin to Lefroy, 23 Feb. 1853; Provincial Secretary to Secretary of State for the Colonies, 28 Feb. 1853, 4–7; 15 Mar. 1853, 26–30. *Journals*, 1852–3, 485, 489, 490; Toronto *Globe*, 1 Mar. 1853
40 Toronto *British Colonist*, 8 Mar. 1853
41 Hodgins, DHE, XI: 204n. Province of Canada, House of Assembly, *Debates* (1853), 2856–7, 3092–3; Hodgins, DHE, X: 304–5, clauses xv–xvi
42 PRO, CO42/591, Public Offices, Vol. 3, 8 Apr. 1853. *Journals*, 1852–3, 1085. Fleming, 'Early Days of the Canadian Institute,' 22; Toronto *British Colonist*, 8 Apr. 1853. "The Observatory," CJ 1/12 (July 1853): 282–4. PRO, MO, Part V, Sabine to Cherriman, 11 Aug. 1853, 299–302; Sabine to Lefroy, 25 Feb. 1853, 290–2. John McCaul quoted in Toronto *British Colonist*, 8 Apr. 1853.

CHAPTER 9 Encompassing the North

1 Chauncey Loomis, 'The Arctic Sublime,' in U.C. Knoepflmacher and G.B. Tennyson, eds., *Nature and the Victorian Imagination* (Berkeley 1977), 99. JEUC (Dec. 1851): 186–7
2 W.E. Knowles Middleton, *A History of the Theories of Rain and Other Forms of Precipitation* (New York 1965), chap. 8; Gisela Kutzbach, *The Thermal Theory of Cyclones* (Boston 1979), 2–3. Toronto *Daily Leader*, 5 and 11 Aug. 1853; Morley K. Thomas, 'A Brief History of Meteorological Services in Canada,' *Atmosphere*, Part I (1839–1930) 9/1 (1971): 3–15. Howard Daniel, *One Hundred Years of International Co-operation in Meteorology* (Geneva 1973), 3
3 J.G. Hodgins, ed., DHE (Toronto 1894), XI: 275; XV: 145. *Canada Gazette* quoted in CJ (June 1855): 269; PRO, MO, Part V, Sabine to Cherriman, 12 July 1855, 311–12; 17 Aug. 1855, 314. Cherriman, 'On the Variation of Temperatures at Toronto,' CJ 2 (Aug. 1853): 14–18
4 Thomas Nattress, 'The Western District Literary, Philosophical and Agricultural Association,' OHS, *Papers and Records* 6 (1905): 81–3; C.F.J. Whebell, 'Two Polygonal Settlement Schemes from Upper Canada,' *Ontario Geography* 1/12 (1978): 85–92
5 CJ 2 (May 1854): 241–5
6 Toronto *Globe*, 29 Sept. 1854; CJ 3 (Feb. 1855): 154

7 CJ 3/17 (Dec. 1855): 405-8; 'Memorandum on the Steps which have been taken by the Educational Department, to establish a system of Meteorological Observations throughout Upper Canada,' 410-11

8 'Report of the Editing Committee on Major Lachlan's Supplementary Remarks,' ibid. 3 (Jan. 1855): 135; 'Meteorological Observations,' ibid. 1/1 (Jan. 1856): 91

9 G.T. Kingston, [Review of] 'The Tenth Annual Report of the Smithsonian Institution for the Year 1855,' ibid., n.s. 2/7 (Jan. 1857): 46; 'On the Employment of the Electric Telegraph for Predicting Storms,' ibid. n.s. 2/9 (May 1857): 179. Charles Smallwood in Montreal agreed: 'On the Cold Term of January, 1859,' CN and G 4/2 (Apr. 1859): 39; 'Report of the Meteorological Committee,' CJ n.s. 3/16 (July 1858): 361-4; see also Hodgins, ed., DHE, XII: 149.

10 Hodgins, 'Memorandum'; J.H. Lefroy, 'Memorandum on the Supply of Instruments for the Canadian Grammar Schools,' JEUC, (Jan. 1856): 4-5; 'Lessons on the Thermometer,' and 'Principles of Barometric Indications,' (Jan. 1858): 11-12. PAO, RG2, Series Q-1, Box 1, 'General Instructions for Making Meteorological Observations at the Grammar School Stations in Upper Canada,' (1857). 'Philosophical Instruments at the Central School, Hamilton,' Hamilton *Spectator*, quoted in JEUC (Feb. 1858): 30; 'Meteorology in Upper Canada,' *The Ottawa Citizen*, 9 Feb. 1856, in ibid. (Feb. 1856): 31

11 'Meteorological Observations-The Education Office and the Grammar Schools,' Barrie *Northern Advance*, in JEUC (Jan. 1858): 14

12 JEUC (May 1862): 67. See also PAO, RG2, Series G-1-A, Vols. 1-3, Inspectors' Reports. 'Meteorological Stations at the Senior County Grammar Schools,' JEUC (Mar. 1861): 38

13 PAO, RG2, Series G-1-A, Vol. 1, Report of William Ormiston, 1858, 561; Vol. 3, 1863, 127. PAO, RG2, Series G-1-E, Box 7, Grammar School Meteorological Reports, Alexander Burdon to Ryerson, 8 Mar. 1858; Burdon to Ryerson, 13 Sept. and 2 Oct. 1858; Burdon to Ryerson, 17 Jan. 1859; Vol. 1, 1866, Burdon to Ryerson, 21 Feb. 1866; Checkley (Barrie) to Hodgins, 10 Apr. 1866; Vol. 3, Burdon to Ryerson, 25 Feb. 1870; Preston (Goderich) to Ryerson, 19 Feb. 1870. PAO, RG2, Series C-1, Vol. 22, Letterbook X, Ryerson to Burdon, 6 May 1858, 1264; Vol. 23, Letterbook Y, Ryerson to Burdon, 26 June 1858, 1722; Ryerson to Burdon, 16 Oct. 1858, 3156. Observations were often taken by headmasters' untrained relatives or pupils; 'Annual Report for 1857' explains the extra £100 per annum granted to senior county grammar schools, in Hodgins, HOP, V: 227

14 PAO, RG2, Series G-1-E, Hodgins to Bradbury (Cornwall) 3 Nov. 1869; Johnston (Windsor) to Ryerson, 24 Feb. 1870; Mulholland (Simcoe) Form D, May 1867: 'we hope soon to see everything green again (of course barring the Fenians.) ... on breaking the last wet-bulb by melting the ice, and taking off the cotton, we became converted to the anabaptist theory, believing in immersion. The extraor-

dinary phenomenon to-day, however, has shaken our faith, and determined us again to put on the cotton'; PAO, RG2, Series G-I-E, Box 7, Checkley (Barrie) to Ryerson, 17 Nov. 1866. Report of Inspector Checkley (1863) Series G-I-A, Vol. 3

15 'Grammar School Meteorological Observations in Upper Canada,' JEUC (Jan. 1860): 2; Hodgins, DHE, XII: 118. 'The Chief Superintendent's Annual Report for 1861,' in Hodgins, HOP, V: 308. 'Explanatory Remarks on the New Grammar School Act,' JEUC (Sept. 1865): 132. Province of Canada, House of Assembly, *Journals*, 1864, App. 7

16 'Grammar School Meteorological Observations,' 2

17 Hodgins, DHE, XII: 292–4. *Journals* (1856), App. 53. Kingston, 'On Deducing the Mean Temperature of a Month,' CJ n.s. 3/13 (Jan. 1858): 5–6; 'On the Magnetic Disturbances at Toronto during the Years 1856 to 1862, Inclusive,' ibid. 8/44 (Mar. 1863): 157–63; 'Remarks on the Temperature Coefficients of Magnets,' *ibid.* 8/46 (July 1863): 280–3; 'Monthly Absolute Values of the Magnetic Elements at Toronto, from 1856 to 1864, Inclusive,' ibid. 10/51 (Mar. 1865): 114–18. '[Notice of Publication of] *Abstracts of Magnetical Observations made at the Magnetical Observatory, Toronto, Canada West*,' ibid. 8/48 (Nov. 1863): 467–8. Kingston, 'The Toronto Observatory,' BAJ n.s. 1 (1863)

18 'Mean Temperature of a Month,' 6. Kingston, 'On the Abnormal Variations of Some of the Meteorological Elements at Toronto and Their Relations to the Direction of the Wind,' ibid. 9/50 (Mar. 1864): 109; 'On the Annual and Diurnal Distribution of the Different Winds at Toronto,' ibid. 9/49 (Jan. 1864): 10–25. Canada, House of Commons, *Sessional Papers* (1872), Vol. 7, Report of the Minister of Marine and Fisheries; (1873), Vol. 3, App. 4; (1874), Vol. 4, App. 8. R.F. Stupart, 'Meteorology in Canada,' RASC, *Journal* 6/2 (Mar.–Apr. 1912): 77

19 Hodgins, DHE, XV: 145, 159. 313; XVIII: 23. Montreal *Gazette*, 25 Sept. 1863. *Journals*, 1860, 246. Margaret Cohoe, 'The Observatory in City Park 1855–1880,' *Historic Kingston* No. 27 (Jan. 1979): 78–91; Hilda Neatby, *Queen's University*, I (1841–1917) (Montreal 1978), 107; 'Kingston Observatory,' reprinted in JEUC (Aug. 1862): 118; 'The Kingston Observatory' (Mar. 1864); 44–5; R.A. Jarrell, 'Origins of Canadian Government Astronomy,' Royal Astronomical Society of Canada, *Journal* 69 (April 1975): 81.

20 *Journals*, 1856, 359; (1858), 537. CJ n.s. (July 1857): 409. Metropolitan Toronto Reference Library, Baldwin Room, W.E. Logan Scrapbook, 28 Mar. 1856. On McGill see Montreal *Gazette*, 20 Jan. 1857; Public lectures in Montreal on the climate of Canada by Dr W.H. Hingston also drew 'large audiences': Annual Report of the NHSM, CN *and* G 4/3 (June 1859): 236. Annual Report in CN *and* G 6/3 (June 1861): 236; *Journals* (1859). 'McGill College Observatory,' *Montreal Transcript* in JEUC (Mar. 1864): 43. Smallwood, Presidential Address, CN *and* G n.s. 2/2 (Dec. 1866): 132

21 Smallwood, 'Meteorology,' ibid. n.s. 4/1 (Mar. 1869): 112-15, 119
22 GSC, 'Report of Progress for the Year 1857 [R of P 1857]'; LP, Box 4, 1207/11, Edward Ashe to Logan, 11 Aug. 1856, 8 Sept. 1856; Ashe, 'On the Employment of the Electric Telegraph in Determining the Longitude of Some of the Principal Places in Canada,' CJ n.s. 4/24 (Nov. 1859): 453-65. 'Report of the Committee appointed to consider a proposition from Lt Ashe, R.N., for the establishment of an Astronomical Observatory at Quebec,' CJ n.s. 2/10 (July 1857): 309; *Journals*, 1857, 105
23 LP, Jack to Logan, 15 Oct. 1856; 24 Feb., 25 Mar., and 5 May 1857. Derek Howse, *Greenwich Time and the Discovery of the Longitude* (Oxford 1980)
24 PAM, *The Nor'Wester*, 19 Feb. and 5 Mar. 1862
25 Daniel Wilson, 'Science in Rupert's Land,' CJ n.s., 7/40 (July 1862): 336-47
26 Alexander von Humboldt, *Kosmos: Entwurf einer physischen Weltbeschreibung* (Stuttgart 1845), I: 345-7. F. Model, 'Alexander von Humboldts Isothermen,' *Deutsche Hydrographische Zeitschrift* 12 (1859): 29-33; Werner Horn, 'Die Geschichte der Isarithmenkarten,' *Petermanns Geographische Mitteilungen* 103/3 (1959): 225-32; A.H. Robinson and Helen M. Wallis, 'Humboldt's Map of Isothermal Lines: A Milestone in Thematic Cartography,' *Cartographic Journal* 4/2 (Dec. 1967): 119-23
27 Georg Forster, *Kleine Schriften*, III (Leipzig 1794), 87. *Kosmos*, 345-6
28 Toronto *Globe*, 24 Mar. 1847
29 *Journals*, 1857, App. 17. GSC, R of P 1845-6. Great Britain, Parliament, House of Commons, *Report of the Select Committee on the Hudson's Bay Company* (1857), 20. On Blakiston, see 'Report on the Interior of British North America,' in John Palliser, *Further Papers Relating to the Exploration of British North America* (London 1860), 40.
30 Abbé J.B.A. Ferland, *Cours d'Histoire du Canada*, 2nd ed. (Quebec 1882), I: 507-8 and App. B; first published 1861. P.F.X. de Charlevoix, *Journal of a Voyage to North America* (London 1761), I: 237, 240. Edward Gibbon, *The History of the Decline and Fall of the Roman Empire*, I (London 1791), 346-7
31 Gibbon, *Decline and Fall of the Roman Empire*, 346-8. J.B. Black, *The Art of History* (London 1926), 143; cf. also chap. 8, note 31.
32 Kenneth Kelly, 'The Changing Attitude of Farmers to Forest in 19th-century Ontario,' *Ontario Geography* 1/8 (1974): 64-7. Catharine Parr Traill, *The Backwoods of Canada* (London 1836), 312. PAO, Crown Lands Department, RG1, Series AI4, Vol. 35, 'Statement prepared,' 23 Mar. 1855, 97
33 J. Sheridan Hogan, *Canada: An Essay to which was awarded the First Prize by the Paris Exhibition Committee of Canada* (Montreal 1855), 53. 'The Climate of Canada,' JEUC (May 1862): 67. 'British American Confederacy,' New York *Evening Post*, reprinted in JEUC (Dec. 1864): 182
34 JEUC (May 1862): 67. Ferland preferred to believe that Canadian temperatures were linked to terrestrial magnetism and the proximity of the north magnetic pole; see

Cours d'Histoire, 510. Great Britain, House of Commons, *Report of the Select Committee*, 17, 189.

35 By the 1870s adverse effects of over-clearance added new subtleties to the arguments; see Kelly, 'Changing Attitudes,' 70–2. But Andrew Thomson, 'The Meteorologist and Local History,' CHR 27/1 (Mar. 1946): 102, still asks the same kinds of questions. *Journals* (1857), App. 25, 'Report of the Commissioner of Crown Lands'

36 Lorin Blodget, *Climatology of the United States and of the Temperate Latitudes of the North American Continent* (Philadelphia 1857). G.S. Dunbar, 'Isotherms and Politics: Perception of the Northwest in the 1850s,' in A.W. Rasporich and H.C. Classen, eds., *Prairie Perspectives*, II (Toronto 1973), 80–101; John Warkentin, 'Steppe, Desert and Empire,' ibid., 102–36; Doug Owram, *Promise of Eden* (Toronto 1980), chap. 3. PRO, MO, Part VI, no. 119, Richardson to Sabine, [1846]; Lefroy and Richardson, *Magnetical and Meteorological Observations at Lake Athabasca and Fort Simpson and at Fort Confidence, in Great Bear Lake* (London 1855). Alexander Morris, *Canada and Her Resources: An Essay, to which, upon a reference from the Paris Exhibition Committee of Canada, was awarded ... second prize* (Montreal 1855), 136–43

37 [H.Y. Hind], 'The Future of Western Canada,' *Canadian Almanac and Repository of Useful Knowledge* (1856), 33–6; 'Our Railway Policy–Its Influence and Prospects' (1857), 30–3; 'The Great North-West' (1858), 28. Hind, *Narrative of the Canadian Red River Exploring Expedition of 1857 and of the Assiniboine and Saskatchewan Exploring Expedition of 1858* (1860; reprint ed. Edmonton 1971) II: 370; Lefroy's observations were used by Hind as 'indisputable testimony' of the effects of warm Pacific winds north of the fifty-eighth parallel: 234, 354, 357. W.L. Morton, *Henry Youle Hind* (Toronto 1980), 26. 'The Interior of North America,' JEUC (Apr. 1858) 56. Cf. Francis Parkman, *The Oregon Trail* (1849; Signet Classic, New York 1978), 14, 34.

38 'Canadian Expeditions to the North-West Territory,' CJ 5/30, (Nov. 1860): 550. (Unsigned review of) H.Y. Hind, *Narrative*, CN and G n.s. 6/1 (Feb. 1861): 68. Blodget, *Climatology*, chap. XIX, 'Climate of the Northwestern Districts,' 529–35. G.T. Kingston, (Review of) Lorin Blodget, *Climatology of the United States*, CJ n.s. 3/13 (Jan. 1858): 28–34. Warkentin, 'Steppe, Desert and Empire,' 109–10

39 Hind, *Narrative*, II: 354. Blakiston, 'Report,' 40

40 Hind, 'Our Railway Policy,' 32; *Journals*, 1857, App. 17. Blakiston, 'Report,' 43, 57. John Palliser, *The Journals, Detailed Reports, and Observations relative to the Explorations of British North America* (London 1863), 264–316; for an assessment of the contributions of Palliser and Hind, see Owram, *Promise of Eden*, chap. 3.

41 Owram, *Promise of Eden*, 40–1. *Journals*, 1857, App. 25, was supplemented in March 1857 by an isothermal map of 'the North West Part of Canada [,] Indian Territo-

ries and Hudson's Bay' (PAO, map B-24). J.E. Hodgetts, *Pioneer Public Service* (Toronto 1955), 118. PAO, Crown Lands Dept., RG1, Ser. AI4, Vol. 35, 23 Mar. 1855, 107. Great Britain, House of Commons, *Report ... on the Hudson's Bay Company*, 12, 18. Toronto *Globe*, 23 Mar. 1857. E.T. Fletcher, 'On the Secular Change of Magnetic Declination in Canada, from 1790 to 1850,' LHSQ, *Transactions* n.s. 3 (1864-5): 137-40. J.H. Lefroy, *Autobiography*, ed. Lady Lefroy (London [1896]) 78. Alexander Morris, *The Hudson's Bay and Pacific Territories* (Montreal 1858), 58, 67, 73, 79-80

42 Gibbon, *Decline and Fall of the Roman Empire*, I: 348-9
43 Traill, *Backwoods of Canada*, 203. Hogan, *Canada*, 53; Morris, *Canada and Her Resources*, 136-7, 143; Morris, *Hudson's Bay and Pacific*, 89. Morris, *Nova Britannia* (Montreal 1858), 49
44 PRO, MO, Part V, Sabine to Michael Faraday, 26 Sept. 1850, 224-5
45 PAC, Sandford Fleming Papers, Vol. 28, Folder 199, Lefroy to Fleming, 2 Mar. 1878. Lefroy, *Autobiography*, 73
46 David B. Wilson, 'Concepts of Physical Nature: John Herschel to Karl Pearson,' in Knoepflmacher and Tennyson, eds., *Nature and the Victorian Imagination*, 208-10. Also J.T. Merz, *A History of European Thought in the Nineteenth Century*, I (Edinburgh 1890), 16
47 J. Morrell and A. Thackray, *Gentlemen of Science* (Oxford 1981), 372
48 Graeme Patterson, 'The Myths of Responsible Government and the Family Compact,' *Journal of Canadian Studies* 12/2 (Spring 1977): 3-16; Ronald Robinson and John Gallagher with Alice Denny, *Africa and the Victorians* (New York 1968), 469
49 Charles R. Tuttle, *Our North Land* (Toronto 1885), 17-18; J.G. Bourinot, 'The Intellectual Development of the Canadian People,' *Rose-Belford's Canadian Monthly and National Review* 5 (Dec. 1880): 633-4
50 BAAS, *Annual Report* (1844), xliv; (1851), xlv. 'Union of the British North American Colonies,' Toronto *British Colonist*, 28 Jan. 1853. Colonel Myers, 'Notes on the Weather at Halifax, N.S., during 1863, with Comparisons of the Temperature of that Place with Some Other Parts of British North America,' NSINS, *Proceedings and Transactions* 1 (1863-6): 71-2
51 *CJ* n.s. 3/14 (Mar. 1858): 103
52 Ibid. 5/25 (Jan. 1860): 111-12, 114
53 Egerton Ryerson, 'Address to County School Conventions,' JEUC (Mar. 1860): 35. *CJ* 10/51 (Mar. 1865): 84

CHAPTER 10 Adventitious Roots

1 Lucille Brockway, *Science and Colonial Expansion: The Role of the British Royal Botanic Gardens* (New York and London 1979)

2 Martin Möbius, *Geschichte der Botanik*, 2nd ed. (Stuttgart 1968), 21-5
3 Keith Thomas, *Man and the Natural World: Changing Attitudes in England 1500-1800* (London 1983), 66, 53, 240. A.G. Morton, *History of Botanical Science* (London 1981), 364-77; Sir James Edward Smith, *A Grammar of Botany* (London 1826), 1-2. On cell theory see *Dictionary of Scientific Biography* XII: 173-7, 240-5.
4 Möbius, *Geschichte der Botanik*, 48; Caroli Linné, *Genera Plantarum*, 8th ed. (Prague 1791), v-vi. R.J. Harvey-Gibson, *Outlines of the History of Botany* (London 1919), 88
5 Möbius, *Geschichte der Botanik*, 44-6; Harvey-Gibson, *History of Botany*, 56; A.M. Lysaght, *Joseph Banks in Newfoundland and Labrador, 1766* (London 1971), 36. James L. Larson, *Reason and Experience: The Representation of Natural Order in the Work of Carl von Linné* (Berkeley 1971), 2, 144-6
6 W.B. Turrill, *Joseph Dalton Hooker: Botanist, Explorer, and Administrator* (London 1963), 10. Philip R. Sloan, 'The Buffon-Linnaeus Controversy,' *Isis* 67/238 (Sept. 1976): 365-75; 'John Locke, John Ray, and the Problem of the Natural System,' JHB 5/1 (Spring 1972): 1-2, 27
7 Möbius, *Geschichte der Botanik*, 48; Harvey-Gibson, *History of Botany*, 24
8 Harvey-Gibson, *History of Botany*, 74; M.A.P. de Candolle, *Théorie élémentaire de la Botanique* (Paris 1813). P.F. Stevens, 'Haüy and A.-P. de Candolle: Crystallography, Botanical Systematics, and Comparative Morphology, 1780-1840,' JHB 17/1 (Spring 1984): 80-1.
9 Stevens, 'Haüy and A.-P. Candolle,' 54-5, 60-2; Agnes Arber, *The Natural Philosophy of Plant Form* (Cambridge 1950; reprint ed., Darien, Conn., 1970), 2, 60. De Candolle, *Théorie élémentaire*, 53; Harvey-Gibson, *History of Botany*, 54
10 Smith, *Grammar of Botany*, xiii, x; 'A Review of the Modern State of Botany with particular reference to the natural systems of Linnaeus and Jussieu,' in his *Memoirs and Correspondence*, II (London 1832), 443.
11 Turrill, *Joseph Dalton Hooker*, 11; De Candolle, *Théorie élémentaire*, 53. Smith, *Memoirs and Correspondence*, 589
12 J.W. Johnson, 'Of Differing Ages and Climes,' *Journal of the History of Ideas* 21/4 (Oct.-Dec. 1960): 470-1. William Jackson Hooker, 'Some Account of a Collection of Arctic Plants found by Edward Sabine,' Linnean Society, *Transactions* 14 (1825): 360-94
13 Morton, *History of Botanical Science*, 432-3; Gareth Nelson, 'From Candolle to Croizart: Comments on the History of Biogeography,' JHB 11/2 (Fall 1978): 273-5. *Barr's Buffon* (London 1810), VII: 27-39; IX: 315-18; Paul L. Farber, 'Buffon and the Concept of Species,' JHB 5/2 (Fall 1972): 259-84
14 Joseph Kastner, *A World of Naturalists* (London 1977), 121-3. Charles Lyell, *Principles of Geology* (London 1832), II: 67, 71; Farber, 'Buffon and the Concept of Species,' 284; Harcourt Brown, 'Buffon and the Royal Society of London,' in *Studies*

and *Essays in the History of Science and Learning* (New York 1947), 145; Lysaght, *Joseph Banks*, 36

15 Ernst Mayr, *Systematics and the Origin of Species*, 2nd ed. (New York 1964), 6. Michel Foucault, *The Order of Things* (London 1970), 144, 162

16 Thomas, *Man and the Natural World*, 52; John Dean, 'Controversy over Classification: A Case Study from the History of Botany,' in Barry Barnes and Steven Shapin, eds., *Natural Order: Historical Studies of Scientific Culture* (Beverly Hills 1979), 212

17 Lysaght, *Joseph Banks*, 36–7; David Mackay, 'A Presiding Genius of Exploration: Banks, Cook, and Empire, 1767–1805,' in Robin Fisher and Hugh Johnston, eds., *Captain James Cook and His Times* (Vancouver 1979), 27

18 Mackay, 'Banks, Cook, and Empire,' 23; H.C. Cameron, *Sir Joseph Banks* K.B., F.R.S.: *The Aristocrat of the Philosophers* (London 1952), 63; Philip Miller, *The Gardeners Dictionary*, 7th ed. (Dublin 1764), I: preface

19 Morris Berman, *Social Change and Scientific Organization* (Ithaca, NY, 1978), 46, 60–1, 99; Agnes Arber, 'Sir Joseph Banks and Botany,' *Chronica Botanica* 9 (1945): 104, mentions Banks's interest in wheat diseases.

20 Harold R. Fletcher, *The Story of the Royal Horticultural Society 1804–1968* (London 1969), 9, 16, 27; William Morwood, *Traveller in a Vanished Landscape* (London 1973), 11; Lysaght, *Joseph Banks*, 37. Royal Horticultural Society of London, *Transactions* V (1821–2): iii; Brockway, *Science and Colonial Expansion*, 71

21 Fletcher, *Royal Horticultural Society*, 100

22 Morwood, *Traveller in a Vanished Landscape*, 21. Royal Horticultural Society of London, *Transactions* V (1821–2): iii, vi

23 David Pearce Penhallow (1854–1910) was actually American born and educated. R.B. Thomson, 'A Sketch of the Past Fifty Years of Canadian Botany,' RSC, *Fifty Years Retrospect* (n.p. 1932), 173; D.P. Penhallow, 'A Review of Canadian Botany from 1800 to 1895,' RSC, *Transactions* Ser. 2, Vol. 3, Sec. 4 (1897): 9

24 Mary Alice Downie and Mary Hamilton, eds., '*And Some Brought Flowers*': *Plants in a New World* (Toronto 1980), xi, xiii. Brian L. Evans, 'Ginseng: Root of Chinese-Canadian Relations,' CHR 66/1 (March 1985): 1–26. D.P. Penhallow, 'A Review of Canadian Botany from the First Settlement of New France to the Nineteenth Century,' RSC, *Transactions*, Sec. 4 (1887): 55

25 Thomson, 'Past Fifty Years of Canadian Botany,' 173. CN and G n.s. 9/1 (1881): 186–7

26 R.P. Stearns, 'The Royal Society and the Company,' *The Beaver*, Outfit 276 (June 1945): 13; R.H.G. Leveson Gower, 'H.B.C. and the Royal Society,' *The Beaver*, Outfit 265/2 (Sept. 1934): 29–33, 66. Glyndwr Williams, ed., *Andrew Graham's Observations on Hudson's Bay 1767–91*, Hudson's Bay Record Society Publications, Vol. 27 (London 1969), 129; E.E. Rich, ed., *James Isham's Observations on Hudson's Bay, 1743*, The Champlain Society, Vol. 12 (Toronto 1949), xxxvi

27 Williams, ed., *Graham's Observations*; Mrs George Bryce, 'Early Red River Culture,' HSSM, *Transactions* no. 57 (1901): 9; University of Toronto Library, James Hargrave Papers, George Barnston to Hargrave, 17 July 1834, mfm reel 133, 825. William Fraser Tolmie, *Physician and Fur Trader* (Vancouver 1963), II, 387; Jean Murray Cole, *Exile in the Wilderness: The Biography of Chief Factor Archibald McDonald 1790–1853* (Don Mills 1979), 175, 209–11

28 John Goldie, 'Description of some new and rare plants discovered in Canada in the year 1819,' *Edinburgh Philosophical Journal* VI (1821–2): 319–33; *Diary of a Journey through Upper Canada and some of the New England States 1819* (privately published n.d.); G.U. Hay, 'John Goldie, Botanist,' RSC, *Transactions* Ser. 2, Vol. 3, Sec. 4 (1897): 125–30

29 W.J. Hooker, 'On the Botany of America,' reprinted in *American Journal of Arts and Science* 9 (1825): 263–4. The original article is in *The Edinburgh Journal of Science* II (1825): 108–29. Hooker was referring also to the impressive body of work by American botanists like John Bartram (1699–1777); Benjamin Smith Barton (1766–1815), whose *Elements of Botany* (1803) was the first botanical textbook published in the United States and who became a patron to both Frederick Pursh and Thomas Nuttall; Amos Eaton (1776–1842), for his *Manual of Botany for the Northern States* (1817); Thomas Nuttall (1786–1859), whose *The Genera of North American Plants, and a Catalogue of the Species* (1818) was the first comprehensive study of the American flora, and for *An Introduction to Systematic and Physiological Botany* (1827); and John Torrey (1796–1873) for his *Flora of the Northern and Middle Section of the United States* (1824). For secondary literature see E. Earnest, *John and William Bartram: Botanists and Explorers* (Philadelphia 1940); E.M. McAllister, *Amos Eaton: Scientist and Educator* (Philadelphia 1941); J.E. Graustein, *Thomas Nuttall, Naturalist: Explorations in America 1808–1841* (Cambridge, Mass., 1967); and the *Dictionary of Scientific Biography*.

30 David Chisholme, 'Review of William Bell, *Hints to Emigrants*,' *Canadian Review and Literary and Historical Journal* 1/2 (1825): 287; 'M.P.S.E.,' 'Hints and Observations on the Natural History of Canada,' *Canadian Review and Magazine* 1/3 (1824–5); 1/4 dealt with zoology, but the journal folded before botany could be discussed in 1/5.

31 William Adams, *Modern Discoveries in North and South America [Flowers of Modern Voyages and Travels]* (London 1824), 57–8. Kenneth Kelly, 'The Evaluation of Land for Wheat Cultivation in Early 19th-century Ontario,' *Ontario History* 72/1 (March 1970): 57–64. Samuel Strickland, *Twenty-Seven Years in Canada West* (London 1853), esp. chap. 13

32 William Sheppard, 'Observations on the American Plants described by Charlevoix,' LHSQ, *Transactions* I (1824–9): 218–19; 'Catalogue of Canadian Plants collected in 1827 & presented to the Literary & Historical Society, by the R.H. the

Countess of Dalhousie,' 1 (1824–9): 255–61. Marjorie Whitelaw, ed., *The Dalhousie Journals*, 3 vols. (n.p. 1978), esp. I: 7. NHSM, *Annual Report* (Montreal 1828). Thomas C. Haliburton, *An Historical and Statistical Account of Nova-Scotia*, II (Halifax 1829; reprint ed. Belleville 1973), 390, 405–13; Henry How, 'The East Indian Herbarium of King's College, Windsor,' NSINS, *Proceedings and Transactions* 4/4 (1878): 369–79

33 John Young, *The Letters of Agricola on the Principles of Vegetation and Tillage* (Halifax 1822; reprint ed. Halifax 1922), 4. See also Melville Cumming, 'The Junius of Nova Scotia,' *Dalhousie Review* 3 (1923–4): 53–60; and J.S. Martell, *The Achievements of Agricola and the Agricultural Societies*, Public Archives of Nova Scotia, *Bulletin* 1/2 (Halifax 1940).

34 Rosalind Mitchinson, *Agricultural Sir John: The Life of Sir John Sinclair of Ulbister 1754–1835* (London 1962), 120–1, 150, 204, 212

35 Young, *Letters of Agricola*, 13–23, 368

36 Gerald S. Graham, *Sea Power and British North America 1783–1820* (Cambridge 1941), 142, 152. R.G. Albion, *Forests and Sea Power* (Cambridge 1926), 232, 241. PAC, G Ser. 65, Upper Canada, Murray to Colborne, Despatch No. 50, 12 Dec. 1829

37 W.J. Hooker, *Flora Boreali-Americana; or, the Botany of the Northern Parts of British North America*, I (London 1833), 39; II (London 1840), 205 and passim for citations of individual collectors. Athelstan George Harvey, *Douglas of the Fir* (Cambridge, Mass., 1947), 152; Penhallow, Canadian Botany from 1800 to 1895,' 10. Hooker's work was seen to 'mark an epoch in North American botany, which could now be treated as a whole'; see Charles Sprague Sargent, ed., *Scientific Papers of Asa Gray* (Boston 1889), I: 325. For Todd see George Simpson, *Journal of Occurrences in the Athabasca Department, 1820–21*, Hudson's Bay Record Society Publications, I (London 1938), 471–2; and Ross Mitchell, 'Early Doctors of Red River and Manitoba,' HSSM, *Papers* Ser. 3, no. 4 (1947): 38

38 Sargent, ed., *Scientific Papers of Asa Gray*, I: 325. Royal Botanic Gardens, Kew, William Jackson Hooker Papers, North American Letters, II, M.Y. Stark, Dundas, to Hooker, 15 Oct. 1841, 437 (573). Sara Eaton, *Lady of the Backwoods: A Biography of Catharine Parr Traill* (Toronto 1969), 58, 79, 108–9. D.E. Allen, *The Naturalist in Britain* (London 1976). Catharine Parr Traill, *The Backwoods of Canada* (London 1836; reprint ed., Toronto 1980), 91–2.

39 NHSM, *Annual Report* (Montreal 1830), (1835); LHSQ, *Transactions* 30 Dec. 1830, preface, iii; William Sheppard, 'Notes on Some of the Plants of Lower Canada' 2 (1830): 39–40. PAC, G Ser. 83 (Upper Canada), Glenelg to Head, 14 Oct. 1837

40 Hooker Papers, N. Am. Letters, I, James Robb, Fredericton, to Hooker, 24 June 1839, 533. John Torrey, Asa Gray, *A Flora of North America*, I (New York 1838–40), xiii. A. Benedict, University of Vermont, 'On the Vegetation of the Ottawa and Some of Its Tributaries,' *American Journal of Arts and Science* 18 (1830): 352;

Hooker Papers, W. F. Macrae, Montreal, to Hooker, 2 Nov. 1839, 319 (439)
41 Allen, *Naturalist in Britain*, 128; Hargrave Papers, Barnston to Hargrave, 1 Feb. 1837, reel 134, 1342; NHSM, *Annual Report* (1830); Montreal *Gazette*, 2 Aug. 1838
42 Montreal *Gazette*, 19 Jan., 15 June 1837; 2 Aug. 1838; 3 Jan., 27 June 1839; 7 Jan. 1840
43 *York Commercial Directory, 1833-34* (York 1833), 136. PAC, RG7, G16C, Vol. 33: 171, Power to Rees, 8 July 1835
44 *Quebec Mercury*, 28 Dec. 1839. Abraham Gesner shared this expansive outlook: 'The distribution of animals and plants has no reference to the arbitrary boundaries of nations and states, and the description of the productions of a single province would apply to almost the wole of the northern part of the great continent'; *New Brunswick* (London 1847), 355.
45 Penhallow, 'Canadian Botany from 1800 to 1895,' 15; William Canniff, *The Medical Profession in Upper Canada, 1783-1850* (Toronto 1894), 570-1; Thelma Coleman, *The Canada Company* (Stratford 1978), 161. Bytown *Gazette*, 10 Nov. 1836.
46 James J. Talman, 'Agricultural Societies of Upper Canada,' OHS, *Papers and Records* 27 (1931): 547, 549-50. Newton Bosworth, *Hochelaga Depicta* (Montreal 1839), 221. Toronto *Patriot*, 23 May, 24 June 1834; Philip F. Dodds and H.E. Markle, comp., *The Story of Ontario Horticultural Societies* (Picton 1973), 5. Special thanks to Mr Ron Fischer for copies of his 'The Development of the Garden Suburb in Toronto,' *Journal of Garden History* 3/3 (1983): 193-207; and 'Park Lots to Home Smith: Evolving Awareness of Toronto's Landscape, 1793-1914,' given to a conference on the Garden History of Southern Ontario, 30 Mar. 1984.
47 Toronto *Patriot*, 23 May 1834; in the same breath Dalton, a former brewer, pointed out that the cultivation of small fruits carried with it the advantage that the farmer could make wine, 'whether for his own consumption or for Sale.'
48 *York Commercial Directory* (1837), 120. Toronto *Patriot*, 3 Jan. 1834
49 Toronto *Patriot*, 14 Jan. 1834
50 Ibid., 25 March 1834. BAC (June 1845): 176; CAg (Jan. 1855): 2-3; (June 1850); 123; (Jan. 1852): 19; (Aug. 1851): 180-1. Upper Canada, Board of Agriculture, *Journals and Transactions* (1851): 181; (1850): 131-43
51 Morwood, *Traveller in a Vanished Landscape*, 81; 'Eureka,' in Toronto *Patriot*, 25 Mar. 1834. Lyell, *Principles of Geology*, II: 35
52 BAC (June 1845): 176; (Oct. 1845): 295; J.G. Hodgins, ed., DHE (Toronto 1894) VI: 30-1, 36-7, 115, 192, 197
53 BAC (Feb. 1845): 64. A farmer, suggested one article, should learn 'the nature and organic structure of each and every plant he wishes to raise'; (Jan. 1846): 25; (March 1846): 90; (Jan. 1845): 12, 5. Kenneth Kelly, 'The Transfer of British Ideas on Improved Farming to Ontario during the First Half of the Nineteenth Century,' *Ontario History* 62/2 (June 1971): 103-11; Fred Landon, *Western Ontario and the American Frontier* (Toronto 1967), 248; Robert Leslie Jones, *History of Agricul-*

ture in Ontario 1613-1880 (Toronto 1946), 355

54 BAC (Jan. 1846): 4; Jones, *History of Agriculture in Ontario*, 138; *Journals and Transactions* (1850), 39; Montreal *Gazette*, 2 Jan., 19 Jan., 2 Nov. 1846; Toronto *British Colonist*, 23 Apr. 1847

55 BAC (Jan. 1845): 18-19. CAg (Dec. 1849): 326-7; (May 1849): 113; Toronto *Globe*, 23 June 1847; G. Elmore Reaman, *A History of Agriculture in Ontario*, I (Toronto 1970), 97

56 BAC (July 1845): 202-3; (Sept. 1845): 263; n.s. 2/5 (May 1846): 129

57 Ibid. (Aug. 1845): 234; (Sept. 1845): 263; (July 1845): 197; Toronto *British Colonist*, 20 June 1845. Hemp and flax cultivation were considered by many farmers as too labour-intensive for the Canadian situation, especially compared to wheat.

58 10 and 11 Vict., c. 61; J.E. Hodgetts, *Pioneer Public Service* (Toronto 1955), 231; BAC (June 1845): 177; (May 1846): 129; (Aug. 1846): 242; *Journals and Transactions*, 29, 31, 60-1. A.E. Byerly, 'Pioneers and Pioneer Days of Fergus,' OHS, *Papers and Records* 30 (1934): 66-9; CAg (June 1849): 141; Gerald M. Craig, ed., *Lord Durham's Report*, Carleton Library No. 1 (Toronto 1963), 78

59 CAg (Jan. 1849): 1-2, 21-2; (Dec. 1849): 334; (July 1849): 184, 196

60 Ibid. (Jan. 1849): 1; (Mar. 1849): 64-5; (Aug. 1849): 214-17

61 Cf. Alfred W. Crosby, *Ecological Imperialism: The Biological Expansion of Europe, 900-1900* (Cambridge 1986).

CHAPTER 11 The Metamorphosed Leaf

1 John Theodore Merz, *A History of European Thought in the Nineteenth Century*, II (Edinburgh and London 1903), chaps. 8-9; Philip F. Rehbock, *The Philosophical Naturalists* (Madison, Wis., 1983), 10-12

2 Hooker Papers, North American Letters, II, Vol. 64, M.Y. Stark, Dundas, to Hooker, 5 Dec. 1848, 373 (436); I, Lefroy to Hooker, 20 Feb. 1852, 265 (315). Catharine Parr Traill, *The Backwoods of Canada* (London 1836, reprint ed., Toronto 1980), 244-5; J.H. Lefroy, 'Presidential Address' to the Canadian Institute, CJ 1/5 (Dec. 1852): 123

3 D.P. Penhallow, 'A Review of Canadian Botany from 1800 to 1895,' RSC *Transactions* Ser. 2, Vol. 3, Sec. 4 (1897): 12-13; Leonard Huxley, *Life and Letters of Sir Joseph Dalton Hooker*, I (London 1918), 366. Edward Forbes, *An Inaugural Lecture on Botany* (London 1843), 6; 'On the Distribution of Endemic Plants,' BAAS, *Transactions* (1845): 67-8; 'On the Connexion between the Distribution of the Existing Fauna and Flora of the British Isles, and the Geological Changes which have affected their area, especially during the epoch of the Northern Drift,' Geological Survey of England and Wales, *Memoirs*, I (1846): 336-402. Rehbock, *Philosophical Naturalists*, 7-10, 159, 191; Louis Agassiz quoted in W.S.M. D'Urban, 'A Systematic List

of Lepidoptera Collected in the Vicinity of Montreal,' CN and G 5/4 (Aug. 1860): 241-66
4 Huxley, Life and Letters, 437, 475
5 Mea Allen, The Hookers of Kew 1785-1911 (London 1967), 155-6; W.B. Turrill, Joseph Dalton Hooker (London 1963), 43; Penhallow, 'Canadian Botany from 1800 to 1895,' 13
6 University of Toronto Library, James Hargrave Papers (mfm), George Barnston to Hargrave, 12 Apr. 1846; 3 Jan. 1852, reel 5, 4917; 30 June 1852, 5056. Gamaliel Bradford, As God Made Them (Boston 1929), 239. Asa Gray, A Manual of the Botany of the Northern United States (Boston 1848)
7 Dorothy F. Forward, The History of Botany in the University of Toronto (Toronto 1977), 4; DP, Dawson to Joseph Perrault, 18 Mar. 1864
8 J.G. Hodgins, ed., DHE, X: 6, 169-70, 286; XI: 65, 173; XIII: 59-60
9 Ibid., V: 63, 72, 190, 144-5; VI: 14
10 CAg (Sept. 1850): 197; (Apr. 1850): 76; (June 1850): 130; Upper Canada, Board of Agriculture, Journals and Transactions (1851): 158-60; NAm, 27 Dec. 1850
11 NAm, 27 Dec. 1850; CAg II/3 (Mar. 1850): 50; (Jan. 1855): 1. Journals and Transactions (1851): 204-5; DHE, IX: 282-3
12 CAg (Apr. 1853): 114; (Feb. 1853): 33; (May 1851): 106; (June 1856): 149; (15 Apr. 1862): 237-9, 283. Journals and Transactions (1855): 572-3; DHE, XV (1860): 170, 213
13 DHE, X: 213-14; A.B. Macallum, 'Huxley and Tyndall and the University of Toronto,' University of Toronto Magazine 2/3 (Dec. 1901): 68-76; another candidate was T. Cottle of Woodstock, who later published 'A List of Birds Found in Upper Canada,' CN and G 4/3 (June 1859): 231-3; Toronto Globe, 14 Apr. 1849
14 DHE, IX: 273, 275-6, 279; IX: 106; George W. Spragge, 'The Trinity Medical College,' Ontario History 58/2 (June 1966): 78
15 Journals and Transactions, Second Annual Report, 1853-4, 315
16 CAg (Feb. 1854): 47-50; also in CJ (Mar. 1854): 207. CAg (Jan. 1855): 20. Forward, History of Botany in the University of Toronto, 4; C.R.W. Biggar, 'The Reverend William Hincks, M.A.,' University of Toronto Magazine 2/9 (June 1901-2): 232; Macallum, 'Huxley and Tyndall and the University of Toronto,' 75
17 John Lindley, Natural System of Botany, 2nd ed. (London 1836), viii; An Introduction to Botany, 2nd ed. (London 1835); The Elements of Botany, Structural, Physiological & Medical (London 1849). R.J. Harvey-Gibson, Outlines of the History of Botany (London 1919), 95; J. Reynolds Green, A History of Botany 1860-1900 (Oxford 1909), 16-17. Rehbock, Philosophical Naturalists, 11
18 On quinarianism see Rehbock, Philosophical Naturalists, 26-8; P.F. Stevens, 'Haüy and A.P. De Candolle: Crystallography, Botanical Systematics, and Comparative Morphology, 1780-1840,' JHB 17/1 (Spring 1984): 73-5; and Mary P. Winsor, Starfish, Jellyfish, and the Order of Life (New Haven 1976), 82-7. The clearest contem-

porary explication was Swainson, A *Treatise on the Geography and Classification of Animals* (London 1835), 196-348. See also Macleay, 'Remarks on the Identity of certain general Laws which have been lately observed to regulate the natural Distribution of Insects,' Linnean Society, *Transactions* 14 (1822), and the articles by Kirby and Vigors in the same volume. For a contemporary criticism see Peter Rylands, 'On the Quinary, or Natural, system of McLeay, Swainson, Vigors, etc.,' *The Magazine of Natural History* 9 (1836): 130-8, 175-83. Hincks, 'On Vegetable Monstrosities,' BAAS, *Annual Report: Transactions of the Sections* (1838): 120; Biggar, 'Reverend William Hincks,' 232
19 Lindley, 'On the Principal Questions at Present Debated in the Philosophy of Botany,' BAAS, *Annual Report* (1833), 27; Günther Schmid, 'Goethes Metamorphose der Pflanzen,' in Johannes Walther, ed., *Goethe als Seher und Erforscher der Natur* (Halle 1930), 205. Agnes Arber, *The Natural Philosophy of Plant Form* (Cambridge 1950), 5, 40-3, 68-70, and chap. 4. J.W. von Goethe, *Die Metamorphose der Pflanzen* (1790), sec. 115-21; William Hincks, 'On an Anomaly of the *Trifolium repens* (white clover) in which the Pedicles of the Flowers were very much elongated, and the Petals and Pistil converted into Leaves,' BAAS, *Annual Report* (1852), 66; 'An Attempt at an Improved Class of Fruits,' CJ n.s. 6/36 (Nov. 1861): 495; Green, *History of Botany*, 12; Rehbock, *Philosophical Naturalists*, 98-9, 113
20 CAg (July 1853); 198; J. Sheridan Hogan, *Canada: An Essay* (Montreal 1855), 26
21 Hogan, *Canada*, 28-9, 35; CAg (July 1853): 215; *Journals and Transactions* (1853): 424
22 Hargrave Papers, mfm reel 135, 20 Apr. and 2 June 1843, 2566-8, 2608; reel 141, Hargrave to George Simpson, 17 Aug. 1843; Hooker Papers, I, Simpson to Derby, 5 Dec. 1843, 558; R. Glover, 'The Man Who Did Not Go to California,' Canadian Historical Association, *Historical Papers* (1975), 98-9
23 CJ (Nov. 1852), 80; (Mar. 1853): 172-6; CAg (Apr. 1852): 117
24 CJ (Apr. 1853): 204-7, 219. Major differences in classificatory systems are described in J.H. Balfour, *A Manual of Botany*, 3rd revised ed. (London and Glasgow 1855), 359-64
25 CAg (Feb.-Mar. 1851): 36-8, 58 60; CJ (Oct. 1852): 60; *Journals and Transactions* (1853): 325, 338
26 'Observations on the Leafing and Flowering of Plants,' CJ (Mar. 1853): 182-3; 'On the Fruiting and Flowering of Plants' (Apr. 1853): 201
27 Dr and Mrs W. Craigie, 'List of Indigenous Plants found in the Neighbourhood of Hamilton,' CJ (Apr. 1854): 222-3. 'Notes and Queries,' ibid. (Sept. 1852): 42
28 'Vegetation of the Frozen Region,' CAg (Aug. 1853): 249-50. 'The Treasures of Our Forests,' CJ (Nov. 1852): 74-5. Samuel Strickland, *Twenty-Seven Years in Canada West* (London 1853), quoted in Marie E. Catlow, *Popular Geography of Plants*, ed. Charles Daubeny (London 1855), 49
29 *Journals and Transactions* (1854): 466

30 Review of 'The Agriculture of the French Exhibition,' CJ n.s. 1/2 (Feb. 1856): 144–6; 'Agricultural Statistics–Their Importance,' CAg (Dec. 1857): 312–14; (Mar. 1856): 57–8
31 CAg (Aug. 1857): 199–202; (Aug. 1856): 205; (June 1857): 145; (Oct. 1856): 278; (Aug. 1857): 199–201; (Sept. 1856): 251. H.Y. Hind, *Essay on the Insects and Diseases Injurious to the Wheat Crops* (Toronto 1857), i. W.S.M. D'Urban of Montreal undertook his own local inventory on the large farm owned by James Logan, known as a 'scientific farmer'; CN and G (July 1857): 161–70; CAg IC/9 (Sept. 1857): 238–40
32 Hind, *Insects and Diseases Injurious to the Wheat Crops*, 26ff; Émilien Dupont, *Essai sur les insectes et les maladies qui affectent le blé* (Montreal 1857), ii. William Hincks, 'Review of H.Y. Hind, *Essay on the Insects and Diseases Injurious to the Wheat Crops*,' CJ n.s. 2/12 (Nov. 1857): 442–3; CAg (Jan. 1858): 24; J.W. Dawson, 'Things to be observed in Canada, and especially in Montreal and its vicinity,' Lecture of the Popular Course of the Montreal Natural History Society, CN and G 3/1 (Feb. 1858): 4
33 'Instructions for Collecting and Preserving Insects,' CN and G (May 1857): 101–3; 'Notes on the Distribution of Insects,' 2 (1857): 41; 'Entomology; No. 1,' 3/1 (Feb. 1858): 25

CHAPTER 12 Fragile Stems, 1857–1863

1 John Hutton Balfour, *Botany and Religion*, 4th ed. (Edinburgh 1882) [first published as *Phytotheology* 1851], 412; *Manual of Botany*, 3rd ed. (London 1855), xi. See also Isaac Bayley Balfour, 'A Sketch of the Professors of Botany in Edinburgh from 1670 until 1887,' in F.W. Oliver, ed., *Makers of British Botany* (Cambridge 1913), 296–8.
2 Balfour, *Class Book of Botany* (Edinburgh 1854), 351; *Botany and Religion*, 290–2
3 *Botany and Religion*, 2–3; *Class Book*, ix
4 *Manual of Botany*, ix; *Class Book*, 705–6, 581
5 *Manual of Botany*, xii; *Botany and Religion*, 20, 413; *Class Book*, 971
6 University of Toronto Library, James Hargrave Papers, mfm reel 134, George Barnston to Hargrave, 9 Jan. 1838, 1403–4; (1846), 3424; 7 Apr. 1848. Royal Botanic Garden, Edinburgh, John Hutton Balfour Papers, II, James Barnston to Balfour, 7 Sept. 1853, 209; 30 Oct. 1854, 210
7 Balfour Papers, Barnston to Balfour, 30 Oct. 1854, 210; 4 Mar. 1855, 211; IV, Dawson to Balfour, 27 July 1856, 29
8 Royal Botanical Gardens, Kew, W.J. Hooker Papers, I, Logan to Hooker, 4 Feb. 1856, 269; Montreal *Gazette*, 12 Jan. 1856; Hargrave Papers, George Barnston to Hargrave, 16 Aug. 1856, 6077
9 Hooker Papers, I, 64, James Barnston to Hooker, 17 Mar. 1856, 16; Hargrave Papers,

323 Notes to pages 221-5

I, George Barnston to Hargrave, 16 Aug. 1856, 6077
10 *Prospectus of McGill College* (1856-7), 10-12. Balfour Papers, IV, J.W. Dawson to Balfour, 29 Dec. 1856, 30; 'Proceedings of the Botanical Society of Montreal,' CN and G, 2/1 (Mar. 1857): 77. Hargrave Papers, I, George Barnston to Hargrave, 16 Aug. 1856, 6077; George Barnston, 'Remarks upon the Geographical Distribution of the Order *Ranunculaceae*, throughout the British Possessions of North America,' CN and G 2/1 (Mar. 1857): 12-20, read to the Botanical Society on 5 Dec. 1856
11 George Barnston, 'Remarks,' 15, 20
12 James Barnston, 'General Remarks on the Study of Nature, with special reference to Botany,' CN and G 2/1 (Mar. 1857): 34-6, 39-40
13 James Barnston, 'Hints to the Young Botanist, regarding the collecting, naming, and preserving of plants,' CN and G 1/2 (May 1857): 127-34
14 Hargrave Papers, I, George Barnston to Hargrave, 16 Aug. 1856, 6077. *Lovell's City Directory* (Montreal 1858), 441; A.N. Rennie, 'Obituary: James Barnston,' CN and G 3/3 (June 1858): 224; Dawson, 'Things to be Observed in Canada,' 7
15 Balfour Papers, Dawson to Balfour, 25 Jan. 1858, 32; *Prospectus of McGill College* (1857-8), 13. James Barnston, 'Introductory Lecture to the Course on Botany,' CN and G 2/5 (Nov. 1857): 337, 339, 343
16 *Prospectus* (1858-9), 16; (1860-1), 17-18. James Barnston, 'Catalogue of Canadian Plants in the Holmes Herbarium, in the Cabinet of the University of McGill College,' CN and G 4/2 (Apr. 1859): 101, 116. A.F. Kemp, 'Some Account of the Herbarium of the Natural History Society of Montreal,' CN and G 7/3 (June 1862): 228; 'Notice of the Natural History Collections of the McGill University,' CN and G 7/3 (June 1862): 221-2. For details of Barnston's personal life see obituary, Montreal *Pilot*, 31 Mar. 1858; Montreal *Daily Transcript*, 8 May 1857. He had recently married Mary Anne McDonald, daughter of Archibald McDonald of the HBC, an acquaintance of W.J. Hooker; Jean Murray Cole, *Exile in the Wilderness* (Don Mills 1979), appendix.
17 J.W. Dawson, 'Things to be Observed in Canada,' CN and G 3/1 (Feb. 1858): 1-2, 8
18 CN and G 3/5 (Oct. 1858): 386; also in CJ n.s. 3/18 (Nov. 1858): 502. George Barnston, 'Remarks on the Geographical Distribution of Plants in the British Possessions of North America, CN and G 3/1 (Feb. 1858): 28-32
19 George Barnston, 'Remarks on the Geographical Distribution of the *Cruciferae* throughout the British Possessions in North America,' CN and G 4/1 (Feb. 1859): 1-12
20 Owen's address, 385. Daniel Wilson, Editorial Committee's Report, CJ n.s. 2/8 (Mar. 1857): 148-9; CN and G 2/5 (Nov. 1857): 399-400
21 CN and G 3/5 (Oct. 1858): 397; 3/1 (Jan. 1858): 39-50. George Barnston, 'Geographical Distribution of the Genus *Allium* in British North America,' CN and G 4/2 (Apr. 1859): 116-21

22 B.H. MacDonald, 'John William Dawson and Nineteenth-Century Palaeobotany,' MA thesis, University of Western Ontario, 1982, 82–3. Balfour Papers, Dawson to Balfour, 27 July 1855, 29; 22 Oct. 1858, 33. G.U. Hay, 'The Scientific Work of Prof. Chas. Fred. Hartt,' RSC, *Transactions*, Ser. 2, Vol. 5, Sec. 4 (1899): 157. J.W. Dawson, 'Recent Geological Discoveries–review of the supplement to the 5th ed. of Lyell's *Manual of Elementary Geology*,' CN and G 2/3 (July 1857): 194. Henry D. Andrews, *The Fossil Hunters: In Search of Ancient Plants* (Ithaca 1980), 186, 199, argues that Dawson's pioneering studies were so strange that they were largely ignored for more than a half century.

23 'Canadian Natural History,' CJ n.s. 3/17 (Sept. 1858): 461–2

24 E.J. Chapman, 'Review of Geological Survey of Canada, Report of Progress for 1857,' CJ 4/22 (1859): 269; Robert Bell, 'On the Natural History of the Gulf of St. Lawrence,' CN and G 6/2 (Apr. 1861): 120–37; GSC, *Report of Progress* (1858), 226–63. PAC, GSC, Directors' Letterbooks, A.R.C. Selwyn to Agnes Fitzgerald, 21 Mar. 1870, 122

25 'Vegetation in the Arctic Regions,' CAg 9/6 (June 1857): 150; Dawson, 'Things to be Observed in Canada,' 8; 'Gleanings in the Natural History of the Hudson's Bay Territories, by the Arctic Voyageurs,' CN and G 2/3 (July 1857): 170; A.F. Kemp, 'The Fresh Water Algae of Canada,' ibid. 3/5 (Oct. 1858): 331–45, 450; ibid. 6/2 (Apr. 1861): 121–37. For the influence of British dredging see Philip F. Rehbock, 'The Early Dredgers: "Naturalizing the British Seas", 1830–1850,' JHB 12/2 (Fall 1979): 293–368; and Dawson, 'On the Sea Anemones and Hydroid Polyps from the Gulf of St. Lawrence,' CN and G 3/6 (Dec. 1858): 402–4.

26 Hofmeister, *Vergleichende Untersuchungen der Keimung, Entfaltung und Fruchtbildung höherer Kryptogamen* (1851); W.H. Hincks, 'Review of Rev. M.J. Berkeley, *Introduction to Cryptogamic Botany*,' CN and G 2/2 (May 1857): 157–9; reprinted in CJ n.s. 3/16 (July 1858): 342–5

27 Smithsonian Institution Archives, HBC Correspondence Collection 1858–69 and undated (mfm), George Barnston to S.F. Baird, 26 Jan. 1860; Robert Campbell to Baird, 6 July 1861; Strachan Jones to Baird, 22 Apr. 1867. Raymond Duchesne, 'Science et société coloniale: les naturalistes du Canada français et leurs correspondants scientifiques (1860–1900),' HSCT *Bulletin: Journal of the History of Canadian Science, Technology and Medicine* 18 (May 1981): 99–139; international aspects of science in the North Atlantic triangle are emphasized in Carl Berger, *Science, God, and Nature in Victorian Canada*, The 1982 Joanne Goodman Lectures (Toronto 1983), 27.

28 Léon Provancher, *Traité élémentaire de botanique* (Quebec 1858), iv

29 Provancher, 'Naturalistes canadiens,' *Le Naturaliste canadien* 5 (1873): 229–31; R. Duchesne, 'La bibliothèque scientifique de l'Abbé Léon Provancher,' *Revue historique de l'Amérique française* 34/4 (Mar. 1981): 536–7

30 Provancher, *Traité*, iii–vii. Ovide Brunet, *Notes sur les plantes recueillies en 1858* [n.p. 1859?], 1. Provancher's critical review of Brunet, *Éléments de Botanique*, in *Le Naturaliste canadien* II/5 (Apr. 1870): 144–9
31 William Hincks, 'Canadian Flora, Neighbourhood of Toronto,' CJ n.s. 3/5 (May 1858): 266–7
32 Hincks, 'Considerations respecting Anomalous Vegetable Structures,' CJ n.s. 3/16 (July 1858): 311–12; Hincks, 'Review of Louis Agassiz, *Contributions to the Natural History of the United States*,' ibid. 3/1 (May 1858): 245. J.G. Hodgins, ed., DHE IV (1859–60), 194; Hooker Papers, I, J.D. Humphreys, Toronto, to Hooker, 26 Dec. 1856, 246
33 DHE VI (1846), 14. H.P. Gundy, 'Growing Pains: The Early History of the Queen's Medical Faculty,' *Historic Kingston* No. 4 (Oct. 1955): 14–16
34 DHE XIII (1856–8): 133, 135–6; 63; XIV (1858–60); 46, 251
35 Gundy, 'Growing Pains,' 22; Hilda Neatby, *Queen's University, I: 1841–1917*, ed. Frederick W. Gibson and Roger Graham (Montreal 1978), 71
36 Hooker Papers, English Letters 1846, 24, Lawson to Hooker, 24 June 1846, 334; 22 Aug. 1846, 335; 11 Aug. 1846
37 Balfour Papers, II, James Barnston to Balfour, 4 Mar. 1855, 211
38 Hooker Papers, Supplementary Foreign Letters, 1865–1900, 218, Lawson to Hooker, 18 Jan. 1860, 173; Balfour Papers, Lawson to Balfour, 22 Sept. 1857
39 CAg, 9/10 (Oct. 1858): 232, 234–5, 237
40 Neatby, *Queen's University*, 77
41 BSC, *Annals*, I
42 Ibid.
43 Ibid.
44 Ibid., Part I (7 Dec. 1860–8 Mar. 1861), 14, cl. 1
45 Ibid., 9–12, 21, 50, 177
46 CAg 12/24 (26 Dec. 1860): 645; CN and G 5/6 (Dec. 1860): 462–6; BSC, *Annals*, 169; LP, Lawson to Logan, 2 Jan. 1861
47 'The Botanical Society of Canada,' trans. [and abridged] by John Machar, ibid., 174–5; BAJ 2 (1861): 379–80
48 'Die Botanische Gesellschaft Canada's,' *Bonplandia* 9/8 (15 May 1861): 114. BSC, *Annals*, 174–5, 379–80; 15 Nov. 1861, 167
49 CN and G 6/4 (Aug. 1861): 331–3; 6/6 (Dec. 1861): 468–9
50 Lawson, 'Botanical Society of Canada–Abstract of Recent Discoveries in Botany and the Chemistry of Plants,' BSC, *Annals*, 8 Mar. 1861; reprinted in CN and G 6/1 (Feb.–Mar. 1861): 70–1; and CAg 13/5 (1 Mar. 1861): 132–3; BAJ 2 (1861): 380; CN and G 6/6 (Dec. 1861): 471; 4/4 (Aug. 1861): 331–3. BSC, *Annals*, 25–33, 54, 110, 181. Dalhousie University Library, Robert Bell Papers, MS2, 381, A-1, Lawson to Bell, 15 Feb. 1861. Sheppard (Shepherd) was a nurseryman who had belonged to the

Botanical Society of Montreal; see p. 221 above and *Lovell's Montreal Directory* (1861), 198.
51 Hooker Papers, Foreign Letters 1865–1900, 218, Lawson to Hooker, 18 Jan. 1860, 173; 'Canadian Botany,' JEUC (Feb. 1860): 32; CAg 12/8 (16 Apr. 1860): 174
52 CN and G 6/1 (Feb. 1861): 58; PAC, John Macoun Papers, MG24 K28, W.J. Hooker to Macoun, 7 Aug. 1861; JEUC (May 1863): 73
53 New Brunswick, House of Assembly, *Journals*, 1860: 6, 121; 1862: 13–14; 1864: 11. Macoun Papers, Hooker to Macoun, 7 Aug. 1861; Hooker Papers, NAm and SAm Letters, Lord Monck to Hooker, 13 Nov. 1862, 82 (106) mfm
54 Hooker Papers, NAm and SAm Letters, Lawson to Hooker, 24 Oct. 1862, 67 (86) mfm
55 Ibid.; Province of Canada, House of Assembly, *Journals*, 1861: 101; (1862): 123. JEUC (Nov. 1862): 168
56 Hooker Papers, NAm and SAm Letters, J.B. Hurlburt to Hooker (1862): 56 (74) mfm; JEUC (May 1863): 73. For the importance to Canada of the exhibition see CAg 13/14 (16 July 1861): 419–21; and 13/19 (1 Oct. 1861): 579–80. Charles Robb, 'Descriptive List of the Principal Canadian Timber Trees,' CJ n.s. 6/21 (Jan. 1861): 28. Léon Provancher, *Flore canadienne*, 2 vols. (Quebec 1862); noted in JEUC (May 1863): 73
57 CN and G 8/1 (1863): 80
58 CN and G n.s. 1/1 (1864): 1
59 Gundy, 'Growing Pains;' 24; Balfour Papers, Lawson to Balfour, 31 Jan. 1862; 7 Feb. 1862; 12 Mar., 13 Mar. 1862
60 Balfour Papers, Lawson to Balfour, 2 Jan., 2 July, 4 Dec. 1863; DHE XVIII (1863–5): 46; PAC, Robert Bell Papers, William Bell to Bell, 20 Oct. 1863; Neatby, *Queen's University*, 102–3
61 CN and G 5/6 (Dec. 1860): 462; 4/2 (April 1859): 144; 5/6 (Dec. 1860): 462; 8/3 (June 1863): 221; 5/3 (June 1860): 234; CAg 14/3 (1 Feb. 1863): 86, JEUC (Mar. 1856): 46; PAC, Robert Bell Papers, Elizabeth Notman Bell to Bell, 1 Apr. 1863; CN and G 8/3 (June 1863): 221
62 Hooker Papers, NAm and SAm Letters, Hincks to Hooker, 18 Mar. 1861, 54 (70) mfm
63 CJ 12/68 (Apr. 1869): 103; 12/71 (Apr. 1870): 354; CN and G n.s. 1/1 (1864): 59
64 CJ n.s. 6/32 (Mar. 1861): 166; Hooker Papers, Hincks to Hooker, 18 Mar. 1861
65 CJ 9/53 (Sept. 1864): 344, 347–8; Hincks in turn arranged William Saunders's collections, CJ n.s. 7/45 (May 1863): 219–38. Brunet, *Énumérations des genres de plantes de la flore du Canada* (Quebec 1864), 3-4
66 Hincks was never completely satisfied with any system of classification; 'Some Thoughts on Classification in Relation to Organized Beings,' CJ n.s. 11/61 (Jan. 1866): 40; also 'Review of Lawson,' 344; n.s. 10/55 (Jan. 1865): 41; n.s. 10/58 (July 1858): 232–42. He studied political economy in the German tradition of Nationalö-

konomie: CJ (Jan. 1861), (May 1862), and (April 1866). During the 1860s he worked with William Saunders and C.S. Bethune towards a fauna of Canada. Hooker Papers, British North American Letters 1865-1900, Vol. 195, Macoun to Hooker, 10 Jan. 1866

CHAPTER 13 Flower and Fruit

1 PAC, Sandford Fleming Papers, Vol. 93, Folder 3, 'The Canadian Flag,' (1895); Vol. 15, Folder 109, J. Fletcher to Fleming, 18 June 1895. Janet Carnochan, 'The Origin of the Maple Leaf as the Emblem of Canada,' OHS, *Papers and Records* 7 (1906): 141-3
2 CAg 13/11 (1 June 1861): 350; see also [David Boyle], 'The Origin of Our Maple Leaf Emblem,' OHS, *Papers and Records* 5 (1904): 22-6.
3 CAg 13/22 (16 Nov. 1861): 679-80; 14/2 (Feb. 1862): 66. In 1888 John Macoun noted that 'Although [*Elodea canadense*] makes a luxuriant growth in summer, it never chokes Canadian streams, as the frosts of winter destroy it'; *Catalogue of Canadian Plants*, Part IV: *Endogens* (Montreal 1888), 1
4 CAg 14/16 (16 Aug. 1862): 495-7; 15/2 (Feb. 1863): 66-7; 12/21 (1 Nov. 1860): 603; 14/24 (16 Dec. 1862): 744; 15/1 (Jan. 1863): 3
5 Ibid. 11/1 (Jan. 1859): 16; 12/19 (1 Oct. 1860): 504; 14/4 (16 Feb. 1862): 109; 14/22 (16 Nov. 1862): 695; 13/3 (1 Feb. 1861): 79; 14/7 (1 Apr. 1862): 217; 13/4 (16 Feb. 1861): 109-11; CN and G 7/2 (Apr. 1862): 102-6; Léon Provancher, *Le Verger canadien* (Quebec 1862), 3
6 Agassiz, *Lake Superior: Its Physical Character, Vegetation, and Animals, compared with those of other and similar regions* (Boston 1850), 10. The Viking image was resurrected by Professor Blytt of Christiania, Norway, an honorary member of the BSC, who requested George Lawson's co-operation in tracing evidence of grape vines indigenous to British North America; see Lawson, 'On the Northern Limit of Wild Grape Vines,' NSINS, *Proceedings and Transactions* 6/2 (1884): 101.
7 CAg 12/13 (2 July 1860): 295-9; 13/4 (16 Feb. 1861): 111; Hurlburt, 'The Climate of Canada as Adapted to the Culture of the Grape,' CAg 13/20 (16 Oct. 1861): 627-8; D.A. McNabb, 'On the Culture of the Vine in the Open Air,' ibid. 14/6 (16 Mar. 1862): 182-4; C. Arnold, 'On Grape Culture,' 15/3 (Mar. 1863): 107
8 J.D. Hooker, *The Botany: The Antarctic Voyage of H.M. Discovery Ships "Erebus" and "Terror"*, II: *Flora Novae-Zelandiae* (London 1853), i-ii, viii, xxii. 'Letter from M.E. Bourgeau, Botanist to Capt. Palliser's British North American Exploring Expedition. Addressed to Sir William Jackson Hooker, 3 March 1859,' Linnean Society of London, *Journal of the Proceedings* (Botany) IV (1860): 1-16. Michael Ruse, *The Darwinian Revolution* (Chicago 1979), 139-41
9 Forbes, 'On the connexion between the existing flora and fauna of the British Isles,

and the geological changes which have affected their area, especially during the epoch of the Northern Drift,' Geological Survey of England and Wales, *Memoirs*, 1 (1846): 336–432; E.J. Chapman's review in CJ n.s. 5/26 (Mar. 1860): 201; J.F. Whiteaves, 'On the Land and Fresh Water Mollusca of Lower Canada, with thoughts on the geographical distribution of animals and plants throughout Canada,' CN and G 6/6 (Dec. 1861): 454, 457; Ruse, *Darwinian Revolution*, 225; Philip F. Rehbock, *The Philosophical Naturalists* (Madison, Wisc. 1983), 12; Michael Ghiselin, *The Triumph of the Darwinian Method* (Berkeley 1969), 39–40; Hooker, *The Botany*, Part III: *Flora Tasmaniae* (London 1860), xxvi; Charles Darwin, *The Origin of Species*, esp. chaps. 12–13 on geographical distribution, and chap. 5 on variation. Jonathan Howard, *Darwin*, Past Masters (Oxford 1982)

10 *Flora Tasmaniae*, xxvi. R.J. Harvey-Gibson, *Outlines of the History of Botany* (London 1919), 23, 116–17; A.F. Kemp, 'A Classified List of Marine Algae,' CN and G 5/1 (Feb. 1860): 30–3

11 Linnean Society, *Transactions* 23 (1862): 251–310; W.B. Turrill, *Joseph Dalton Hooker* (London 1963), 108–9, 148–9; Hooker's article was reprinted in CN and G six years later: n.s. 3/5 (June 1867–8): 325–62; Hooker, 'Outlines,' 253–4, paraphrased by Asa Gray, 'Dr. Hooker's Distribution of Arctic Plants,' in C.S. Sargent, ed., *Scientific Papers of Asa Gray*, I (Boston 1889), 124. John Macoun, *Autobiography of John Macoun* (Ottawa 1922), 43

12 J.W. Dawson, 'Review of Hooker's Outlines of the Distribution of Arctic Plants,' CN and G 7/5 (Oct. 1862): 335, 341–3; see ed. note.

13 Dawson, *Alpine and Arctic Plants: A Lecture to the* YMCA, *Montreal* (Montreal 1862), 10, 14, 16, 19; printed as 'Notes on the Flora of the White Mountains, in its Geographical and Geological Relations,' CN and G 7/2 (Apr. 1862): 81–102; Francis Darwin, ed., *More Letters of Charles Darwin*, I (London 1903), no. 356, Darwin to J.D. Hooker, 4 Nov. [1862], 468; nos. 144, 358. For negative reactions to Darwin's theory see A.B. McKillop, *A Disciplined Intelligence* (Montreal 1979) and Carl Berger, *Science, God, and Nature in Victorian Canada*, The 1982 Joanne Goodman Lectures (Toronto 1983).

14 Dawson's review, CN and G 5/2 (Apr. 1860): 100–19. Harland Coultas, 'Origin of our Kitchen-Garden Plants,' reprinted in CN and G n.s. 2 (1865): 41–2

15 Hooker Letters, Supplementary Foreign Letters, Vol. 218, Lawson to Hooker, 31 Mar. 1864, 174

16 Lawson, 'On the Flora of Canada,' NSINS, *Proceedings and Transactions* 1/2 (1864): 75; Dawson, 'Review,' 343; CN and G n.s. 1/5 (1864): 378; Lawson, 'Notice of the Occurrence of Heather,' NSINS, *Proceedings and Transactions* 1/3 (1865): 30–5; 'Notes on Some Nova Scotian Plants,' 4/1 (1875): 168. Carl H. Lindroth, *The Faunal Connections between Europe and North America* (Stockholm 1957), 135–43,

explains five criteria of an introduced species: historical, geographical, ecological, biological, and taxonomic.

17 R. Sturton, 'A Few Thoughts on the Botanical Geography of Canada,' LHSQ, *Transactions* n.s. Parts 1–4 (1862–6): 99, 102, 108; John Sommers, 'Introduction to a Synopsis of the Flora of Nova Scotia,' NSINS, *Proceedings and Transactions* 4/1 (1875): 181–3

18 Sommers, 'Notes on Nova Scotia Compositae–Asters,' ibid. 4/3 (1877): 240; 'A Contribution towards the Study of Nova Scotian Mosses,' ibid. 4/4: 363

19 Hooker Papers, NAm and SAm Letters, mfm, Lawson to Hooker, 31 Mar. 1864, 68 (87)

20 PAC, Robert Bell Papers, I, William Bell to Bell, 20 Oct. 1863; Queen's College Board of Trustees to Bell, 2 Nov. 1863; Bell to Sir William Logan, 4 Nov. 1863 (copy in LP); George H. Frost to Bell, 24 Nov. 1863; Robert Bell (Bell's uncle) to Bell, 22 Mar. 1864; Bell to Trustees of Queen's University and College, 9 Feb. 1864; Vol. 49, 'Extract from the written statement,' 30 [sic] Feb. 1864; LP, Logan to Bell, 1 Feb. 1864; Balfour Papers, Dawson to Balfour, 29 Apr. 1864, 36

21 PAC, Robert Bell Papers, 'Extract,' 30 [sic] Feb. 1864. Hincks, 'Review of George Lawson, "Synopsis of Canadian Ferns and Filicoid Plants",' CJ 9/53 (Sept. 1864): 348. W.E. Billings, 'List of Plants found growing as indigenous in the neighbourhood of Prescott, C.W.,' CN and G 6/1 (Feb. 1860): 14; Judge Logie, 'On the Flora of Hamilton and Its Vicinity,' ibid. 6/4 (Aug. 1861): 276–8; Queen's University Archives, A.T. Drummond Papers, John Bell to Drummond, 13 Aug. 1862; Brunet to Drummond, 23 Dec. 1863. Archives du Séminaire de Québec, Louis-Ovide Brunet Papers, 113/106, Drummond to Brunet, 28 Dec. 1863; Balfour Papers, Lawson to Balfour, Jan. 1863. Drummond also began a correspondence with Asa Gray at Harvard.

22 Drummond Papers, Brunet to Drummond, 1864; Drummond, 'Observations on Canadian Geographical Botany,' CN and G 1/6 (1864): 405–13

23 D.A.P. Watt, 'Review of *Ferns: British and Foreign* (John Smith),' CN and G 3/2 (Dec. 1866): 150. Drummond Papers, Watt to Drummond, 30 Nov. 1864

24 CN and G n.s. 2 (1865): 79; Watt, 'A Provincial Catalogue of Canadian Cryptogams,' ibid. 2/5 (Oct. 1865): 390–404; Brunet Papers, 114/45, T.S. Hunt to Brunet, 11 Feb. 1865; Watt, 'Book Notices–*Acadian Geology* (J.W. Dawson, 2nd ed.),' CN and G 3/5 (June 1868): 400

25 Brunet, *Catalogue des Plantes canadiennes dans l'herbier de l'Université Laval* (Quebec 1865), 5–7; *Histoire des Picea qui se rencontrent dans les limites du Canada* (Quebec 1866); reprinted in CN and G 3/2 (Dec. 1866); 'Michaux and His Journey in Canada,' trans. T.S. Hunt, ibid. 1/5 (1864): 325–7; *Notice sur le musée botanique de l'Université Laval* (Quebec 1867), 4–5, 13; *Annuaire de l'Université Laval* (1867–

8), 67–76. *Catalogue des végétaux ligneux du Canada pour servir à l'intélligence des collections de bois économiques envoyées à l'exposition universelle de Paris, 1867* (Quebec 1867), esp. 7

26 Dalhousie University Library, George Lawson Papers, MS 2, 381 A-I, Lawson to Dr John Bell, 27 Nov. 1865. Drummond Papers, Watt to Drummond, 9 Jan. 1866; Lawson to Drummond, 15 Apr. 1867

27 Brunet Papers, 114/100, Watt to Brunet, 3 Feb. 1866; 114/51, 23 Feb. 1865; 114/49, 22 Feb. 1865; 114/29, John Macoun to Brunet, 19 June 1866. Drummond Papers, Macoun to Drummond, 5 Jan. 1866; 31 Mar. 1866; 6 Oct. 1866. Macoun, 'A Catalogue of the Carices Collected,' CN *and* G 3/1 (Feb. 1866): 56–63. Hooker Papers, British North American Papers (BrNAm), Vol. 195, Macoun to Hooker, 7 Aug. 1866, 272–3

28 Watt, 'Filices Canadenses,' CN *and* G 3/5 (June 1868): 402. Lawson, 'On the Flora of Canada,' NSINS, *Proceedings and Transactions* 1/2 (1864): 75. Drummond, 'Some Statistical Features of the Flora of Ontario and Quebec, and a comparison with those of the United States Flora,' CN *and* G 3/6 (Dec. 1868): 430. Lawson, 'On the Ranunculaceae of the Dominion of Canada and of Adjacent Parts of British America,' ibid. 4/4 (Dec. 1869): 407; also in NSINS, *Proceedings and Transactions* 2/4 (1869): 17–51; Drummond Papers, Lawson to Drummond, 14 Apr. 1869; John Bell, 'The Plants of the West Coast of Newfoundland,' CN *and* G 4/3 (Sept. 1869): 256–63; Hooker Papers, BrNAm Letters, Watt to Hooker, 9 Apr. 1866, 326

29 Drummond, 'The Distribution of Plants in Canada in Some of Its Relations to Physical and Past Geological Conditions,' CN *and* G 3/3 (May 1867): 161–77; 'The Introduced and the Spreading Plants of Ontario and Quebec,' ibid. 4/4 (Dec. 1869): 377–88; 'Some Statistical Features,' 431

30 John Macoun, *Autobiography* (Ottawa 1920), 32; Drummond Papers, Macoun to Drummond, 8 Feb. 1868; Agassiz, *Lake Superior*, 139–40

31 Hooker Papers, BrNAm Letters, Macoun to Hooker, 8 Oct. 1867, 274; 7 Mar. 1868, 275. Macoun, *Autobiography*, 42–5. DP, 2211/23, ref. 83, Watt to Dawson, June 1869

32 Macoun, *Autobiography*, 44, 197–8; Agassiz, *Lake Superior*, 153; Lawson, 'On the Flora of Canada,' 77. Macoun Papers, 'Notes of a Trip to Lake Superior during Summer of 1869,' 30; Macoun, 'Plants New to Canada,' CN *and* G 4/3 (Sept. 1869): 362. Hooker Papers, BrNAm Letters, Macoun to Hooker, 7 Aug. 1866, 272–3; Drummond Papers, Macoun to Drummond, 4 Mar. 1871

33 G.F. Matthew, 'On the Occurrence of Arctic and Western Plants in Continental Acadia,' CN *and* G 4/3 (June 1869): 165; Lawson, 'Botanical Society of Canada–Hector on the Areas of Botanical Distribution throughout the central part of British North America,' 6/5 (Oct. 1861): 395–7. 'The Profits of Red River Farming,' CAg 12/4 (Feb. 1860): 77; 12/15 (1 Aug. 1860); 'Lake Superior Region,' 14/22 (16 Nov. 1862): 689; W.O. Buell (Perth) to ed., 15/8 (Aug. 1863): 294–5. Alexander

Morris, *Nova Britannia* (Montreal 1858), 29, 66. Bernard R. Ross, 'An Account of the Botanical and Mineral Products, useful to the Chipewyan tribes of Indians, inhabiting the McKenzie River District,' CN and G 7/2 (Apr. 1862): 133; Victor Hopwood, 'William Fraser Tolmie: Natural Scientist and Patriot. A Review Article,' *BC Studies* 5 (Summer 1970): 45–51

34 Schultz suffered heart palpitations while a student and was advised by his physician to find more relaxing activities to occupy his time. PAM, J.C. Schultz Papers, MG12 E3, Joseph Hackett to Schultz, 24 Mar. 1859; Henry McKenney to Schultz, 22 [?] 1859. Schultz, 'On the Botany of the Red River Settlement and the Old Red River Trail,' BSC, *Annals* 1 (1861): 26–30; reprinted in CAg 13/3 (1 Feb. 1861): 67–8. Murray Campbell, 'Dr. J.C. Schultz,' HSSM, *Papers* 3/20 (1963–4): 7–12; Dougald McDougall, 'The History of Pharmacy in Manitoba,' ibid. 3/2 (1954): 23–4. Lawson, 'Synopsis,' 268; Lawson, 'Botanical Science–Record of Progress,' CN and G n.s. 1/1 (1864): 1

35 T.C.B. Boon, 'The Institute of Rupert's Land and Bishop David Anderson,' HSSM, *Papers* 3/17 (1960–1): 92-III; George Bryce, 'Worthies of Old Red River,' HSSM, *Transactions* 48 (1896): 12. PAM, *The Nor'Wester*, 5 Mar. 1862; PAM, J.J. Hargrave Papers, (1871), 220–1. Smithsonian Institution Archives, HBC Correspondence, George Barnston to S.F. Baird, 12 May 1862

36 Hooker Papers, BrNAm Letters, Macoun to Kew, 3 Nov. 1874, 275; 23 Feb. 1875, 278; Watt to Hooker, 22 Dec. 1874, 355; George Barnston to Hooker, Sept. 1875, 183; Macoun to Hooker, 5 Sept. 1876, 279; 18 July 1877, 282

37 Fleming Papers, Vol. 33, Folder 233, Macoun to Fleming, 20 May 1878. Macoun, 'The Capabilities of the Prairie Lands of the Great North-West, as shown by their Fauna and Flora,' Ottawa Field-Naturalists Club, *Transactions* 1/2 (1880): 38. Macoun Papers, Sept. 1879; Hooker Papers, Macoun to Hooker, 5 Nov. 1885, 288. W.A. Waiser, 'Rambler: Professor John Macoun's Career with the Geological Survey of Canada, 1882–1912,' PH D thesis, University of Saskatchewan, 1983; see also his 'A Willing Scapegoat: John Macoun and the Route of the C.P.R.,' *Prairie Forum* 10/1 (Spring 1985): 65–82. J.W. Dawson too believed that it was 'undoubtedly the duty of those whose scientific studies show them the grandeur of this great question and the nature of the practical results of its solution, to aid in every way that they can the progress towards an unobstructed highway through our territory from the Atlantic to the Pacific'; CN and G n.s. 8/5 (1878): 3.

38 Hooker Papers, Macoun to Hooker, 14 Apr. 1884, 284; James Macoun to Hooker, 11 Apr. 1894, 265; 14 Dec. 1875, 267; 13 Jan. 1896, 268. DP, 909B/40, ref. 2, J.D. Hooker to Dawson, 7 Oct. 1874; G.M. Dawson, 'Notes on the Distribution of Some of the More Important Trees of British Columbia,' CN and G n.s. 9/6 (1881): 321–31

39 Drummond Papers, Lawson to Drummond, 14 Apr. 1869. DP, 22II/22 ref. 67, Lawson to Dawson, 22 Dec. 1869. Lawson, 'Ranunculaceae' (1869); 'Descriptions of

the Plant Species of Mysotis,' CN and G 4/4 (Dec. 1869): 398-407; 'North American Laminariaceae,' ibid. 5/1 (Mar. 1870): 99-101; 'Monograph of Ericiaceae of the Dominion of Canada and adjacent parts of British America,' NSINS, *Proceedings and Transactions* 3/1 (1872): 74-80; 'On Canadian Species of Rubi and Their Geographical Distribution,' ibid. 3/4 (1874): 364-6; 'On the Canadian Species of the Genus Melilotus,' ibid. 6/3 (1885): 180-90; *The Fern Flora of Canada* (Halifax 1889). Jacques Rousseau and William G. Dore, 'L'Oublié de l'histoire de la science canadienne–George Lawson, 1827-1895,' in G.F.G. Stanley, ed., *Pioneers of Canadian Science* (Toronto 1966), 75. Lawson, 'Notes for a Flora of Nova Scotia-Part I,' NSINS, *Proceedings and Transactions* 8/1 (1892): 84

40 Lawson, 'Remarks on the Flora of the Northern Shores of America,' RSC, *Transactions* 5, sec. 4 (1887): 207-8. Dalhousie University Library, Robert Bell Papers, Lawson to Bell, 11 Oct. 1883; 12 Mar. 1887; 18 Dec. 1886; 28 Mar. 1887

41 James Fletcher, 'Notes on the Flora Ottawaensis, with special reference to the introduced plants,' Ottawa Field-Naturalists Club, *Transactions* 2/1 (1884): 29. Lawson, 'On the Present State of Botany in Canada,' RSC, *Transactions* ser. 1, vol. 9, sec. 4 (1891): 19. J[ohn] B[ell], 'Home Botany,' CN and G n.s. 7/2 (1875): 125-7. Drummond, 'Review of Edward Tuckerman, *Genera Lichenum*,' ibid. n.s. 7/1 (1872): 54-5

42 DP, George Barnston to Dawson, 19 June 1871; Hooker Papers, Barnston to Hooker, 3 Apr. 1868, 174. Smithsonian Institution Archives, HBC Correspondence, George Barnston to S.F. Baird, 26 Jan. 1860. CN and G n.s. 6/2 (1872): 222

43 Brunet refused to contribute to Provancher's *Le Naturaliste canadien*; Provancher, for his part, attacked Brunet's *Éléments de botanique* as 'un ouvrage où la science est méconnue, dont la diction est des plus vicieuses, et où la grammaire est horriblement maltraitée'; 2/5 (Apr. 1870): 145.

44 Bell, 'Home Botany,' 126. John Sommers in Nova Scotia agreed; Henry How, 'Additions to the List of Nova Scotian Plants,' NSINS, *Proceedings and Transactions* 4/3 (1877): 320-1. CN and G n.s. 8/8 (1878): 447; DP, 2211/60, ref. 62, Robert Bell to Dawson, 9 Apr. 1878; 2211/61, ref. 9, 16 May 1878

45 Macoun Papers, W.J. Hooker to Macoun, 7 Aug. 1861; Judith Dean Godfrey, 'Notes on Hepaticae collected by John Macoun in southwestern British Columbia,' *Canadian Journal of Botany* 55/20 (15 Oct. 1977): 2600-4. Macoun, *Catalogue*, Part I, v; D.P. Penhallow, 'A Review of Canadian Botany from 1800 to 1895,' RSC, *Transactions* Ser. 2, Vol. 3, Sec. 4 (1897): 25

46 Gray, 'Characteristics of the North American Flora,' in *Scientific Papers*, II: 261.

47 Charles Clarke, *Sixty Years in Upper Canada* (Toronto 1908), 57; J.C. Dent, *The Canadian Portrait Gallery*, IV (Toronto 1881), 147. NAm, 'Is the Earth Full of Seeds?' 23 Oct. 1850; 'Vegetable Instinct,' 30 Oct. 1850; 'Social Relations-The Daisy,' 8 Nov. 1850; 'Why Is the Garden More Fertile than the Field?' 6 Dec. 1850;

Notes to pages 258-60

'The Mysteries of a Flower,' 22 Sept. 1853; 'Crowther the Botanist,' 15 Sept. 1853

48 R.H. Lowie, *The History of Ethnological Theory* (London 1937), 20; Canadian Institute, *Archaeological Report* (1899) 'An Old Letter about the Origin of the Indians,' 164-5; P.R. Sloan, 'The Idea of Racial Degeneracy in Buffon's *Histoire naturelle*,' in *Studies in Eighteenth-Century Culture*, III (Cleveland 1973), 293-322. M.T. Hodgen, *Early Anthropology in the Sixteenth and Seventeenth Centuries* (Philadelphia 1964), chaps. 7-8; R.E. Bieder, 'Albert Gallatin and the Survival of Enlightenment Thought in Nineteenth-Century American Anthropology,' in T.H.H. Thoresen, ed., *Toward a Science of Man* (The Hague 1975), 97; M. Harris, *The Rise of Anthropological Theory* (New York 1968), 84-5. For much the same reason, while the idea of 'national character' and its varieties was widely accepted, environmental and geographical explanations, like those of Bodin and Montesquieu, gave way to social and historical explanations like those of David Hume: J.J. Honigmann, *The Development of Anthropological Ideas* (Homewood, Ill., 1976), 99-100. C.M. Hinsley, *Savages and Scientists* (Washington 1981), 21. J.C. Pritchard, 'On the various methods of research which contribute to the advancement of ethnology and of the relations of that science to other branches of knowledge,' BAAS, *Annual Report* (1847): 231, 236; before the BAAS created a section for ethnology in 1846, anthropology was included with botany and zoology.

49 'Climatic Influences,' CAg (Feb. 1854): 57-8, preferred the theories of Edouard Desor (1811-82), a Swiss geologist and archaeologist who accompanied Louis Agassiz to America in 1847 and returned home after falling out with Agassiz. Before departing, Desor explored the mining regions of Lake Superior; his reports were subsequently published in Swiss and German journals. See *La grande Encyclopédie* (Paris, n.d.), XIV: 263; Larousse, *Grand Dictionnaire universel du XIXe Siècle* (Paris, 1870), VI: 575. Gerald Killan, 'The Canadian Institute and the Origins of the Ontario Archaeological Tradition, 1851-1884,' *Ontario Archaeology* 34 (1980): 3-5. Dr Hodgkin and Richard Cull, 'A Manual of Ethnological Inquiry; being a series of questions concerning the Human Race, ... adapted for the use of travellers and others in studying the Varieties of Man,' BAAS, *Annual Report* (1852?), 243-52

50 James Bovell, *Outlines of Natural Theology for the Use of the Canadian Student* (Toronto 1859), 578-81, 434-9. Bovell quoted Wallace, 'On the Natural History of the Aru Islands,' *Annals and Magazine of Natural History* 20/121 (Dec. 1857): 480-2, which tested Wallace's evolutionary hypothesis first enunciated in 2nd ser. 16 (1855): 143-53. Bovell also criticized similar tendencies in Charles Darwin's *The Voyage of the 'Beagle'*. He shortened Darwin's work without inserting ellipses but was certainly not referring to Erasmus Darwin, as one author states; cf. A.B. McKillop, *A Disciplined Intelligence* (Montreal 1979), 76.

51 Wilson, *Prehistoric Man*, II (London 1862), App. B, 425-6, 478. A typical specimen

334 Notes to pages 260-3

of the 'New England' physiognomy was, for Wilson, Thomas Sterry Hunt, chemist to the Geological Survey of Canada.
52 Ibid., I: 17, 245
53 Wilson, 'Race Head-Forms and Their Expression by Measurements,' CJ n.s. 12/65 (Apr. 1869): 269; 'Hybridity and Absorption in Relation to the Red Indian Race,' CJ n.s. 14/87 (Mar. 1875): 432, 463. In 1836 Sir Francis Bond Head had described the fate of Indians in British North America using plant comparisons: 'Even where the Race barely lingers in existence, it still continue[s] to wither, droop, and vanish before us like Grass in the Progress of the forest in Flames'; quoted in *Canada and Its Provinces*, V (Toronto 1914), 338. In 1866 John Macoun compared the fate of native Canadian plants to that of the Indian: 'species by constant cultivation become so highly *civilized* that like the Anglo-Saxon they can adapt themselves to any soil ... Now on the other hand our native weeds are destroyed by cultivation like the Indians[;] if you try to civilize them you destroy them'; Hooker Papers, BrNAm Letters, Macoun to Hooker, 7 Aug. 1866; cf. Agassiz, *Lake Superior*, 10 and Wilson 'Hybridity,' 464-6
54 W.H. Thompson, 'The Physique of Different Nationalities as ascertained by inspection of Government recruits,' reprinted in CJ n.s. 9/50 (Mar. 1864): 130
55 'The Analogy between Plants and Animals,' CAg 14/5 (1 Mar. 1862): 141-2; Agassiz, *Lake Superior*, 152. Hooker to Darwin, 2 Nov. 1862, no. 355 in Francis Darwin, I: 466. Hooker abandoned these forays into anthropological theory only when he discovered that the German embryologist K.E. von Baer (1792-1876) at Königsberg had preceded him along these lines; see nos. 356-7, 467-9.
56 PAM, Schultz Papers, no. 106, Mair to Schultz, 23 Apr. 1868
57 'The Pines' was reprinted in JEUC (Nov. 1864): 168-9; it was collected in Mair's *Dreamland and Other Poems* [1865] (Toronto 1974), 11-16. Mair also published anonymous nature studies in Canadian periodicals; see Norman Shrive, *Charles Mair: Literary Nationalist* (Toronto 1965), 19-21. Schultz Papers, no. 35, Mair to Schultz, 14 May 1866; *The Nor'Wester*, 22 Jan. 1862. In 1855 Mair's plans to enter the HBC were foiled when relatives interfered because he was too young; Shrive, *Charles Mair*, 20.
58 PAM, *The Nor'Wester*, 22 Jan. 1862; W.L. Morton, 'Two Young Men, 1869: Charles Mair and Louis Riel,' HSSM, *Transactions* 3/30 (1973-4): 34
59 Loren Eiseley, *Darwin's Century* (New York 1958), 10-11. Shrive, *Charles Mair*, 5-8. Dent, *Canadian Portrait Gallery*, 132-3n; Clarke, *Sixty Years in Upper Canada*, 56
60 Carl Berger, *The Sense of Power: Studies in the Ideas of Canadian Imperialism, 1867-1914* (Toronto 1970), 53; 'The True North Strong and Free,' in Peter Russell, ed., *Nationalism in Canada* (Toronto 1966), 3-26. Robert Grant Haliburton, *The Men of the North and Their Place in History* (Montreal 1869), 2

61 T.C. Haliburton, *An Address on the Present Condition, Resources and Prospects of British North America* (Montreal 1857), 13
62 Haliburton, *Men of the North*, 1-2
63 Mair Papers, Haliburton to Mair, 14 July 1869; Schultz Papers, no. 87, Mair to Schultz, 7 Dec. 1867
64 Haliburton, *Men of the North*, 1, 10; *Speech to the Ottawa Literary and Scientific Society on the Young Men of the New Dominion* ([Ottawa?] 1870). Mair Papers, Haliburton to Mair, 14 July 1869
65 Lawson, 'On the Present State of Botany in Canada,' RSC, *Transactions* (1891): 18-19
66 Ibid., 19-20
67 Rev. E.H. Ball, 'The Indigenous Ferns of Nova Scotia,' NSINS, *Proceedings and Transactions* 4/1 (1875): 147
68 NAm, 3 Jan. 1851
69 'Nebenhoukha' to editor, Toronto *Daily Telegraph*, clipping in PAO, William McDougall Papers, Vol. I, 131, mfm; Ruth Bleasdale, 'Manitowaning: An Experiment in Indian Settlement,' *Ontario History* 66/3 (Sept. 1974): 157. McDougall led a chequered trans-partisan political career that earned him the reputation of an 'impossible partisan' among contemporaries. He remains a quintessential 'frontier radical' to most historians: R.C. Brown, ed., *Upper Canadian Politics in the 1850s*, Canadian Historical Readings No. 2 (Toronto 1967).
70 Mary B. Hesse, *Models and Analogies in Science* (Notre Dame 1966), 161-3. William Whewell, *The Philosophy of the Inductive Sciences*, I (London 1847), 494

CONCLUSION

1 James Guillet, 'Nationalism and Canadian Science,' in Peter Russell, ed., *Nationalism in Canada* (Toronto 1966), 221
2 See for example J.D. Bernal, *Science in History*, 4 vols. (London 1969), II: *The Scientific and Industrial Revolutions*, 518ff; John Herman Randall, Jr, *The Making of the Modern Mind*, revised ed. (Cambridge, Mass., 1940), 345-6; Henry Steele Commager, *The Empire of Reason* (New York 1978).
3 C.J.S. Bethune and J.M. Jones, 'Nova Scotian *Lepidoptera*,' Nova Scotian Institute of Natural Science, *Proceedings and Transactions* 2/3 (1869): 78; J.T.H. Connor, 'Of Butterfly Nets and Beetle Bottles: The ESC, 1863-1960,' HSTC *Bulletin: Journal of the History of Canadian Science, Technology and Medicine* 6/3 (Sept. 1982): 162
4 Henry Moyle, Brantford to ed., CAg, 4/1 (Jan. 1852): 24-6; 6/1 (Jan. 1854): 10-11; 3/6 (June 1851): 123
5 George Basalla, 'The Spread of Western Science,' *Science* 156 (5 May 1967): 611-22; Roy MacLeod, 'On Visiting the "Moving Metropolis": Reflections on the Archi-

tecture of Imperial Science,' *Historical Records of Australian Science* 5/3 (1982)
6 Arthur H. Williamson, 'Scotland, Antichrist and the Invention of Great Britain,' in *New Perspectives on the Politics and Culture of Early Modern Scotland* (Edinburgh [1982]), 51. This source was given to me by John H. Noble.
7 F. Kenneth Hare, 'The Re-Exploration of Canada,' *The Canadian Geographer* 1 (1954): 85-8
8 Marlene Shore, 'Post-Darwinian Biology and the Metropolitan Thesis: The Impact of Science on Historical Thought,' paper presented to the Graduate History Colloquium, University of Toronto, 17 Mar. 1986; see also Shore, *The Science of Social Redemption: McGill, the Chicago School, and the Origins of Social Research in Canada* (Toronto 1987). On the 'nation-building' school, see Carl Berger, *The Writing of Canadian History* (Toronto, 1976; revised ed. 1986). For scientific roots of the Frontier thesis, see William Coleman, 'Science and Symbol in the Turner Frontier Hypothesis,' *American Historical Review* 72/1 (Oct. 1966): 22-49; Ray A. Billington, *The Genesis of the Frontier Thesis* (San Marino, Calif., 1971), chap. 5. The classic expression of Harold Innis's economic determinism is *The Fur Trade in Canada* (Toronto 1930), esp. 393. J.M.S. Careless, 'Frontierism, Metropolitanism, and Canadian History,' CHR, 35/3 (Sept. 1954): 1-21. D.G. Creighton's Laurentianism in *The Commercial Empire of the St. Lawrence* (Toronto 1937) echoed J.W. Dawson, that 'Others, ... beside practical merchant men, must regard with intense interest the curious way' life in British North America was dominated by 'our great river'; see his unsigned review in CN and G, 3 (1858): 394
9 Mary Shelley, *Frankenstein; or, the Modern Prometheus* ([1816]; Signet Classics, New York 1965), x

Note on Sources

Space limitations preclude the addition of a full-scale bibliography; this brief note is intended to indicate some of the major archival and primary sources consulted and can only touch on the wealth of secondary material available on this subject. References to material on more specialized questions are cited in the notes.

British archives are rich in material relating to science in Canada during the Victorian age: the Royal Botanical Gardens at Kew, London, house the extensive collections of Sir William Jackson Hooker, Sir Joseph Dalton Hooker, and their successors; the Royal Botanic Garden at Edinburgh has the John Hutton Balfour Papers; these collections include correspondence from British North American naturalists. The Sir Charles Lyell Papers, at the University of Edinburgh Library, contain letters from J.W. Dawson and others. The National Museum of Wales at Cardiff has letters from Sir William Logan in the Sir Henry De la Beche Papers. Collections of the directors-general of the Geological Survey of Great Britain are at the Institute of Geological Sciences; and the Lyon Playfair and A.C. Ramsay Papers are at the Imperial College of Science and Technology, London. The Public Record Office in London has valuable materials in the vast collections of the Metrological Office (some microfilm manuscripts concerning the Magnetic Survey are at the PAC), the War Office, the Foreign Office, and the Colonial Office. The Geological Society of London also has some letters concerning geology in British North America.

In the United States, the Smithsonian Institution has both the Joseph Henry and the Spencer Baird Papers on microfilm, as well as correspondence with Hudson's Bay Company fur traders during the 1850s and 1860s. The James Hall Papers are at the New York State Library, Albany, and contain correspondence with William Logan; a check of the Charles Upham Shepard Papers there failed to turn up evidence that Shepard had been offered the directorship of the Geological Survey of Canada prior to Logan's appointment.

In Canada, the Public Archives in Ottawa (Public Records Division) has a large collection of Directors' Letterbooks and other material of the Geological Survey of Canada; the Robert Bell and Sir Sandford Fleming Papers and minute-books of the British North American Mining Company are in the Manuscripts Division. There is also a small set of John Macoun Papers, but his later papers are at the National Museum of Man. Background information can be found in government records at PAC, including Letterbooks of governors general and provincial secretaries, and correspondence of civil secretaries; the Upper Canada State Papers, Series G and Q; papers of Sir John Colborne, Earl of Dalhousie, Lord Durham, Lord Sydenham, Sir Charles Bagot, and Sir John A. Macdonald. The Molson Company of Montreal kindly gave permission to consult its collections at PAC.

The Public Archives of Ontario has records of the Crown Lands Department, the Department of Education, and the Canada Company, all of which provide useful insights. The Thomas Fisher Rare Book Library at the University of Toronto has such journals as the *Canadian Naturalist and Geologist*, the *Canadian Journal of Science, Industry and Art*, the *Annals* of the Botanical Society of Canada, the *British American Journal*, the *Canadian Agriculturist*, and the *British American Cultivator*, as well as a wide range of other primary sources on Victorian science. The Science and Medicine Library at the University of Toronto has the *Transactions* of the Nova Scotian Institute of Natural Science, *Le Naturaliste canadien*, and many other journals. The Baldwin Room at Metropolitan Toronto Reference Library has the journals and scrapbooks of Sir William Logan, annual reports of the Montreal Mining Company, *The Canadian Review and Magazine*, and other useful pieces of Canadiana. The United Church Archives in Toronto has the Egerton Ryerson and J.G. Hodgins Papers and some published material such as Ryerson's speeches.

McGill University Archives has large and well-organized collections of Sir J.W. Dawson and Sir William Logan Papers, and a small set of Robert Bell correspondence. The Blacker Wood Rare Book Library at McGill has minute-books of the Natural History Society of Montreal. The Archive du Séminaire de Québec has papers of Abbé L.-O. Brunet; and those of Abbé Léon Provancher are at the Archive du Séminaire de Chicoutimi.

Dalhousie University Library in Halifax has small collections of Robert Bell and George Lawson Papers; the Centre for Newfoundland Studies at Memorial University also has some letters of Robert Bell. Queen's University Archive has a useful collection of A.T. Drummond Papers.

In addition to private papers and periodical publications, newspapers offer a wealth of information on attitudes to science in the context of other issues and problems of the day. The best starting-point is perhaps the Montreal *Gazette* because of its longevity, but a wide political and regional selection provides the best overview. Almost as important for filling in gaps and identifying individuals are the *Dictionary of Canadian Biography* and its British, American, and scientific counterparts; local histories are often indispensable for the specialized information they provide. One study that combines the best qualities of biographical, local, and scientific approaches is Gerald Killan, *David Boyle: From Artisan to Archaeologist* (Toronto 1983).

Secondary sources in the history of science in Canada have only recently built upon the foundation set by H.M. Tory, ed., *A History of Science in Canada* (Toronto 1939); W.S. Wallace, ed., *Centennial Volume 1849-1949* of the Royal Canadian Institute (Toronto 1949); and G.F.G. Stanley, ed., *Pioneers of Canadian Science* (Toronto 1966). General introductions are Trevor H. Levere and Richard A. Jarrell, eds., *A Curious Field-Book: Science and Society in Canadian History* (Toronto 1974) and its companion volume, B. Sinclair, N. Ball, and J.O. Peterson, eds., *Let Us Be Honest and Modest: Technology and Society in Canadian History* (Toronto 1974). An official and elaborately illustrated history of surveying and mapping in Canada is Don W. Thomson, *Men and Meridians*, 2 vols. (Ottawa 1966). The official history of the Geological Survey of Canada is the indispensable *Reading the Rocks* (Ottawa 1974) by Morris Zaslow. Vittorio M.G. de Vecchi, 'Science and Government in Nineteenth-Century Canada,' PH D thesis, University of Toronto, 1978, focuses on the post-Confederation period.

Several studies of the intellectual history of Victorian Canada offer informative insights: W.L. Morton, ed., *The Shield of Achilles* (Toronto 1968); Carl Berger, *The Sense of Power* (Toronto 1970); and Doug Owram, *Promise of Eden* (Toronto 1980). A.B. McKillop, *A Disciplined Intelligence* (Montreal 1979) and Carl Berger, *Science, God, and Nature in Victorian Canada*, The 1982 Joanne Goodman Lecture (Toronto 1983), emphasize the negative reception by some Canadians of Darwin's theory of evolution by natural selection.

For background on the political history of the period: J.M.S. Careless, *The Union of the Canadas: The Growth of Canadian Institutions 1841-1857* (Toronto 1967); W.L. Morton, *The Critical Years: The Union of British North America 1857-1873* (Toronto 1964); Donald Creighton, *The Road to Confederation* (Toronto 1964); P.B. Waite, *The Life and Times of Confederation* (Toronto 1962).

On the place of science in Victorian culture, Susan Faye Cannon, *Science in Culture: The Early Victorian Period* (New York 1978) is a highly suggestive contribution. Two older works are still helpful: John T. Merz, *A History of European Thought in the Nineteenth Century*, 4 vols. (Edinburgh 1890); John Herman Randall, Jr, *The Making of the Modern Mind* (Cambridge, Mass., 1940). J.D. Bernal, *Science in History*, 4 vols. (London 1969) and Daniel J. Boorstin, *The Discoverers* (New York 1983) place developments in science into their larger historical context. Social attitudes to natural history are documented in David Elliston Allen, *The Naturalist in Britain: A Social History* (London 1976) and his *The Victorian Fern Craze* (London 1969); and Keith Thomas, *Man and the Natural World: Changing Attitudes in England 1500-1800* (London 1983). Robert M. Young, *Darwin's Metaphor: Nature's Place in Victorian Culture* (Cambridge 1985); John C. Greene, *Science, Ideology, and World View* (Berkeley 1981); and Michel Foucault, *The*

Order of Things: An Archaeology of the Human Sciences (New York 1970) offer incisive insights into relationships between science and ideology. Along these lines Morris Berman, *Social Change and Scientific Organization* (Ithaca, NY 1978) documents the utilitarian tradition of the Royal Institution. Among the most recent works are William H. Goetzmann, *New Lands, New Men: America and the Second Great Age of Discovery* (New York 1986), which discusses the contribution of American efforts to world exploration; Alfred W. Crosby, *Ecological Imperialism: The Biological Expansion of Europe, 900–1900* (Cambridge 1986); and Stephen Jay Gould, *Time's Arrow, Time's Cycle: Myth and Metaphor in the Discovery of Geological Time*, The Jerusalem-Harvard Lectures (Cambridge, Mass., and London 1987).

Index

'A.B.' *See* Baddeley, Frederick Henry
Acadian Geology (Dawson) 90
Agassiz, Louis 87, 242, 251-2
'Agricola.' *See* Young, John [pseud. Agricola]
Agricultural Association of Upper Canada 207
agriculture 88; and chemistry 4, 198; and mining 27; and science 149, 150-1, 190; improvement of 194-5, 199-200, 202, 215-16, 230; crisis (1840s) 200-2 (1850s) 215-16; education 206-7
Allan, George W. 132, 155
Allan, William 23, 132
American Association for the Advancement of Science (AAAS) 100-1, 143
American Fur Company 70
American Philosophical Society 119, 141, 143
American Sault Ste Marie Company 72
Anderson, David (bishop of Rupert's Land) 98
annexationism 74-5
anthropology 4, 258-61; and botanical analogy 257, 259, 261, 267, 271, 333n53
Arago, François 148

Arctic: Victorian public's interest in 161
Armour, Andrew H. 33
Armour, Robert, Jr 33
Armour, Robert, Sr 33
Ashe, Edward 169
astronomy 121-2
Audubon, John James 126-7
aurora borealis 122, 129, 135, 138-9, 157, 303n20
Aylwin, Thomas 40-1, 66-7

Bache, Alexander Dallas 123, 143
Bacon, Francis 14, 117, 147, 262
Baconianism 4, 14, 37, 46, 49, 145, 146, 185, 190; and incremental science 22, 37, 126, 135, 145-6
Baddeley, Frederick Henry 16, 26, 28-9, 39
Badgley, William 33, 48
Bagot, Sir Charles 41-2, 46-8 *passim*
Baldwin, Robert 63, 142, 155, 207
Balfour, John Hutton 218-19, 220, 229-30, 237, 256, 273
Banks, Sir Joseph 189-90
Barnston, George 219, 220, 221, 224-5, 227, 252, 256, 265
Barnston, James 184, 230, 245, 252; early

career of 219–20; botanical activities in Montreal 220–2; appointed to chair of botany at McGill College 222–3; legacy of 223
Barnston, John George 245
Barrie Northern Advance 165
Bayfield, Henry Wolsey 120, 124
Beadle, D.W. 242
Bédard, Thomas 227
Beddoes, Thomas 34
Bell, Rev. Andrew 84–5, 99
Bell, John 255–6
Bell, Robert (Carleton Place) 84
Bell, Robert (GSC) 84, 106–7, 108, 226, 234, 247–8, 255
Bell, Robert (*Ottawa Citizen*) 85–6, 90
Bethune, Rev. Charles 272
Bigsby, John Jeremiah 17–18
Billings, Elkanah 86, 101, 249; and botany 214–15
biogeography. *See* botany: and plant geography
Blakiston, Thomas 171, 175
Blodget, Lorin 151, 173, 174, 251
Bonnycastle, Sir Richard Henry 16, 26, 122
Bonpland, Aimé de 188, 233
Bonplandia 233–4
Bos, Abbé J.B. du 172
botanic medicine. *See* Thompsonian medicine
Botanical Club of Canada 265
Botanical Society of Canada (BSC) 230–5, 236–7, 238, 247, 248, 250, 262, 265; purposes of 231, 234; international reception of 233–4
Botanical Society of Edinburgh 218
Botanical Society of Montreal 220, 221, 223
botany: and Canadian development 184, 198, 257, 260–1, 265–6, 266–8, 271; and Canadian public opinion 184, 189, 197, 201, 203, 208, 211, 230–1, 271; inventorial organizational problems 184–5, 250, 255–7, 265; branches of 185; and Linnaean classification 185–7 *passim*; and natural systems of classification 186–7, 189, 205, 209–10; and comparative plant morphology 187; and early exploratory voyages 188, 189, 190, 192; and 'devolution' of man in North America 188, 193, 258–61, 266; and plant geography 189, 205, 206, 214, 219, 223–5, 231, 232, 237, 243–7, 251, 252, 253–4, 257, 261, 265–6, 271; economic importance of 189, 191, 197; early study in British North America 190–2; and the Hudson's Bay Company 192, 197, 212; arguments for a Canadian botanical inventory 193–6 *passim*, 216–17; and Natural History Society of Montreal 193–4, 196, 197, 252, 256; and agricultural improvement 194–5, 199–200, 202, 215–16, 230; in the United States 196–7; and botanical gardens in British North America 198, 233, 234, 238, 249–50; and religion 199, 200–1, 204, 218, 221, 258–9; and transmutation of species 199–200, 219; and the agricultural press 200–3 *passim*; and plant disease 200–1, 215–16; and provincial agricultural associations 201–2; taught at Canadian universities 206–11 *passim*, 216–17, 218–19, 220, 222, 223, 228–9, 230, 237–8, 247, 264–5; and historical consciousness 225, 265; and the Geological Survey of Canada (GSC) 226, 249–50; and the Botanical Society of Canada (BSC) 230–5; and Canadian government funding 235–7, 249, 265; and

national symbolism of plants 240-1; and Canadian fruit culture 241-3; and Charles Darwin 244-6 *passim*, 265; and expansionism 219, 224, 252-5 *passim*, 262. *See also* cell theory; palaeobotany; quinarianism; teratology
Boulton family 131
Bourgeau, M.E. 244, 252-3
Bovell, James 259
Bowmanville scandal (1858) 95-6
Brewster, David 121, 153
British American Cultivator 200, 201, 202
British American Journal of Medical and Physical Science 71, 74, 77, 137, 138, 139, 199, 233
British American League 75, 96
British American Mining Company 67
British Association for the Advancement of Science (BAAS) 39; and international chain of stations and observatories 119, 126, 134-5
British imperialism 272, 295n18; and geology 16, 17, 19, 39, 48-9, 52-3, 65, 91, 93, 105, 110; and physical science 116, 119, 177-8, 272; and agriculture 148, 201-2; and botany 184
British North American Company 71
Brondgeest, John T. 33, 40
Brontes canadensis 61
Brown, George 72-3, 96-7, 98, 155
Brown, James 133
Brown, Robert 192
Brunet, Abbé Louis-Ovide 227, 228, 239, 248, 249-50, 256
Buckland, George 202-3, 207, 208, 209, 215
Buckland, Rev. William 14, 39, 42, 45, 48
Buffon, Comte Georges Louis Leclerc de 186, 188-9, 258, 259
Buffon's Law 188-9, 224

Burke, Joseph 212, 229
Burwell, Mahlon 24-5, 29, 120
By, John 16, 21
Bytown Gazette 198

Cameron, Malcolm 77
Campbell, Sir Alexander 107, 233
Canada and Her Resources (Morris) 173
Canada Company 23-4; Huron Tract 19-21, 23-6 *passim*
Canada First movement 104; and the myth of the north 262-4, 266
Canada Land Company. *See* Canada Company
Canadensium plantarum (Cornut) 191
Canadian Agriculturist 202-3, 209, 233, 241, 257
Canadian Institute, Royal 93, 115, 154-5, 157, 158, 159, 160, 162-3, 164, 165, 169, 170, 173, 213, 214, 234, 238
Canadian Journal of Industry, Science and Art 100, 155, 159, 164, 212, 214, 226, 238
Canadian Naturalist and Geologist 7, 103, 169, 221, 222, 223, 224, 225, 233, 238, 249, 253
Canadian Pacific Railway 73, 83; proposed 73-4
Canadian Review and Magazine 18, 31, 193
Canadian Shield 49, 52, 55, 251, 270. *See also* Pre-Cambrian geology
Canadian Statesman (Bowmanville) 91-2
Candolle, Augustin-Pyramus de 187, 188, 189, 205, 209-10, 211
Carpenter, W.B. 103
Carthew, John 25, 26
Cartier, Sir George Étienne 8, 79
Catalogue des plantes (Brunet) 249
Catalogue of Canadian Plants (Macoun) 257

Cathcart, Charles Murray, second Earl 67, 136
Cauchon, Joseph 95, 96
cell theory 185
Chapman, E.J. 99–100
Charlevoix, Pierre François Xavier de 171
Cherriman, J.B. 155, 160, 162, 164
Chippewa nation. See Ojibway nation
Chisholme, David 18–19, 30, 31, 32, 33, 35, 38–9, 48; support for William Logan 42; advocates a Canadian botanical inventory 193
Christie, A.J. 17, 49–50
Clarke, Charles 266
Clear Grit Party 266–7
climatology 115, 144, 162, 163, 167, 168, 242; and theory of climatic progress 53, 122–3, 140, 156, 171–3, 174–5; and epidemics 123, 136–8; and H.Y. Hind 150–1, 155, 168, 173–6 passim, 213; and territorial expansion 171, 174–6, 178, 179; and isotherms 173–5 passim; and palaeobotany 225–6; role in human variation 258–61, 264, 266; and terrestrial magnetism 311n34
Climatology of the United States (Blodget) 151, 173, 174
Cloutier, J.B. 227
coal 29, 42–3, 49, 54, 66, 80–1, 110, 111, 112; in Nova Scotia 8, 13, 25, 44, 57, 64–5, 106, 108; sought in Canada 27, 29, 56–7, 58–9, 61–2, 64, 95–6; in New Brunswick 39, 57; in Gaspé 57, 58, 60, 61, 64; and Canadian economic stability 74; Bowmanville scandal (1858) 95–6; British reserves 105–6; in Newfoundland 106–7
Coal Question (Jevons) 106
Cochrane, William 194

Colborne, Sir John 23–5 passim, 121, 197, 198
Colebrooke, Sir William 135
Comparative View of the Climate of Western Canada (Hind) 150–1
Compositae 247
Confederation: and scientific metaphors 6–7; and inventory science 8–9; and GSC 105, 108–9, 110–12, passim; and terrestrial magnetism 180; and botanical inventory 253–5; and Victorian science 270
Cook, Capt. James 118, 171, 188, 189, 190
copper 13, 54, 56, 59, 61, 65–6, 72, 75, 80, 98, 110; Bruce Mines 82
Corn Laws 190, 201
Cornut, Jacques 191
Couper, William 216–17, 218
Cours d'histoire du Canada (Ferland) 172–3
Craigie, William 137, 163, 214, 242
Creighton, Donald 274
Croft, Henry H. 135, 155, 200
Cruciferae 224
cryptogams 186, 187, 197, 223, 226–7, 229, 244, 249, 250, 256

Dade, Charles 122–3
Dalhousie, George Ramsay, ninth Earl of 35, 122, 194
Dalhousie, Lady 193, 195
Dalhousie University 238, 248
Dalton, Thomas 24, 25, 30, 198, 199
Darwin, Charles 103, 204, 244–6 passim, 261, 265, 266, 268
Darwin, Erasmus 183
Davy, Sir Humphry 35
Dawn of Life (Dawson) 104
Dawson, George Mercer 254
Dawson, Sir John William 44, 57, 88,

90, 95, 96, 99, 100–1, 102–5 passim, 107, 108, 111, 168–9, 216, 221, 223, 234, 247, 273, 295n17; and palaeobotany 206, 225–6; and plant geography 223–4; and variation in species 245–6
Dawson, William McDonell 175
De la Beche, Sir Henry 14, 45, 48, 53, 54–5, 60, 83–4, 88, 206, 273
Delisle, Auguste 227
De Magnete (Gilbert) 117
De philosophiae naturalis principia mathematica (Newton) 117
Desor, Edouard 333n49
Douglas, David 190–1, 195
Dove, Wilhelm 170
Draper, William 97, 155, 180
Drew, John 165
Drummond, Andrew 248
Drummond, Andrew Thomas 233, 234, 248–9, 250, 251, 252, 255
Drummond, Thomas 195
Dufrenoy, M. 82
Duncombe, Charles 26, 33
Dunkin, Christopher 79
Dunlop, Robert Graham 27–8, 29, 33, 39
Dunlop, William ('Tiger') 20–2, 33, 39, 58–9, 63, 198; and the Huron Tract 20–1, 23; lobbies for government support of a geological survey 22
Dupont, Emilien. *See* Provancher, Abbé Léon
D'Urban, W.S.M. 226
Durham, John George Lambton, first Earl of 34–7, 46, 177, 202; and Natural History Society of Montreal 35–6; *Report on the Affairs of British North America* 37

Edinburgh Botanical Club 218
electromagnetism 117, 123, 138

Elements of Botany (Gray) 196
Elgin, James Bruce, eighth Earl of 146, 152, 156, 159, 202
Ellice, Edward 19, 98
Emmons, Ebenezer 58
Entomological Society of Canada 271
entomology 4, 210, 215–16, 271–2
Eozoon canadense controversy 51, 103–4
epidemics: and climatology 123, 136–8
Espy, James Pollard 116, 140
Essai sur la géographie des plantes (Humboldt and Bonpland) 188
expansionism 53–4, 96–9, 271; in uniformitarian geology 53–4, 107–9; in geological maps 90, 108; and meteorology 155, 171, 174–6, 178–9; and climatology 171, 174–6, 178, 179; and botany 219, 224, 252–5 passim, 262

Family Compact 131–4 passim, 145, 146, 152, 178
Faraday, Michael 117, 123, 177
Fauna Boreali-Americana (Richardson) 210
Ferland, Abbé J.B.A. 172–3
Ferrier, James 82
Fillmore, Millard 143
Fleming, Sir Sandford 83, 93, 155, 240, 253–4
Fletcher, James 217
Flora Americae septentrionalis (Pursh) 191, 196
Flora Boreali-Americana (Michaux) 191
Flora of North America (Torrey and Gray) 196, 206
Flore canadienne (Provancher) 236
Forbes, Edward 205–6, 243, 244, 245, 248, 251
Forster, Georg 171
Fothergill, Charles 22, 198

Franklin, Sir John 161, 195
Fredericton, N.B.: observatory 121-2, 135, 169
Freemasonry, Montreal 45-6
French Canadians: and science 6, 33-4, 87, 95, 153, 227
Fruit Growers' Association of Upper Canada 241-2

Galt, Sir Alexander Tilloch 101, 102, 105
Galt, John 19-21 *passim*
Gaspé Fishing and Coal Mining Company 64
Gauss, Carl Friedrich 116, 119, 177
Genera plantarum (Jussieu) 186-7
Geological Society of London 15-16, 60, 98
Geological Survey of Canada (GSC): origins 16; public interest in 83, 85, 91-2; and railway surveys 83; public criticism of 86-7; funding crisis of 1861 101-2; and Confederation 105, 108-9, 110-12 *passim*; and palaeobotany 206; and botany 226, 249-50. *See also* Logan, Sir William Edmond
geology, conceptual development 13-14, 43; and Scottish Enlightenment 14-15; institutionalized 15; uniformitarianism 43-4; expansionist outlook 53-4; Palaeozoic systems 55; Carboniferous 55. *See also* coal; copper; GSC; geosyncline concept; Laurentian series; Logan, Sir William Edmond; Lyell, Sir Charles
Geology of Canada (Logan) 52, 102, 103, 105
geomagnetism. *See* magnetism, terrestrial
geosyncline concept 100
German Magnetic Association 119
Gesner, Abraham 14, 38-9

Gibbon, Edward 172, 176
Gibson, David 26, 33
Gilbert, William 117
ginseng 191, 192, 253
Glackmeyer, Louis-Edouard 227
Glenelg, Charles Grant, Baron 196
Goethe, Johann Wolfgang von 211
Goldie, John 192, 195
Gordon, Alexander 198-9, 243
Gosford, Archibald Acheson, second Earl of 121
Graham, Andrew 192
Graham, Robert 195-6
Grand Manitoulin Island Improvement Company 107
Gray, Asa 196, 206, 211, 245, 246, 257, 273
Great Exhibition (London, 1851) 78-82; organization of 79; Canadian preparations 79-80; Canadian mineral exhibit 80-1, 82; reception of Canadian exhibit 81-2
Green, Anson 147-8
Griffin, Frederick 38
Guyot, François 142

Hagerman, Christopher 132
Hagerman, Mary Jane ('Polly') 132
Haliburton, Robert Grant 263-4
Haliburton, Thomas Chandler 263
Halifax *Acadian Recorder* 194
Hall, Archibald 137-8, 139
Hall, James 58, 85, 90, 100, 273
Hamilton Gazette 26-7
Hamilton Spectator 165-6
Hansteen, Christopher 118
Harris, John 120
Harrison, S.B. 67
Harvey, Sir John 135
Haüy, Abbé René-Just 187

Index

Head, Sir Edmund Walker 89-90, 93, 169, 236
Head, Sir Francis Bond 26, 28, 29-30, 121, 140, 196
heather (*Caluna vulgaris*) 246
Hector, Sir James 97, 105-6, 230, 252-3
Henry, Joseph 116, 121, 138, 142-3, 164, 174
Henry, William 127-8, 129
Herschel, Sir John 134-5
Hill, George 216
Hincks, Sir Francis 72, 76, 77, 90, 152
Hincks, Thomas Dix 152
Hincks, Rev. William 152, 208, 213, 216, 224, 228, 242, 271; appointed professor of natural history, University of Toronto 209; intellectual rigidity of 209, 210, 211, 216-17, 256; influenced by John Lindley 209-10, 211, 227, 239; and quinary classification 210; interest in teratology 210-11, 227; philosophical idealism of 227; and the Botanical Society of Canada (BSC) 238-9, 248
Hind, Henry Youle 97, 100, 109, 155, 159, 167, 200, 208, 216, 225, 252, 253, 262, 271; as teacher 149-50; as publicist for scientific education 150-2; and climatology 150-1, 155, 168, 173-6 passim, 213; and botany 213
Histoire des Celtes (Pelloutier) 172
Histoire naturelle (Buffon) 186
History of England from the Invasion of Julius Caesar (Hume) 172
History of the Decline and Fall of the Roman Empire (Gibbon) 172
Hitchcock, Edward 32, 47
Hodder, Edward M. 212-13
Hodgins, John George 147, 165
Hofmeister, Wilhelm 226
Hogan, John Sheridan 172, 176, 211-12
Holmes, Andrew Fernando 32-3, 36, 38, 40, 47-8, 194, 195, 222, 224
Holmes, Benjamin 40, 41
Holton, Luther 102
Honeyman, David 108
Hooker, Sir Joseph Dalton 189, 204, 218, 237, 254, 256, 273; and arctic flora 205, 244-6, 261, 262; and palaeobotany 206; theories of plant distribution 243-6, 251, 255, 261, 265; influenced by Charles Darwin 244-6 passim, 261
Hooker, Sir William Jackson 184, 189, 197, 205, 257; encourages botanical investigation in British North America 192-3, 220-1; and *Flora Boreali-Americana* 195, 197, 212; and George Lawson 229, 235, 236; plans update of BNA floras 235-7, 239
Hope, Thomas Charles 15
Horae entomologicae (Macleay) 210
Horan, Rev. Edouard 87
Houghton, Douglass 59-60, 65
How, Henry 108
Hudson's Bay Company (HBC) 5, 68, 70, 71, 72, 73, 96-8, 163, 170, 253, 267; and meteorology 123, 137, 143, 170; and Lefroy's journey (1843-4) 127-8, 130; and botany 192, 197, 212
Humboldt, Alexander von 116, 118-19, 148, 151, 170, 171, 173, 179-80, 188
Hume, David 172
Hunt, Thomas Sterry 77, 88, 103, 110, 221, 333n51
Hurlburt, J.B. 200, 241, 242-3, 257
Huron Tract. *See* Canada Company
Huronian formations 100. *See also* Laurentian series
Hutton, James 13, 15, 43
Huxley, Thomas 103, 208

Industrial Exhibition (New York, 1853) 82-3

industrialization 6, 27, 35, 74, 96; in Scotland 15; ideology of 34, 37, 269-70
Innis, Harold 274
Institute of Rupert's Land 170, 253
inventory science 52, 83, 269-74 *passim*; tradition 4-5; and utilitarianism 5-6; and Confederation 8-9. See also natural history; science
iron ore 49, 59, 80-1, 91, 99
Isbister, Alexander Kennedy 98
isodynamic ovals 127, 128

Jack, William Brydone 121-2, 169-70
Jameson, Robert 15
Jardin des plantes (Paris) 191
Jarvis, W.B. 67
Jevons, William Stanley 106
Johnston, J.F.W. 203
Jones, Thomas Mercer 23-4
Journal (Charlevoix) 171
Journal of Education for Upper Canada 147, 150, 151, 153, 154, 155, 174
Jukes, Joseph Beete 14, 48, 106
Jussieu, Antoine Laurent de 186-7
Jussieu, Bernard de 186-7, 205

Kalm, Peter 191
Kelly, William 122, 172
Kemp, A.F. 221, 226
Kennedy, William 98
Kew Gardens (London) 190, 235, 236, 237, 247, 250, 257, 273
Killaly, Hamilton H. 40, 71-2
Kingston, George Templeman 162, 164-5, 167-8, 170, 173, 174; and telegraphic storm warning system 165, 168, 169
Kingston (Ont.): early interest in a geological survey 28
Kingston Chronicle and Gazette 28

Lachlan, Robert: advocates chain of meteorological stations 163-4; and storm alarm system 164
Lake Huron: and early geology 17
Lake Superior: search for copper 59-60, 65-72
Lake Superior (Agassiz) 251-2
Langton, John 83, 89, 146-7, 155, 168
Laplace, Pierre Simon, Marquis de 150
Laurence, Abbott 143
Laurentian series 88, 99, 100, 102, 104, 108, 109. See also Canadian Shield
Lawson, George 184, 228, 252, 253, 256, 262, 263, 265, 272; early career of 229-30; and Sir William Hooker 229, 235, 236; association with Queen's University 230, 237-8; and the Botanical Society of Canada (BSC) 230-4, 238, 239, 247, 250; moves to Dalhousie University 238, 248; and Nova Scotia flora 246; plans a series of volumes on Canadian flora 255; proposes Botanical Club of Canada 265
Lefroy, Sir John Henry 98, 204-5, 212, 270, 272; and Edward Sabine 125, 126, 128-30 *passim*, 132, 140-3 *passim*, 146, 157, 158, 160, 177-8; appointed to conduct magnetic survey 125-6; and John J. Audubon 126-7; magnetic survey through HBC territory (1843-4) 127-8, 140, 173, 177; plans for global overland survey 129; plans for second journey to the Arctic north-west 129-30; and Toronto Magnetic and Meteorological Observatory 130, 135-6, 140-3 *passim*, 156-8, 160, 177-8; and Toronto society 131-4; marries Emily Robinson 132; conservative ideals of 133-4; and University of Toronto 135-6, 141; and ozone disease theory 138; and aurora

borealis 138–9; and Egerton Ryerson 143–4, 146, 172; and relationship of science to education 145–6; and H.Y. Hind 150–1; and storm alarm system 153, 157; and the Royal Canadian Institute 155, 158; and thermometric registers 155–6; and theory of climatic progress 156, 173; and territorial expansion 175–6

Leitch, William 231–2, 233, 238

Lindley, John 209–10, 211, 226–7, 239

Lindsay, William Lauder 228–9

Linnaeus, Carolus 185–7 *passim*, 189

Linnean Society (London) 190, 210, 244

Literary and Historical Society of Quebec (LHSQ) 36, 38, 40, 121, 122, 158, 246; and botany 193, 196

Literary and Philosophical Society of Newcastle-upon-Tyne 35

Literary and Philosophical Society of Upper Canada (Toronto) 22

Locke, John (American magnetician) 127

Locke, John (English philosopher) 186

Logan, Hart 19, 42

Logan, James 19, 45, 46, 67, 69

Logan, Sir William Edmond 5, 14, 17, 19, 41, 169–70, 171, 216, 247, 270, 272, 273; theories of *in situ* origins of coal 42–3, 54; early career of 42–3; and Charles Lyell 44, 62, 88, 107; attempts to secure position of provincial geologist 44, 45, 46, 48; and Montreal business connections 45, 68; and Freemasonry 45–6; appointed provincial geologist 48; work on Pre-Cambrian formations 49, 55–6; phases of career 52, 77, 93, 94; personal qualities of 52–3, 62; geological vision of 53; expansionist outlook of 53–4, 97, 105, 107, 108, 109; business sense of 54; interest in copper 54, 56, 59, 61, 65–6, 110; and Sir Henry De la Beche 54–5, 60, 83–4, 88; and public opinion 56–60 *passim*, 62, 64, 76, 83, 92; planning the survey 56–7; and the search for coal formations 56–7, 58–9, 60, 61–2, 64–5, 66, 95–8, 105, 110, 111; first 'Report of Progress' (1843) 57–60; and search for iron ore deposits 59–60; 'Report of Progress' (1844) 61; eccentric personal habits of 62–3; and governmental financial support for the GSC 63–4, 77, 93, 94, 101–2; duties as provincial geologist 64, 89; and Montreal Mining Company 68–70 *passim*; and the Great Exhibition (London, 1851) 78–82 *passim*; and Select Committee (1854) 85–9 *passim*; and Universal Exposition (Paris, 1855) 90–1; and geological map of British North America 90; named *Chevalier* of the Legion of Honour 91; receives knighthood 92; awarded Wollaston Medal, Geological Society of London 92; and the Hudson's Bay Company 97; and AAAS meeting (Montreal, 1857) 100–1; *Geology of Canada* (1863) 102; and *Eozoon canadense* controversy 103–4; and Newfoundland geological survey 108; and Nova Scotia survey 108; and New Brunswick survey 109; awarded Royal Society (London) Medal 109; and Confederation 109, 112; retirement of 109–10; legacy of 110–12 *passim*; and meteorology 139–40; and Royal Canadian Institute 155; and botany 220, 226, 233

Logie, Alexander 236, 241, 242

London Guardian 235

Loomis, Elias 116, 141, 143

Lyell, Sir Charles 14, 43–4, 55, 62, 88, 89,

103, 104, 107, 205, 225, 244, 246, 265; and botany 188, 189, 199

McCaul, John 136
Macaulay, James Buchanan 131
Macaulay, John 30
McCord, J.S. 33, 38, 40, 123, 140, 163
McCulloch, Michael 46-7
Macdonald, Sir John A. 67, 93, 233, 236
Macdonald, John Sandfield 102
Macdonell, Allan 70-1, 72, 75, 98; proposes transcontinental railway 73-4
McDougall, William 202-3, 207, 257-60, 262-3, 266-7
Macfarlane, James 30
McGee, Thomas D'Arcy 7, 52, 74, 111, 185
McGill, Peter 46, 67
McGill, Robert 207
McGill University 194; and botany 206-7, 218, 220, 222, 223
McGillivray, Simon 19
Macintosh, John 26, 33
Mackenzie, Alexander 128
Mackenzie, William Lyon 26, 27, 30, 33
Macleay, William Sharp 210
MacNab, Sir Allan 77, 83
McNab, James 195
Macoun, John 235, 239, 250, 251-2, 254; botanical tours of 252, 253-4; and expansionism 252-3; as publicist for the Canadian north-west 254; botanical reputation of 256-7
Magdalen Islands 75-6
magnetism, animal 121
magnetism, terrestrial 177, 270-1; study introduced to Canada 116; and international co-operation 116, 117, 119, 126; and navigation 116-17; and electricity 117; and Alexander von Humboldt 118-19; Canadian popular interest in 120-1, 130, 136-7, 139, 153, 161-2, 177, 178; and Confederation 180; and climatology 311n34
Magnetismus der Erde (Hansteen) 118
Mair, Charles 262, 264
Manitoulin Island 107
Manual of the Botany of the Northern United States (Gray) 206, 222, 228, 239
maple leaf: as symbol of Canadian unity 240-1
Marmora Iron Company 82
Mécanique céleste (Laplace) 150
Men of the North (Haliburton) 263-4
metamorphic processes (geology) 43, 55, 88, 103
Metamorphose der Pflanzen (Goethe) 211
Merritt, William Hamilton 40, 77
Mesmer, Franz Anton 121
Metcalfe, Sir Charles 57, 64, 136
Meteorological Society of Mannheim 117
meteorology: study introduced to Canada 116; and international co-operation 116, 117, 119, 123, 140, 141-3, 179-80; early studies 117-18; and weather forecasting 118, 162, 167, 178; Canadian popular interest in 121, 130, 136-7, 139-40, 153, 161, 165-7 *passim*, 178, 270-1; Fredericton observatory 121-2, 135, 169; and Literary and Historical Society of Quebec 122; and Natural History Society of Montreal 123; and the Hudson's Bay Company 123, 137, 143, 170; Isle Jésus observatory 138, 168-9; and grammar school meteorological stations 143-4, 152-3, 160, 162, 164, 165-7, 172; and territorial expansion 155, 171, 174-6, 178-9; and Robert Lachlan 162-4; Kingston observatory 168; Quebec observatory 169; and

Confederation 180. *See also* climatology

Michaux, François André 191

mining 27; and mineral development 80

Moffatt, George 63, 67, 69, 75

Molson, John, Sr. 25, 46

Molson family 45, 106

Monck, Sir Charles Stanley, fourth Viscount 236

Montreal Gazette 35, 42, 44, 48, 69

Montreal Herald 69

Montreal Horticultural Society 198

Montreal Literary Club 263

Montreal Mining Company 67-70 passim, 71, 82

Morin, A.N. 84, 85, 159

Morris, Alexander 96, 99, 173, 176

Morris, J.H. 240

Morris, William James 225

Mowat, Sir Oliver 180

Murchison, Sir Roderick 14, 47, 48, 49, 55, 93

Murray, Alexander 45, 58, 60, 77, 97, 106, 108; and meteorology 139-40

Murray, Anthony 45

Murray, Sir George 45

Murray, Robert 136, 155

museums: NHSM 31, 32, 36; geological 60, 63-4, 79, 283n41; British Museum 196; educational 207; Botanical Society of Montreal 221; agricultural 241; Laval University botanical 249

Naphegyi, Gabor 209

Narrative of the Canadian Red River Exploring Expedition (Hind) 174

Nation 104, 107-8, 111

nation-building 6, 8-9, 25, 75, 85-6, 90, 96, 110-11, 180, 269-70; mythology 52, 104-5, 176, 262, 266; and scientific inventory 154; and variation 260-1; and Canadian historians 273-4

native peoples 258-61. *See also* Ojibway nation

natural history: tradition 14, 192, 269, 271-2. *See also* White, Gilbert *and individual sciences*

Natural History of Selborne (White) 14, 196, 231

Natural History Society of Montreal (NHSM) 31-3, 83, 100-1, 158, 163, 226, 238; 'Indian Committee' 5, 31-2; and Lord Durham 35-6; and Literary and Historical Society of Quebec 38; and Lord Sydenham 39-40; support for William Logan 46-8 *passim*; meteorological interests of 123; and botany 193-4, 196, 197, 252, 256

Natural Philosophy (Wesley) 147

Natural Theology (Paley) 14

nature, changing perceptions of: from mechanic to organic models 4, 7; historical 9, 43, 225, 265, 271. *See also* trees, settlers' antipathy to

Naturphilosophie 210, 211

Neilson, John 40

Neptunists. *See* Wernerian geology

New Atlantis (Bacon) 117

Newton, Sir Isaac 117, 119, 176, 177

Newtonianism 4, 7, 73, 117, 270

New York Evening Post 173

New York State Geological Survey 58

North American (Toronto) 257

north magnetic pole 115, 118, 119, 178

North West Company 192

northern man, myth of 99, 176, 179, 262-4

Notman, William 104

Nova Britannia (Morris) 96

Nova Scotia, Board of Agriculture 194

Nova Scotian Institute of Natural Science 263

Oersted, Hans Christian 117
Ojibway nation 19, 71
onion, wild (*Allium*) 225
On the History and Natural Arrangement of Insects (Swainson) 210
On the Origin of Species (Darwin) 4, 204, 244, 245, 265
Ordnance Geological Survey of Great Britain 14
Oregon Trail (Parkman) 174
Ottawa Citizen 85-6, 93
Ottawa Valley 253; early geological surveys 17; and expansionist interests 98-9; and botany 197, 198, 205
'Outline of the Natural History and Statistics of Canada' (Rae) 41
Outlines of Natural Theology for the Use of the Canadian Student (Bovell) 259
Owen, Richard 224
Owen, William Fitz-William 120
ozone: and etiology of disease 138

palaeobotany 205-6, 225-6, 245-6
Paley, William 14
Palliser, John 97, 168, 175, 176, 252
Parent, Etienne 6, 33, 66-7
Parkman, Francis 174
Pelloutier, Simon 172
Penhallow, D.P. 264-5
Pennant, Thomas 5
Percival, Anne Marie 195
Perth, Upper Canada 17, 82; as centre for expansionism 98-9
petroleum 107
phanerogams 186, 187
Phillpotts, George 71
philosophical idealism. *See* Forbes, Edward; Hincks, Rev. William; Goethe, Johann Wolfgang von; quinarianism
philosophical naturalists. *See* philosophical idealism
Philosophy of Storms (Espy) 140
Phytologist 233
Plutonism. *See* Vulcanism
Poe, David Allen. *See* Watt, David Allen Poe
Portraits of British Americans (1865-8) 104
potato blight 200-1
Practical Meteorology (Drew) 165
Pre-Cambrian geology 49, 55, 56
Prehistoric Man (Wilson) 260-1
Prince, John 66-7
Principles of Geology (Lyell) 43, 262; and botany 188, 189
progress: material 7, 111-12; climatic 53, 122-3, 140, 156, 171-3, 174-5
Provancher, Abbé Léon [pseud. Emilien Dupont] 169, 216, 227, 242, 256
Province of Canada, Board of Agriculture 207-8, 209, 212
Province of Canada, House of Assembly: select committee (1841) supports funds for a geological survey 40; financial support for the Geological Survey of Canada (GSC) 63-4, 77, 93, 94; and licences for mineral exploration 66-8; and petitions for a canal at Sault Ste Marie 72; select committee on the Magdalen Islands (1852) 76; motion to reprint Logan's reports 76; select committee (1854) 84-9 *passim*; Act of Continuance (1956) 93, 94; budgetary restrictions (1861) 101-2; petition for an astronomical observatory at York [Toronto] 120-1; support for the Toronto Magnetic and Meteorological

Observatory 142-3, 157-60 passim, 168; support for other Canadian observatories 168-9, 179-80. See also Upper Canada, House of Assembly
Provincial Exhibition (Montreal, 1850) 80
Provincial Normal School, Toronto 149-50
Pursh, Frederick 191, 221, 234, 256

Quebec and Lake Superior Mining Association 70-1
Quebec Gazette 39
Quebec Mercury 197
Quebec Morning Chronicle 81, 159-60
Queen's University: and botany 206-7, 218, 228-9, 230, 237-8, 247
Quesnel, F.-A. 40
quinarianism 210

Rae, John 23, 41, 197
Rae, John (British explorer) 98
Ramsay, A.C. 109
Ramsay, Hew 33
Ranunculaceae 221
Rawson, Rawson W. 126
Rees, William 22, 197, 198
Réflections sur la Poésie et la Peinture (du Bos) 172
Report on the Affairs of British North America (Durham) 37
Report on the Geology of Massachusetts (Hitchcock) 32
Rhine River: early climate compared to modern Canada 156, 171-2
Richardson, James 226
Richardson, Sir John 98, 127, 129, 173, 195, 210, 243
Riddell, Charles James Buchanan 120, 123-5

Rideau Canal 16, 17
Rideau Valley: early geological surveys 17
Riel, Louis 267
Robb, James 38, 90, 196
Robinson, Emily Merry 132
Robinson, Sir John Beverley 131, 133, 152, 155, 163
Robinson, W.B. 71, 72
Robinson family 131-2
Roche, Alfred 97, 98
Ross, Donald 106
Ross, Sir James Clark 119
Rottermund, E.S. Comte de 86-7, 95
Roy, David 227
Roy, Thomas 26, 280n32
Royal Botanic Garden (Edinburgh) 218
Royal Botanical Gardens (London). See Kew Gardens
Royal Canadian Institute. See Canadian Institute, Royal
Royal Engineers 41; early Canadian geological observations 16. See also Baddeley, Frederick Henry *and* Bonnycastle, Sir Richard Henry
Royal Horticultural Society (London) 190
Royal Institution (London) 190
Royal Society of Canada 265
Royal Society of London 109, 117, 119, 130
Ryerson, Rev. Egerton 136, 155, 168, 174, 178, 180, 240-1, 267; promotes grammar school meteorological stations 143-4, 152-3, 160, 162, 164, 165-7, 172; supports physical science in Ontario's educational curriculum 146-8 *passim*, 152; encourages agricultural education 148-9, 150, 207; and the Family Com-

pact 152; and organized scientific inventory 153-4

Sabine, Sir Edward 120, 124, 139, 188, 212; early career of 118; and Alexander von Humboldt 118-19; plans chain of British imperial observatories 119, 135, 176-7; and Toronto Magnetic and Meteorological Observatory 125, 130, 134-5, 140-3, 157, 158, 177-8; and J.H. Lefroy 125, 126, 128-30 passim, 132, 140-3 passim, 146, 157, 158, 160, 177-8; and isodynamic ovals 127, 128; and University of Toronto 135-6, 141
Sabine, Joseph 190
St Cyr, Napoléon 227
Sarrazin, Michael 191
Sault Ste Marie ship canal 71-3
Saunders, William 217, 241, 249
Schleiden, J.M. 185
Schönbein, C.F. 138
Schultz, Sir John Christian 170, 233, 253, 262, 264, 267
Schwann, T.A.H. 185
science: and Victorian Canadians 3-9 passim, 269-74 passim; and the 'geographical' tradition 4; and Confederation 6-9, 180; French Canadians and 6, 33-4, 87, 95, 153, 227; and social order 6, 30, 37, 133-4, 145-6, 152, 153; and economic progress 34, 51, 65, 88; and Canadian popular opinion 147; and agriculture 149, 150-1, 190; and a unified world-view 177. See also inventory science
Scotland: contributions to geological theory 14-15; and Canada 273
Scottish Enlightenment 4, 15, 110; and geology 14-15; and economic backwardness 15

Sedgwick, Rev. Adam 14, 48, 55
Seemann, Berthold 233-4
Selwyn, A.R.C. 109, 254
Semper, Gottfried 290n8
Shea, Sir Ambrose 106
Shelley, Mary 161, 274
Shepard, Charles Upham 47
Shepherd, Forrest 68-9, 70
Shepherd, George 221, 234
Sheppard, Harriet 195
Sheppard, William 36, 195
Silliman's Journal 44
Simcoe, John Graves 133
Simpson, Sir George 67, 68, 70, 71, 92, 97, 176, 243; and meteorology 123; and John Henry Lefroy 127, 129
Sinclair, Sir John 4-5, 194, 200
Smallwood, Charles 138, 163, 168-9, 222; and botany 214
Smith, James 63
Smith, William ('Strata') 13
Smithsonian Institution 142, 143, 156, 158, 163, 212, 227
Sommers, John 246-7
species, transmutation of 199, 218
Spragge, William 172
Statistical Account of Scotland (Sinclair) 4-5, 194
statistics 4, 5, 24
Stewart, Frances 196
Stewart, John 237, 238
stigmaria 42
Strachan, John (bishop of Toronto) 22, 132, 134, 146, 153, 155; interest in natural philosophy 133
Strickland, Samuel 214
Sturton, R. 246
Sullivan, Robert Baldwin 171
surveys 4-5; geological: in New Brunswick and Newfoundland (1830s) 4, 14,

38; W.E. Logan 5, 56–7; American (1820s) 13; salt (1790s) 14; Lake Huron (Bigsby) 17–19; general: NHSM 'Indian Committee' (1828) 5; Huron Tract 20–1; Upper Canada (Rae) 23; north shore of Lake Huron (Carthew) 24–6; canal: Rideau (1820s) 16; institutional: Durham's 36; magnetic: Lefroy's 127–8; agricultural 207–8. See inventory science: tradition; Pennant, Thomas; Sinclair, Sir John

Swainson, William 210

Sydenham, Charles Poulett Thomson, first Baron 37–41 passim, 47

Systema naturae (Linnaeus) 186

Taché, Sir E.P. 6–7
Taché, J.C. 85, 90, 91
Taylor, Fennings 104
telegraph, magnetic 121, 123
teratology 210–11, 227
Théorie élémentaire de la botanique (Candolle) 187
Thom, Adam 33
Thompsonian medicine 208–9
Times (London) 91
Toronto Albion of Upper Canada 26
Toronto British Colonist 64–5, 150, 159–60, 180
Toronto Globe 69, 74, 82, 98, 163
Toronto Horticultural Society 198
Toronto Leader 167
Toronto Magnetic and Meteorological Observatory 4; early petition for 120–1; established 124; and University of Toronto 124, 135–6, 141, 142, 157, 162, 168; and Edward Sabine 125, 130, 134–5, 140–3, 157, 158, 177–8; and J.H. Lefroy 130, 135–6, 140–3 passim, 156–8, 160, 177–8; government support of 142–3, 157–60, 168; and Canadian public opinion 143, 177–8, 180

Toronto Mechanics' Institute 21–2, 161
Toronto Normal School 207
Toronto Patriot and Farmer's Monitor 24, 198
Torrey, John 196, 206
Tour in Scotland 1769 (Pennant) 5
Tournefort, Joseph Pitton de 191
Traill, Catharine Parr 51, 172, 176, 183–4, 195–6, 205, 241, 265
Traité de la minéralogie (Haüy) 187
Traité élémentaire de botanique (Provancher) 227
Treadwell, C.P. 212
Treatise on the Geography and Classification of Animals (Swainson) 210
trees, settlers' antipathy to 126, 172
Tupper, Sir Charles 108
Tuttle, Charles R. 179

uniformitarian geology 43–4, 53–4, 55, 107–9, 110, 112
United Grand Lodge of Freemasons 45
United States: and Canadian science 13, 56, 60–1, 85, 100–1, 116, 123, 126, 127, 140–3, 151, 196, 206, 272–3. See also Bache, Alexander Dallas; Espy, James Pollard; Hall, James; Henry, Joseph; Loomis, Elias
Universal Exposition (Paris, 1855) 90–1
University College, Toronto. See University of Toronto
University of King's College, Toronto. See University of Toronto
University of Toronto 120, 141, 158, 168; and Toronto Magnetic and Meteorological Observatory 124, 135–6, 141, 142, 157, 162, 168; and botany 200, 206–11 passim, 216–17

Upper Canada, House of Assembly: and 1832 petition by the Literary and Philosophical Society of Upper Canada for a geological survey 22; and memorial from Dr John Rae (1832) 23; and select committee to plan for a provincial geological survey (1836) 26–8, 33. *See also* Province of Canada, House of Assembly
utilitarianism 5–6, 110, 185

Van Cortlandt, Edward 17
Vankoughnet, Philip M. 215–16
variation: magnetic 117, 119; in species 221, 247; and natural selection 244
Vegetable Kingdom (Lindley) 209–10
viniculture 242–3
Vulcanism 43

Wakefield, Edward Gibbon 121
Wallace, Alfred Russell 259
Watt, David Allen Poe 221, 226, 249, 250, 252
Wellington, Arthur Wellesley, first Duke of 16
Werner, Abraham Gottlob 13

Wernerian geology 15, 17, 19, 20–1, 30, 43
Wesley, John 147, 154
Western District Agricultural and Horticultural Society 198
Weston: search for coal 62
Whewell, William 79
White, Gilbert 14, 126, 196
Williamson, James 229, 234
Wilson, Sir Daniel 180, 260–1, 263, 264, 266
Wilson, James 17, 67, 82
Winder, William 213

York Literary and Philosophical Society. *See* Literary and Philosophical Society of Upper Canada (Toronto)
York Mechanics' Institute. *See* Toronto Mechanics' Institute
Young, John 84
Young, John [pseud. Agricola] 194, 200, 263
Younghusband, Charles Wright 125, 131

Zaslow, Morris 53
zoology 4, 127, 129, 188, 210, 321n22, 326n66